Geothermal Energy: Concepts, Models and Development

Geothermal Energy: Concepts, Models and Development

Editor: Hunter Cox

www.murphy-moorepublishing.com

Murphy & Moore Publishing,
1 Rockefeller Plaza,
New York City, NY 10020, USA

Visit us on the World Wide Web at:
www.murphy-moorepublishing.com

ISBN: 978-1-63987-257-2 (Hardback)

Cataloging-in-Publication Data

Geothermal energy : concepts, models and development / edited by Hunter Cox.
    p. cm.
Includes bibliographical references and index.
ISBN 978-1-63987-257-2
1. Geothermal resources. 2. Power resources. 3. Geothermal engineering. I. Cox, Hunter.

TJ280.7 .G46 2022

333.88--dc23

Table of Contents

Permissions

List of Contributors

Index

Preface

Energy which determines the temperature of matter is called thermal energy. Geothermal energy is the thermal energy that is generated and stored inside the Earth. It is generated from the continual heat loss due to radioactive decay of materials. The high temperature and pressure in the Earth's interior causes some rocks to melt and solid mantle to behave plastically. Water from geothermal springs can be used for a variety of purposes such as bathing, heating and generating electricity. Geothermal power is cost-effective, reliable, sustainable and environmentally friendly. Geothermal energy is an upcoming field of science that has undergone rapid development over the past few decades. The ever growing need of advanced technology is the reason that has fueled the research in this field in recent times. Scientists and students actively engaged in this field will find this book full of crucial and unexplored concepts.

The information shared in this book is based on empirical researches made by veterans in this field of study. The elaborative information provided in this book will help the readers further their scope of knowledge leading to advancements in this field.

Finally, I would like to thank my fellow researchers who gave constructive feedback and my family members who supported me at every step of my research.

Editor

Modeling coupled convection and carbon dioxide injection for improved heat harvesting in geopressured geothermal reservoirs

Tatyana Plaksina[1*] and Christopher White[2]

*Correspondence: tatiana_plaxina@yahoo.com
[1] Department of Petroleum Engineering, Texas A&M University, 907 Richardson Hall, College Station, TX 77843, USA
Full list of author information is available at the end of the article

Abstract

Geopressured geothermal saline aquifers are an abundant low-enthalpy geothermal energy resource available in many coastal regions including the US Gulf of Mexico. In such geographic areas thick geopressured sandstones (up to several hundred meters thick) hold tremendous geothermal heat with conservative estimates of gross extractable energy approximately 0.2 EJ per cubic kilometer of the formation. Additionally, widespread geopressure in sedimentary deposits of the Gulf region preserves favorable petrophysical properties of unconsolidated sandstones such as high porosity and permeability, thus, enhancing productivity and economics of potential heat harvesting projects. In this study we investigate the potential of a typical geopressured reservoir in the US Gulf coast to deliver commercial quantities of geothermal heat with the possibility of simultaneous supercritical CO_2 sequestration into the same formation. Specifically, we focus on numerical simulation study of heat extraction from a model based on the Camerina A sand of South Louisiana. In our numerical experiments, we consider both theoretical and practical implications of combining a traditional heat harvesting method with supercritical CO_2 injection. Moreover, this study pays specific attention to the effect of natural convection due to the formation's tilt and uneven heating at the reservoir boundaries and its impact on the forced convection due to geofluid withdrawal. The numerical simulation results suggest that introduction of supercritical CO_2 might have an observable positive effect on the ultimate heat recovery and that a strategic injection/production well placement might further enhance density-driven flows inside the geothermal formation.

Keywords: Geothermal energy, Geopressured brines, Saline aquifers, Natural convection, Forced convection, Carbon dioxide sequestration, TOUGH2

Background

Geothermal systems hold abundant and carbon-free thermal energy for potential electricity generation, space heating, and air-conditioning. For example, the subsurface potential of the US contains approximately 170,000 EJ (1 EJ = 10^{18} J) of energy (MIT 2006). One such energy source readily available in many coastal regions across the globe is geopressured saline sedimentary aquifers. Geopressured aquifers are usually undercompacted, brine saturated, porous, and permeable sandstone formations that have anomalously high pore pressures and reservoir temperatures over 100 °C. Among all geothermal systems, geopressured fields are considered a medium- and low-grade (or low-enthalpy) geothermal resource that occupies large subsurface areas in coastal regions (Esposito and Augustine 2011). The US states of Louisiana and Texas are examples of geographic locations where geopressured systems occur frequently and occupy the areal extent of more than 145,000 km^2 (MIT 2006).

Several technical obstacles may render development of most coastal geopressured systems sub-commercial. These low-enthalpy systems require drilling multiple injection wells for improved heat sweep because they have lower heat content and thermal efficiency. Costly pressure maintenance programs and surface handling of withdrawn geofluids may make these reservoirs unattractive for commercial development (Freifeld et al. 2013). In addition to these problems, withdrawal of geothermal fluids might cause land subsidence due to compaction in the producing geologic formation unless the produced geofluid is re-injected into the reservoir or shallower formations (Gustavson and Kreitler 1979). As a result, pilot commercial projects exploit only those sites that have anomalously high geothermal gradients and strong water drives—the so-called "low-hanging fruit" of the tremendous resource.

In deep sedimentary basin geothermal production techniques are usually divided into two main categories: coproduced fluids and geopressured geothermal extraction. The first development strategy uses thermal energy of hot water coproduced with oil and gas. This type of geothermal resource is confined to existing hydrocarbon fields at depths between 4 and 6 km (MIT 2006). In the US, the annual volume for coproduced hot water is approximately 33 billion barrels which is equivalent to 3000 MW based on geofluid temperature of 100 °C (Curtice and Dalrymple 2004). Geothermal projects of the second category, geopressured brines, are independent from oil and gas production and develop thermal potential of deeply buried aquifers. According to the latest conservative estimates, the Northern Gulf of Mexico (GOM) basin stores raw thermal energy of about 46,000 EJ (White and Williams 1975).

Although coproduced and independent geothermal reservoirs seem to be very similar for modeling purposes, their initial states (and thus, model initialization methods) differ. This research demonstrates benefits of initializing numerical models as quiescent systems with a proper temperature distribution (quiescent period is a period during which the reservoir experiences no injection or production, and it might range from 100 to millions of years). The thermal profile of coproduced geothermal reservoirs, however, is distorted by oil and gas production, and initialization with the quiescent period would not provide the correct temperature distribution for a coproduced project. This study focuses on geopressured brines, and therefore, we assume no forced convection (due to injection or withdrawal of geofluid) prior to heat harvesting.

Despite the fact that geothermal heat extraction is still considered a marginally profitable energy industry, a number of scientists have already proposed several strategies to make geothermal projects more economically attractive. More specifically, Ganjdanesh et al. (2015) have investigated how the energy cost could be reduced with capturing and storing CO_2 inside the geothermal formation. National agencies such as the US Geological Survey (USGS) and Department of Energy (DoE) also propose favorable conditions and location for such project that include vast subsurface areas of the GOM (Warwick et al. 2014; Goodman et al. 2011; Nicot 2008). Several prominent studies have proposed to utilize sequestered CO_2 as a secondary fluid to deliver the heat from reservoir's hot lower boundary (Randolph and Saar 2011; Salimi and Wolf 2012; Randolph et al. 2013; Adams et al. 2014). This interest in using CO_2 in geothermal projects stimulated further research on the behavior of the supercritical greenhouse gas in the subsurface conditions. In this context, for our investigation the most interesting works on CO_2 behavior in geologic formations include the numerical study of CO_2 flow under non-isothermal conditions by Singh et al. (2011), the investigation of supercritical CO_2 injection into a deep saline aquifer by Vilarrasa et al. (2013), and the study of the dynamics inside the CO_2 geologic storage by Pool et al. (2013). All these works provide a solid foundation and expectation of how supercritical CO_2 plume should behave inside a geothermal system.

This study builds on this foundation and offers an investigation of a new method for improved heat recovery from low-enthalpy geopressured aquifers by combining the effects of natural and forced convection and density-driven effects of CO_2 injection. Particularly, we demonstrate the advantages of characterizing a natural convection pattern inside a tilted or flat geothermal formation that might help place injection/production wells strategically to enhance subsequent heat extraction by the coupled convection. This approach allows for better geothermal resource estimation, potentially improved economics and a selection of a more efficient production arrangement. Because the current study might be particularly beneficial to the US GOM region, in our numerical experiments we use models with petrophysical and thermodynamic properties of typical GOM geopressured formations. Additionally, this study discusses heat harvesting simultaneous with small scale CO_2 sequestration, the way to avoid land subsidence with re-injection of the produced geofluid into the same formation and the effect of the formation dip on the ultimate heat recovery. One South Louisiana aquifer, the Camerina A, is a central example for this study used for a more detailed investigation of an optimal geothermal production scenario.

Although the economic feasibility evaluation is not a part of this study, because of constant changes in the carbon tax agenda, high volatility on energy markets, and varying abilities to fund renewable projects, we can outline a possible life cycle of such geothermal reservoir for engineering and energy production purposes. A geopressured reservoir could be brought down to pressure suitable for economic injection of the supercritical gas (like in the case of Camerina reservoir the upper portion of which has been depleted during oil production). However, to prevent quick subsidence of mostly unconsolidated sediments, the geofluid must be re-injected into the shallower layers. At this stage of the production cycle both thermal and dynamic energy can be used for electricity generation purposes. Once overpressure is depleted, the project can be categorized as a carbon dioxide sequestration project and corresponding financial benefits can be applied to offset high compression costs. However, at this point in the life cycle all withdrawn geofluids must be returned into the formation after heat harvesting to maintain reservoir pressure and help mixing the supercritical gas with the brine. In addition to creating density-driven convection, this mixing is important in controlling the gas plume and ensure caprock integrity as discussed later (Islam et al. 2013; Shukla et al. 2010; Karimn-ezhad et al. 2014; Wang et al. 2015).

Methods

In this section we introduce necessary theoretical background on convection in flat and tilted porous media which are analogous to brine saturated sandstone geothermal reservoirs used in this study as well as describe the numerical simulation model used to obtain thermal energy production data. Additionally, we provide the description of the model initialization process and the design of experiment that helps identify the most significant parameters affecting the energy output.

Convection in flat and inclined porous media

Natural convection may occur in a fluid-saturated porous medium subjected to non-uniform heating. The Rayleigh number describes the energy and mass transfer (Horne 1975).

$$Ra = \frac{k\rho^2 c\gamma \Delta Tgh}{\mu K} \tag{1}$$

In Eq. 1, γ is the thermal expansivity of the fluid, ρ is fluid's density, k is permeability of the porous medium, g is the acceleration due gravity, c is fluid's specific heat, h is the height of the system's square cross-section, ΔT is a change in temperature, μ is the fluid's viscosity, and K is the average thermal conductivity of the fluid and the rock matrix. The value of the Rayleigh number indicates if conditions favorable for convection have been reached. Specifically, when Ra exceeds $4\pi^2$ due to non-uniform heating and/or compositional heterogeneities (for instance, changes caused by mixing with gas or salt), the investigated system becomes unstable and convection cells begin to form (the critical value of $4\pi^2$ is derived from a solution for a system with an infinite horizontal dimensions, uniform petrophysical properties, and constant top and bottom temperatures (Nield and Bejan 2006)). Based on the same study, a typical fluid density change in naturally convecting systems is about one percent.

Although convection in flat systems is of interest in some development cases as highlighted in Zhang et al. (2014), in the GOM region many hot saline aquifers are dipping. Dip in such geologic formations can be local or sustained for the entire length of the reservoir. Tilted geopressured formations are particularly common around salt structures that cause deformation of adjacent sand deposits as well as their anomalously high temperature. The base case of this study, the Camerina A sand, is a dipping system due to its proximity to the Gueydan salt dome (Smith and Reeve 1970). Therefore, a more careful examination of natural convection in inclined reservoirs is needed.

For tilted systems a modified definition of the Rayleigh number can be used as in Eq. 2 (Nield and Bejan 2006):

$$Ra = \frac{k\rho^2 c\gamma \Delta TgL \sin\theta}{\mu K} \tag{2}$$

Here the height of the system h is replaced by the length multiplied by sine of the dip $Lsin\theta$. This subtle change in the formula also affects calculation of the critical Rayleigh number. For inclined systems it becomes $4\pi^2 sin\theta$ implying that in dipping reservoirs convection starts to dominate conduction at relatively small Ra values (Nield and Bejan 2006). Apart from the reservoir dip and non-uniform heating, salt dissolution and precipitation have been found to promote thermohaline convection in GOM sediments (Hanor 1987).

Simulation model and setup of numerical experiments

The Northern GOM basin is a prolific geopressured geothermal province with many thick sandstone saline aquifers. To build a realistic numerical model for subsequent flow simulations, we use the properties of a hot saline aquifer in the GOM coast of Louisiana suitable for heat extraction. The Camerina A sand of South Louisiana is selected as a prototype for the simulation model (Plaksina 2011; Gray 2010).

The Camerina A sand is a Late Oligocene deposit identified near the Gueydan salt dome in Vermillion Parish, LA (Fig. 1) at an approximate true vertical depth of 4300 m

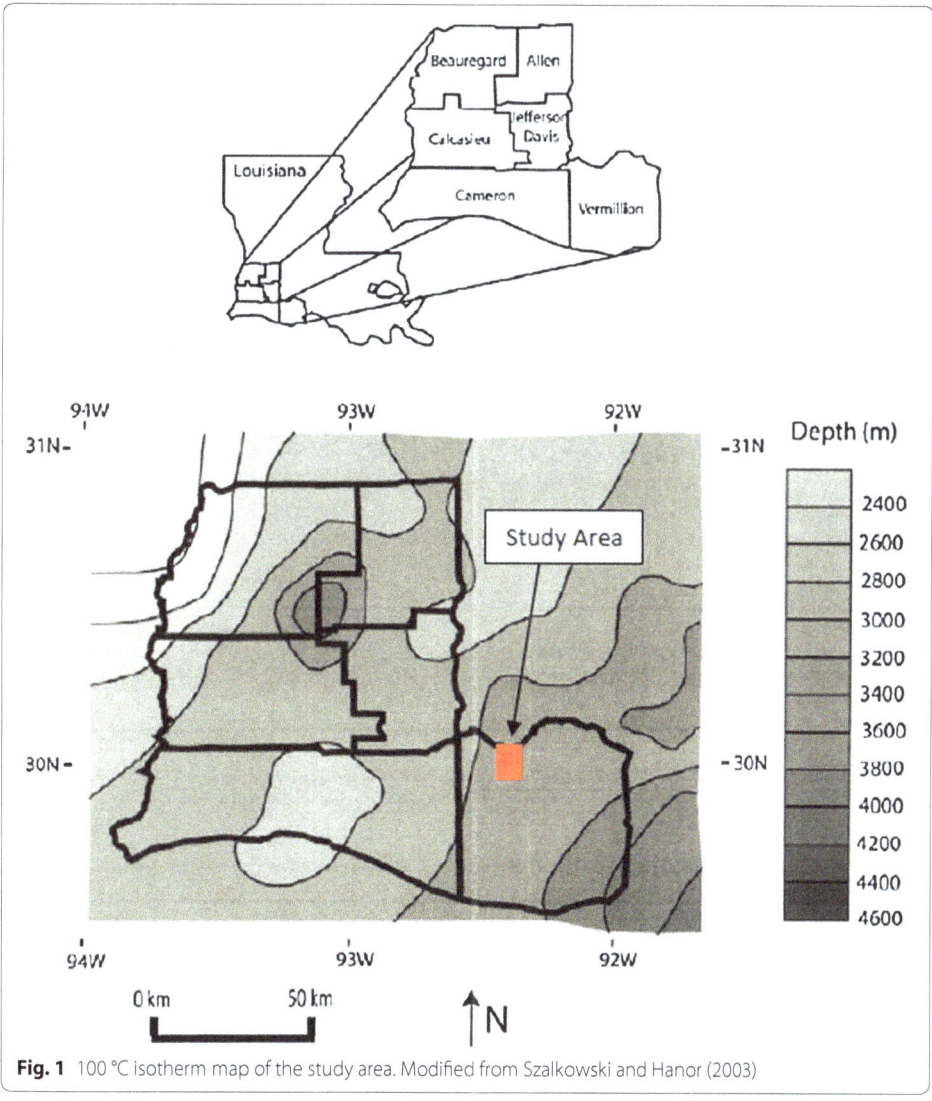

Fig. 1 100 °C isotherm map of the study area. Modified from Szalkowski and Hanor (2003)

(Gray 2010). Its depositional environment is a delta front to distributary mouth bar and it is a part of a marine transgressive sequence. The sand's average thickness is approximately 100 m, permeability is approximately 200 md, and its porosity varies between 9–31 percent. The Camerina A sand is a dipping aquifer with varying dips ranging between 1.2–28°. The corrected formation temperature is approximately 140 °C with a geothermal gradient of approximately 29 °C/km and estimated formation pressure is over 80 MPa (Gray 2010). Other relevant properties used for 2D TOUGH2 simulations are summarized in Table 1. The sand formation has all properties relevant to the objectives of this study such as dip, anomalously high formation temperature suitable for electricity generation, substantial thickness, and areal extent. Therefore, its properties can be used for the base case simulation model in our investigation of an optimal heat harvesting scenario with zero mass net withdrawal.

Although surface facilities and energy conversion is not in the scope of this study, it is important to note that the Camerina A sand is located in a geographic area where relatively low geothermal gradients are expected. Thus, one of possible methods to convert the heat energy of the aquifer's fluid into electricity is an organic binary cycle for electricity generation (MIT 2006).

Model initialization and design of experiments

To see the benefits of initializing geomodels with proper geothermal gradient and running the quiescent period, a number of 2D simulations were performed with varying geometries and petrophysical properties (Pruess et al. 1999). Because one of the objectives of this study is to find parameters that influence heat recovery the most, a factorial design was used to vary the factors of interest, and for each case the Rayleigh number and its critical value were calculated (tabulated in Additional file 1: Appendix B). To encompass a wide range of geometric and petrophysical properties found in the Northern GOM basin, two levels of permeability (100 and 1000 md) and thickness (100 and 200 m), and three levels of dip (0, 2 and 15°) were used in the design of experiments (Ewing et al. 1984).

All simulation models have three layers of rock: the top and the bottom ones are impermeable bounding layers with infinite heat capacity to imitate multiple formations

Table 1 Reservoir properties for 2D simulation runs

Property	Value	Units
Initial pressure	3.45×10^7	Pa
Initial average temperature	135	°C
Porosity	0.20	–
Matrix compressibility	2.0×10^{-8}	1/Pa
Injection water enthalpy	3.0×10^5	J/kg
Rock density	2600	kg/m^3
Wet rock heat conductivity	2.0	W/m °C
Reservoir length	4000	m
Reservoir width for 2D run	100	m
Salinity	0	ppt
Geothermal gradient	29	°C/km

with constant temperatures above and below the reservoir and the middle layer is the permeable sandstone reservoir (Fig. 2). An initialization script assigns temperature value to each grid block (including those of the bounding layers) according to the chosen geothermal gradient (e.g., the gradient of the Camerina A sand). Although salinity (or more generally, composition) is another factor that might influence convection and, thus, the ultimate heat recovery, in this study we focus exclusively on significance of geometric parameters of the geologic system and rock properties (Hanor 1987).

The formation pressure and geothermal gradient (Table 1) are lower than they would be expected in the geopressured zone in the GOM coast. Gray (2010) suggests that the Camerina A sand, which is a typical sand deposit in the zone, has a geothermal gradient of 29 °C/km and initial formation pressure over 80 MPa. In our numerical investigation, however, the value of the initial formation pressure is lowered for two reasons. First, the solutions obtained with the equation of state (EWASG) in the simulator of choice (TOUGH2) become less stable and reliable at high pressures. To our knowledge, this numerical problem has not been solved to date, thus, these extreme pressure conditions are still out of numerical reach even though the researchers are on track to address this problem for extra deep reservoirs (Zhang et al. 2011). Because of this gap in current numerical tools, we decided to resort to a common in such situation approach and work in the range of pressures that provide reliable solutions. Nevertheless, we emphasize that the actual reservoirs of such temperatures and pressures are found much deeper. Second, we conduct a comparative study of production cases with and without CO_2 injection to analyze the effect of density-driven convection. To achieve this, it is instrumental to keep the numerical test conditions as similar as possible. However, at high initial pressures which become even higher, downdip simulation of CO_2 injection with current software tools is impossible for reasonable injection pressure range. Thus, to compare energy output from the simulated systems in which only CO_2 injection rate is varying (0

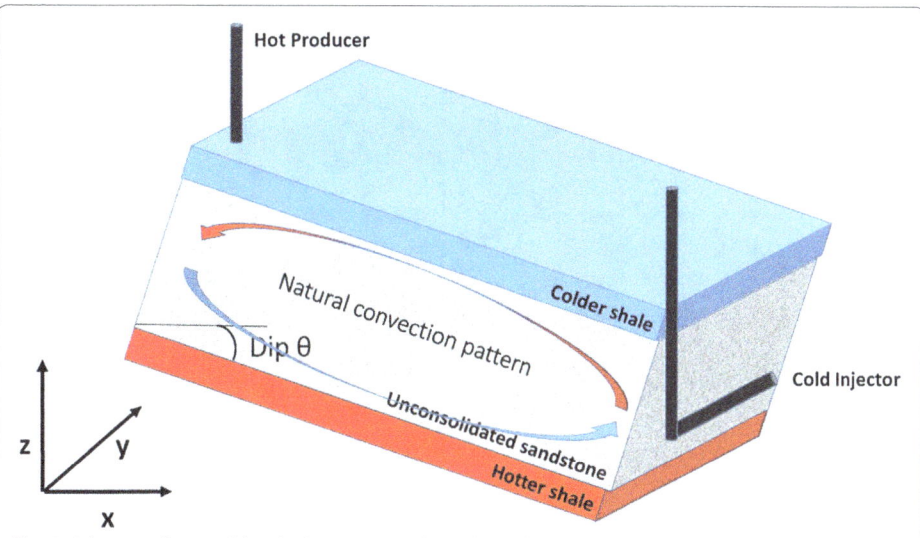

Fig. 2 Schematic figure of the tilted geopressured geothermal reservoir with one hot water producer well and one cold water injector well. The geologic systems dip at the angle θ. *Three layers* correspond to the upper colder layers of shale, the middle porous unconsolidated sandstone with high permeability, and the lower hot layers of shale. The natural convection takes cold fluid downdip and sends hot fluid updip. The injector helps create displacement of hot fluid updip

for no CO_2 injection and 10^{-4} kg/s for the case with CO_2 injection), the initial formation pressure is kept at 34.5 MPa. This pressure value was derived from the experiments with the simulator and the range of stability of the equation of state.

As for the land subsidence problem usually associated with geothermal development, we propose to re-inject the withdrawn geofluid into the same formation (in the case of extreme overpressure, however, injection into a shallower formation might be considered at the initial production stage). This strategy is common place for geothermal projects because it helps solve several problems. First, keeping net zero withdrawal prevents compaction of the unconsolidated sandstone and preserves favorable petrophysical properties. Second, cold water re-injection is an important component in setting the forced convection in motion and keeping the supercritical gas plume from rising toward the top of the reservoir. Final, re-injection of the geofluid back into the formation relieves some costs associated with surface handling of large volumes of geothermal fluids (MIT 2006).

The final aspect of the modeling process that is essential for understanding the dynamics of the simulated system is the production arrangement with a CO_2 injection well. Schematic Fig. 3 shows that the design with the CO_2 well is slightly different from that without CO_2 injection. Because caprock integrity is one of the biggest concerns in any CO_2 sequestration project, we disperse the gas plume by placing the CO_2 injection well at bottom of the reservoir at approximately middle of the model. This arrangement allows us to keep the supercritical gas sufficiently far away from the hot water producer and sufficiently close to the descending cold water front from the cold water injector. As Anchliya (2009) demonstrated, strategic placement of injection wells allows to dynamically control the process of CO_2 sequestration and ensure the gas plume containment

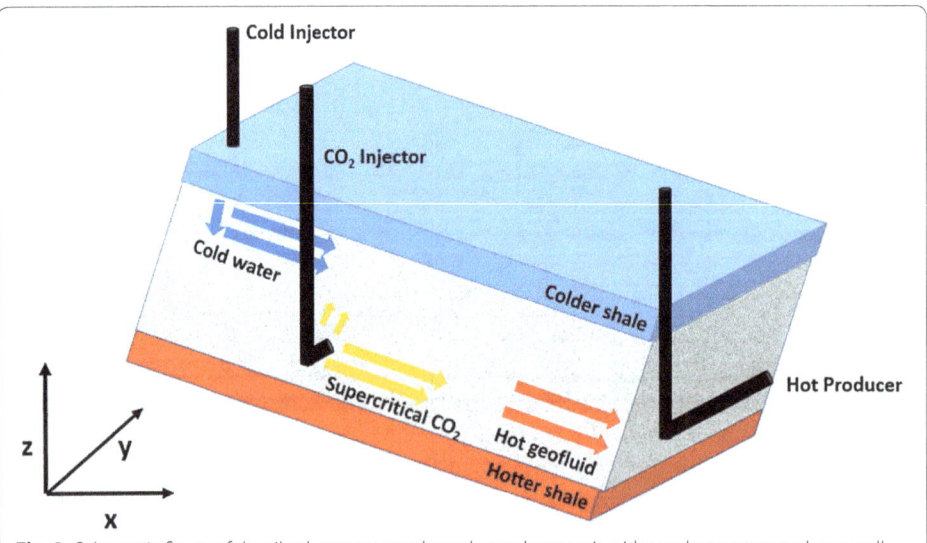

Fig. 3 Schematic figure of the tilted geopressured geothermal reservoir with one hot water producer well, one cold water injector well, and CO_2 injection well. The producer well withdraws hot geofluid downdip because in this production arrangement the cold water injector plays double role: it returns cold geofluid to maintain the formation pressure and prevents the supercritical gas plume which is lighter than formation fluid before mixing to rise to the caprock. Cold water drives the gas plume downdip and enhances its mixing with the formation fluid

within the reservoir. It also helps maintain the integrity of the caprock because only small concentrations of the supercritical gas eventually rise to the top of the reservoir.

Results

Natural convection modeling

The ability to predict heat transport (such as conduction or convection) is valuable for successful production planning, but not sufficient for adequate geothermal resource estimation and developing an optimal production strategy. In addition to the Rayleigh number, the engineer needs to know the approximate shape of the natural convection pattern, the span of the quiescent period during which natural convection stabilizes, and the effect of the bounding layers on the reservoir's temperature profile. This section provides the results of modeling these three aspects.

Nield and Bejan (2006) outline problems with computing and examining convection patterns in inclined porous media and conclude that in general tilted systems tend to have unicellular convection pattern until very high Ra are achieved. A coarse grid simulation with the properties listed in Table 1 reveals a kilometer-scale natural convection loop, with hot fluid being convected along the top of the reservoir (Fig. 4). Another important aspect in characterization of natural convection is duration of the quiescent period during which a geothermal system reaches stable temperature distribution. For computational efficiency, the quiescent period should be the shortest time span after which no significant change in the convection pattern occurs. To establish this time period, we consider cases with moderate Ra's and estimate the time span after which the temperature and aqueous phase flow in each grid block change negligibly. Cases two and eight from the experimental design (Additional file 1: Appendix B) are suitable for this

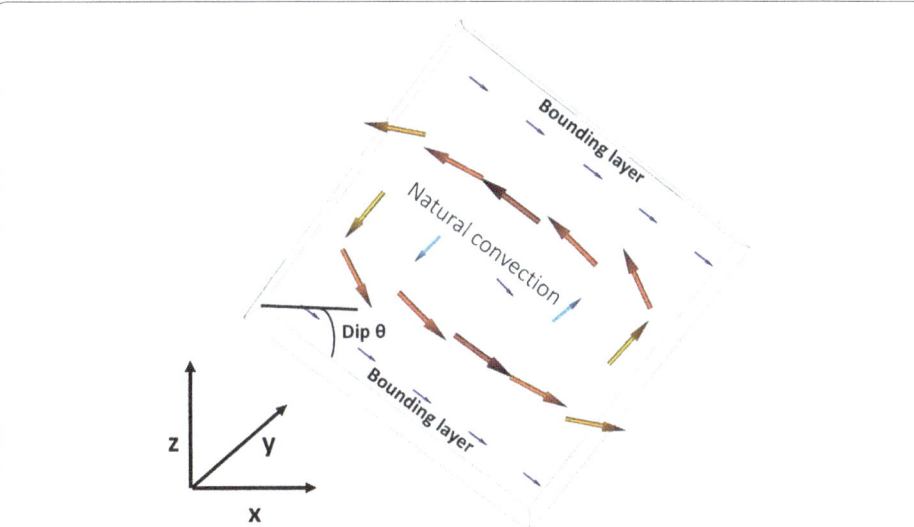

Fig. 4 Vector plot of aqueous phase flow (kg/s) for 5 × 1 × 5 blocks (500 m in x-direction, 100 m in y-direction, and 500 m in z-direction) model dipping at 45° with top and bottom bounding layers of infinite heat capacity. Pattern snapshot is taken after 1000 years of quiescent period, modeled in TOUGH2, and visualized with PetraSim software. Heat is conducted into the bounding layers high in the reservoir, and into the reservoir at greater depths. Length and color of the vectors reflect the magnitude of mass transfer. The range is from short vectors with *cold colors* to long ones with *warm colors*. For this image, the range is from 0 to 0.00003 kg/s

analysis and were run for 1 million years as quiescent systems. Additional file 1: Appendix C provides details of how the quiescent period of 1000 years is obtained with available software tools. Because the mean of difference of aqueous phase flow values between 1000 and 10,000 years comprise less than 0.1 percent of the mean of initial aqueous phase flow values in both cases, the change after 1000 years is not significant and can be neglected. We get the same results if we consider temperature instead of aqueous phase flow. These results are omitted for brevity.

To demonstrate how the natural convection pattern stabilizes, the previously used coarse model with $5 \times 1 \times 5$ grid blocks (5 grid blocks in the x-direction, 1 grid block in the y-direction, and 5 grid blocks in the z-direction with each grid block having dimensions of 500 m $\times$ 100 m $\times$ 500 m) is kept quiescent for 1 million years. Comparison of vector plots of aqueous phase flow after 1 s, 1000 years, and 1 million years show little visual or quantitative difference after 1000 years (Fig. 5).

The final aspect of natural convection modeling that requires discussion in this section is the effect of the bounding layers on the reservoir's temperature profile. Impermeable colder top and hotter bottom bounding layers with infinite (or very large) rock heat capacity simulate low layers permeability (for instance, shale) that allow heat, but not mass transfer in and out of the reservoir. The bounding layers with large heat capacity supply the porous medium with constant heat fluxes that produce evenly spaced contour lines in the temperature profile (Fig. 6). After 1000 years quiescent period, temperature contours appear nearly horizontal (without vertical exaggeration shown on the figures), and their slight curvature is due to natural convection in the reservoir's geofluid (Fig. 7). When the bounding layers are not included, and the model is run for the same quiescent period of 1000 years, the range of temperatures decreases, and with lesser natural convection, the temperature contours are nearly planar and simpler in structure (Fig. 8).

Even though the reservoir has uniform petrophysical properties, the range of temperatures without bounding layers is 6 percent less than in the previous case and the contours are not equally spaced. Because the bounding layers give a wider, evenly spaced, and more realistic temperature profile (realistic in a sense that any GOM geopressured aquifer is bound by other formations that conduct heat in and out of the reservoir), all production cases discussed below are initialized and kept quiescent for 1000 years. The last illustration in this section (Fig. 9) shows the interdependence of the Rayleigh number and variance of temperature with time. The plot confirms that the systems with high Ra have higher variability in temperatures due to more vigorous convection and that stabilization time for the geomodels with uniform properties is relatively short. Because the graph has nearly horizontal trend (in other words, additional 10-, 100-, etc., fold increases in quiescent time do not change the temperature distribution), 1000 years is sufficient duration for the quiescent period.

Geothermal production and CO_2 injection modeling

Using the experimental design (Additional file 1: Appendix B), we generated three sets of simulations: (1) twelve geofluid production cases initialized without the proper geothermal gradient and quiescent period, (2) twelve geofluid production cases with natural convection in-place at the time of heat extraction, and (3) a set of twelve cases with simultaneous geofluid production and CO_2 injection, initialized with natural convection.

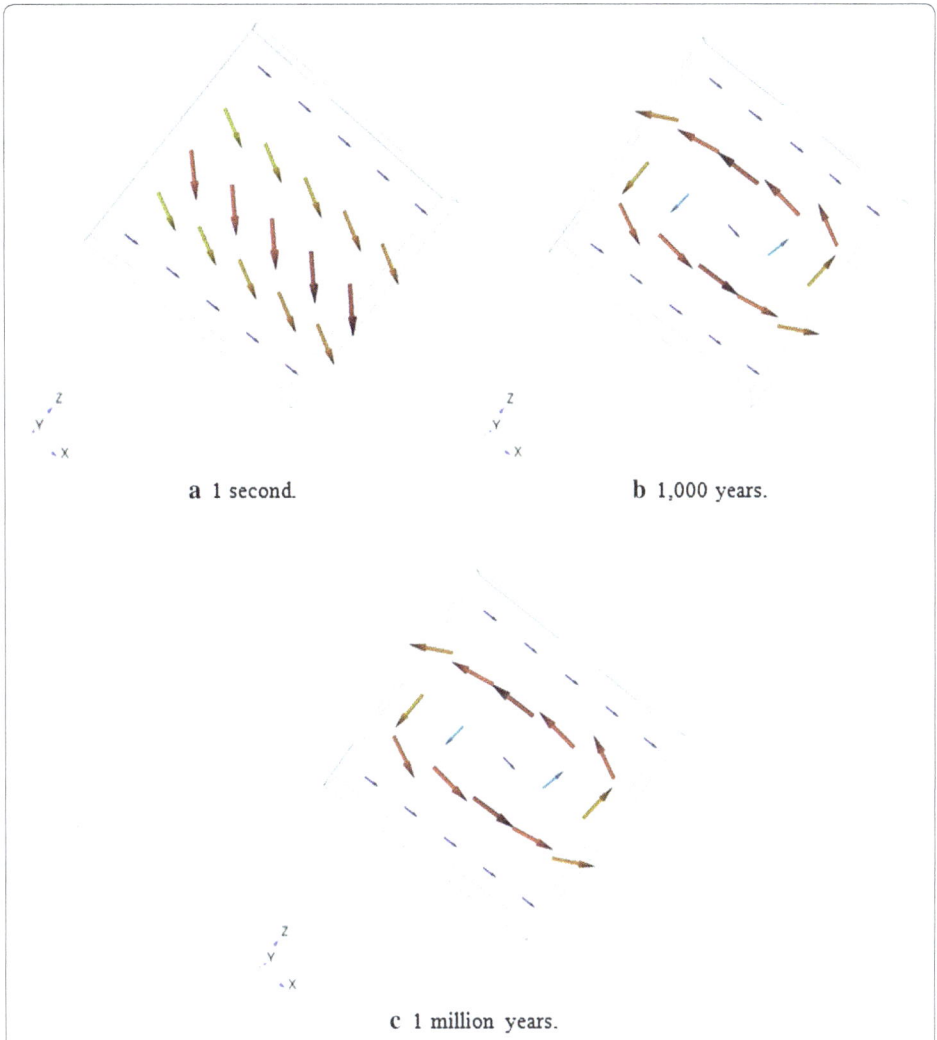

a 1 second.   **b 1,000 years.**

c 1 million years.

Fig. 5 Vector plots of aqueous phase flow (kg/s) for a model with 5 × 1 × 5 blocks (500 m in the *x*-direction, 100 m in the *y*-direction, and 500 m in the *z*-direction) with top and bottom bounding layers of infinite heat capacity. Pattern snapshots are taken after **a** 1 second, **b** 1000 years, and **c** 1 million years of the quiescent period. Visually the pattern stabilizes after 1000 years

The response of interest for all simulation runs is energy extracted, E, after 10, 20, and 30 years of production. These results along with the factors are merged into one dataset and imported into statistical modeling software (R Team 2013).

To focus on the most important factors, the dataset is split into subsets by time (10, 20, 30 years) and flow rates (0.2, 2, 20 kg/s) and inspected for correlation. Correlations between energy outputs for 10 and 20 years of production and 20 and 30 years of production are 0.999 and 0.998, respectively. Correlations between subsets split by the production flow rate are 0.999 and 0.997 for 0.2–2 kg/s and 2–20 kg/s, respectively. Therefore, it is possible to reduce the number of factors by analyzing only one subset with energy output after 10 years of hot water production at a flow rate of 0.2 kg/s. All significant factors found for this subset are significant for the entire dataset.

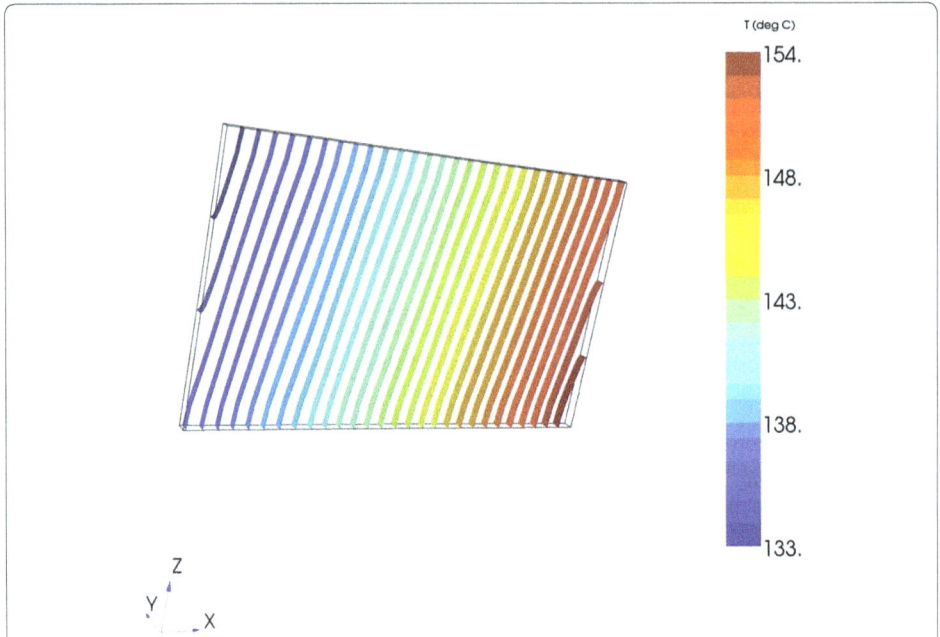

Fig. 6 Temperature profile for a 50 × 1 × 20 grid blocks model (50 blocks in the x-direction, 1 in the y-direction, and 20 in the z-direction) with each grid block having dimensions 80 m × 100 m × 5 m model with the bounding layers, the vertical gradient of 18 ‰/km and 15° dip after quiescent period of 1000 years (plotted with the 20-fold exaggeration in the z-direction). The systems experiences natural convection. Presence of the bounding layers that supply constant heat influx into the reservoir produces equally spaced temperature contour lines expected for a medium with uniform petrophysical properties

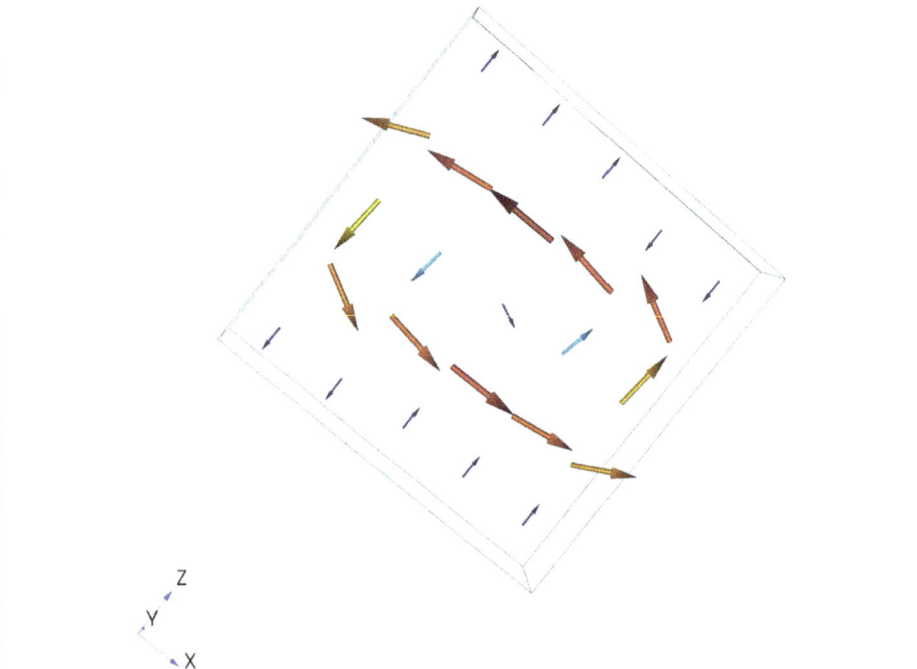

Fig. 7 Vector plot of heat flow for the model with 5 × 1 × 5 blocks (each block has 500 m in the x-direction, 100 m in the y-direction, and 500 m in the z-direction) model with bounding layers and 45° dip. The snapshot is taken after 1000 years and demonstrates heat flow out of the reservoir on the top and into on the bottom. Length and color of the vectors reflect the magnitude of heat transfer. The range increases from short vectors with *cold colors* to long ones with *warm colors*. For this snapshot, the range is from 0.06 to 1.68 W per m²

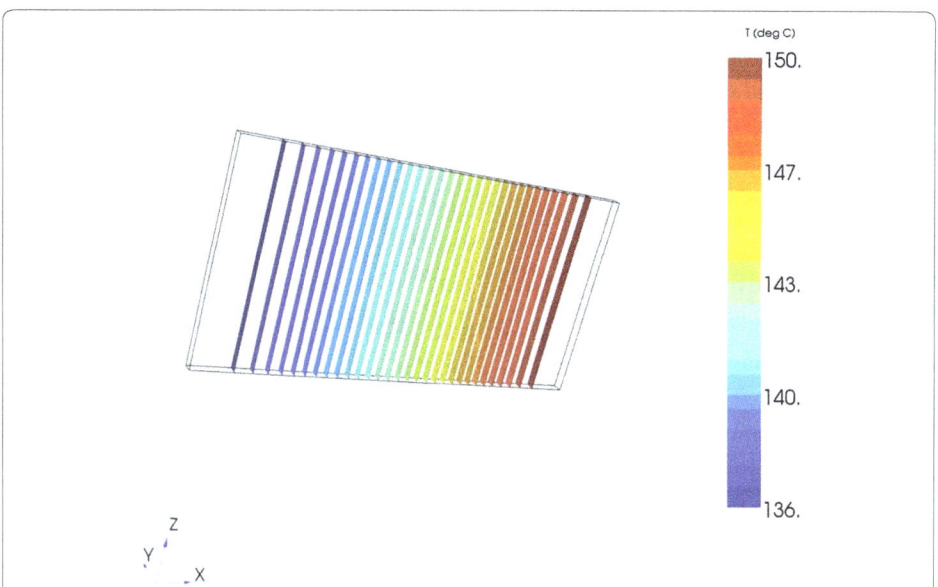

Fig. 8 Temperature profile for a model with $50 \times 1 \times 20$ grid blocks (50 blocks in the x-direction, 1 in the y- direction, and 20 in the z-direction) with each grid block having dimensions of 80 m $\times$ 100 m $\times$ 5 m with the vertical gradient of 18 ‰/km and 15° dip initialized without the bounding layers (plotted with the 20-fold exaggeration in z-direction). The temperature contour lines are not equally spaced and the range of temperatures is less than in the case with the bounding layers due to absence of heat influx from the bounding layers

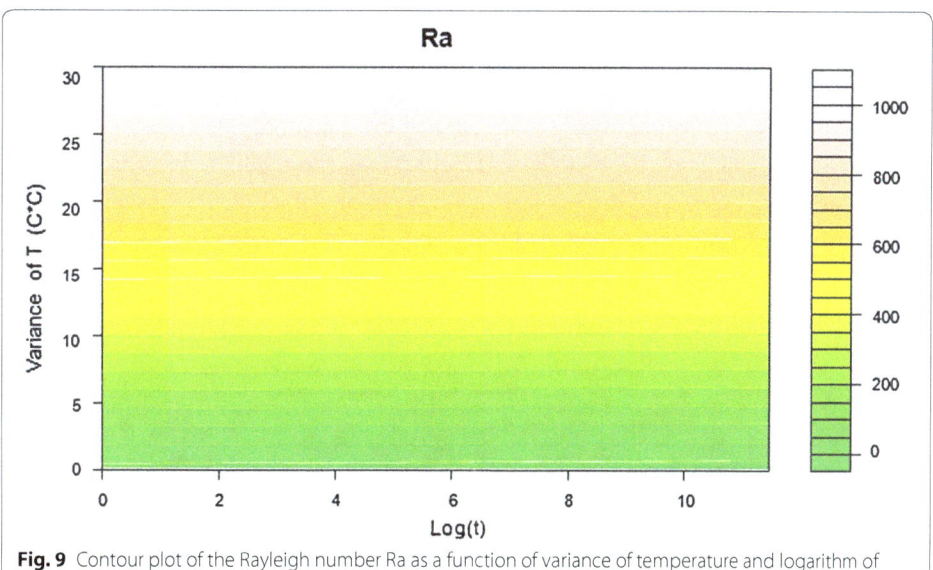

Fig. 9 Contour plot of the Rayleigh number Ra as a function of variance of temperature and logarithm of time. The higher Ra correspond to higher variances, reflecting the greater range of temperature in systems with high Rayleigh numbers. The *linear color scale* gives the Ra values

A stepwise regression run and a subsequent ANOVA test show that dip, convection, and their product are the most significant factors:

$$E = \beta_0 + \beta_D D + \beta_C C + \beta_{CD} CD \tag{3}$$

where C is a boolean variable indicating whether the simulation was run with natural convection initialization, D is dip in degrees, and E is gross energy output in joules; β's are the regression coefficients. Even though this model is a result of the analysis that tries to fit all possible combinations of the factors and outputs the best fit, the multiple R^2 is relatively low (0.57), indicating poor fit. Nevertheless, an ANOVA test confirms that the two factors identified by the stepwise regression are the most significant (Additional file 1: Appendix C).

Previously, we have established that one factor that affects energy output is initialization with natural convection. A subset of cases initialized without and with natural convection (no CO_2 added) was used to compute the difference in energy output, E. Prior to the analysis, the expectation was that flat systems would not be significantly affected by initialization. Tilted systems, however, should exhibit increasing difference in energy output. This should occur due to natural convection initialization that causes a wider temperature range and, thus, a higher enthalpy of the produced geofluid. The contour plot (Fig. 10) corroborated the expectation and previous research (Nield and Bejan 2006). Indeed, energy outputs from the systems with zero dips are virtually unaffected by the initialization (Eq. 4). Meanwhile, the upper portion of the contour plot corresponding to high dips shows relative difference in energy outputs from convection and no convection cases of about five percent. The simulation results suggest that cases with significant natural convection always recover more heat.

$$\Delta E = \frac{2(E_{conv} - E_{no})}{(E_{conv} + E_{no})} \tag{4}$$

Natural convection modeling in this study emphasizes the importance of the bounding layers and the quiescent period. Although the top and bottom bounding layers have the greatest effect on temperature profile in a quiescent system (due to the areal extent of these layers), it would be interesting to investigate the impact of side bounding layers in future research. One potential benefit of modeling a side bounding layer is the ability to incorporate a salt dome with its heat fluxes. However, introduction of additional bounding layers (or heat sources) will impact the duration of the quiescent period. Therefore, a

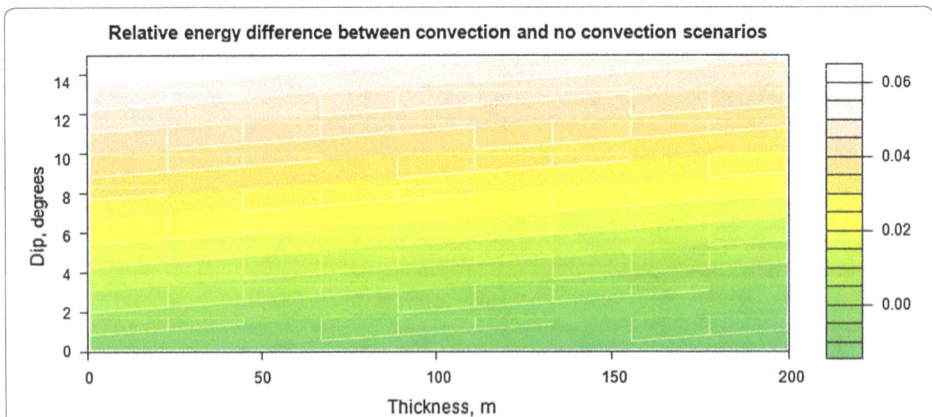

Fig. 10 Contour plot of relative increase in energy recovery for the cases initialized without natural convection compared to the cases with natural convection, 2(Econv − Eno)/(Econv + Eno). The systems are produced at 0.2 kg/s for 10 years. The systems with convection included always have a higher heat recovery. Dip has a greater impact on relative increase in recovery than thickness of a system, for the ranges of parameters considered. However, thinner systems are more affected by convection

more thorough analysis with a new experimental design might be necessary to establish the time span after which convection in such complex system becomes stable.

In addition to the initialization with natural convection, it is important to establish whether strategic placement of a horizontal CO_2 injection well has a positive impact on energy recovery. Based on previous theoretical and experimental research on CO_2 sequestration and the fact that the supercritical gas injector can be spatially isolated from the heat extraction well, we expect thermal energy recovery to be comparable or better than in cases without CO_2 injection (Farajzadeh et al. 2007; Kneafsey and Pruess 2010). Figure 11 confirms this expectation and demonstrates that differences in energy outputs (analogous to Eq. 4) are small but positive, indicating better performance of the cases with CO_2 injection. Here again, cases with higher dip yield higher energy recovery.

Discussion

Can CO_2 sequestration be done simultaneously with heat extraction, without impairing hear recovery? The simulation runs (Fig. 11) show that injection of small amounts of supercritical CO_2 away from the geofluid producer and injector is beneficial. Because water displacement (or forced convection) is the dominant component in coupled convection (natural and forced), we conclude that increased energy output in the cases with CO_2 injection is due to additional displacement rather than gas dissolution and subsequent density-driven convection. This conclusion, nevertheless, should not undermine further attempts to simultaneously harvest geothermal heat and sequester CO_2 with higher injection rates at which the mentioned effects might become more pronounced. The sequestration rate could be increased to match those in major CO_2 sequestration projects (NETL 2008). The choice of the low rate of 10^{-4} kg/s per meter of the horizontal well was dictated by the necessity to compare against the same production arrangements in different geologic systems (ten-fold difference in permeability, high and zero dips) and does not mean that the rate of 10^{-4} kg/s per meter of the horizontal well is the upper limit for each sedimentary geothermal aquifer. For 1000 md permeable and 200 m thick systems, the CO_2 injection rate could have been much higher, but would cause rapid

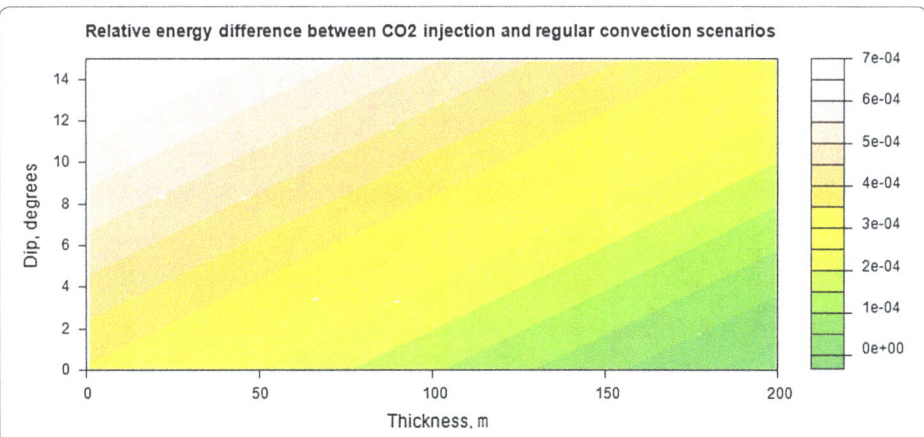

Fig. 11 Contour plot of relative increase in energy recovery for the cases with CO2 injection compared to regular production. The systems are produced at 0.2 kg/s for 10 years. CO2 injection rate is 10^{-4} kg/s in the 100 m out-of-plane thickness of the 2D model (10^{-6} kg/s m). Both dip and thickness influence the recovery. The change in recovery is small but positive; CO2 injection does not impair thermal recovery for these cases

pressure buildup in lower permeability reservoirs (and may lead to stability problems in equation of state). Therefore, the next step in research is to determine whether aquifers with thermodynamic and petrophysical properties favorable for CO_2 sequestration can also be prolific geothermal systems.

In addition to these findings, we established that efficient dynamic control over the supercritical gas plume can be established with judicious choice of the locations for both CO_2 injection and cold water injection wells. The visual evidence that the supercritical plume has not reached the upper portion of the reservoir and that the injected gas was driven downdip and mixed with the geofluid until very low concentrations is provided in Figs. 12, 13, 14. Figure 12 shows the onset of the rising gas plume after the first year of CO_2 injection. At this point, the cold re-injected water from the cold injector well has not reached the rising plume. Figure 13 gives the CO_2 concentration surface at the end of the simulation period. As we anticipated, the cold water helped drive the high concentration plume downdip while mixing and dispersing it. The cross-sectional planes of this CO_2 concentration surface in Fig. 14 further illustrate that the upper portion of the reservoir has very small concentrations of the supercritical CO_2. Thus, it is possible to prevent the uncontrolled rise of the gas plume, ensure caprock integrity, and still produce higher thermal energy output.

Conclusions

In this study, we conducted a numerical investigation of the effects of the coupled convection and CO_2 injection on heat extraction from sedimentary geothermal aquifers. The analysis showed that there were certain benefits in characterizing natural convection pattern prior to heat harvesting, because the knowledge of the convection pattern

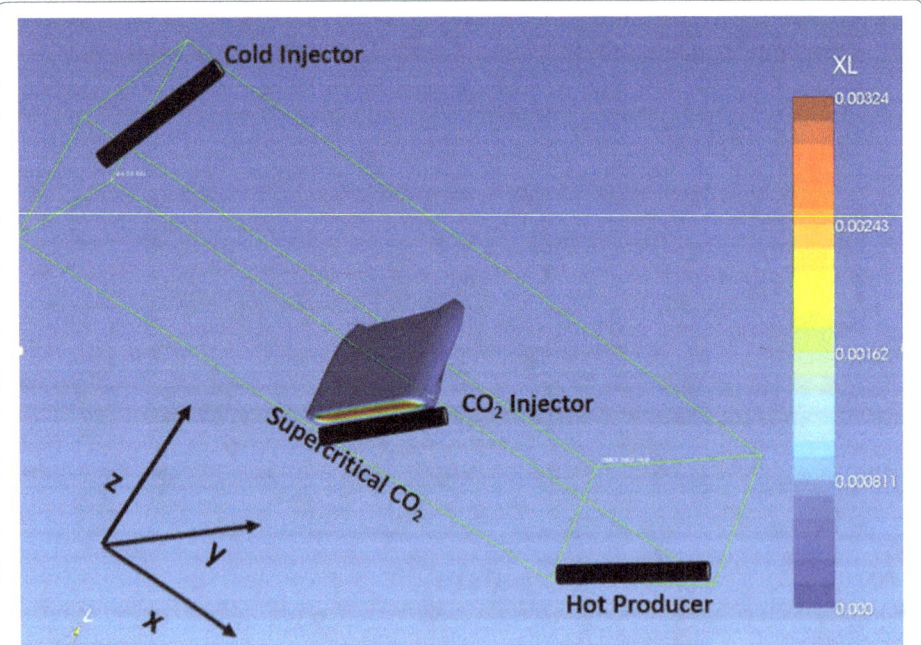

Fig. 12 Fraction of the supercritical CO_2 after 1 year of gas injection. The view from the *bottom* of the reservoir (visualized with PetraSim)

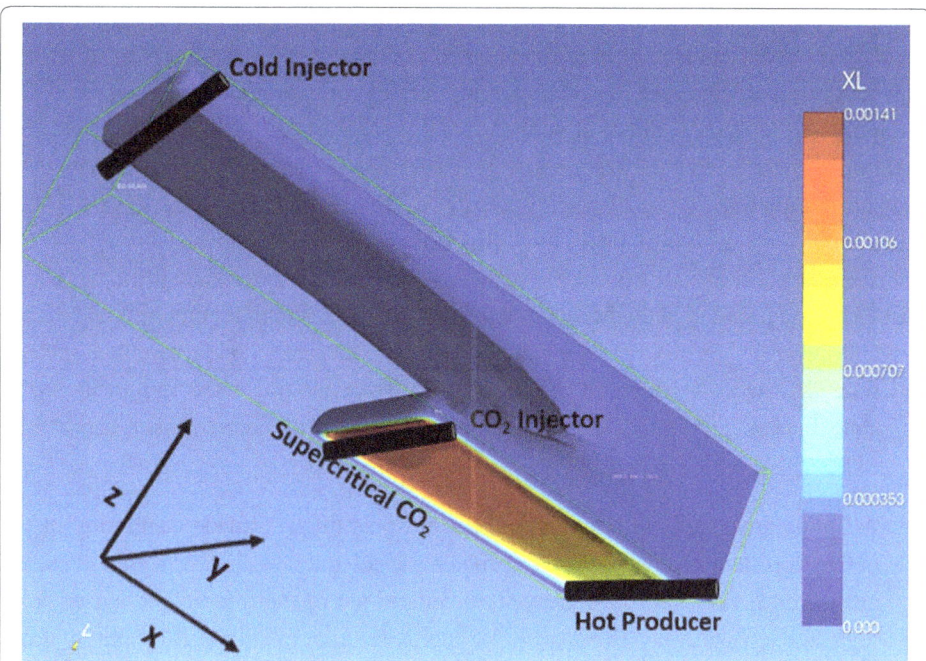

Fig. 13 Fraction of the supercritical CO_2 after 20 year of gas injection. The view from the *bottom* of the reservoir (visualized with PetraSim)

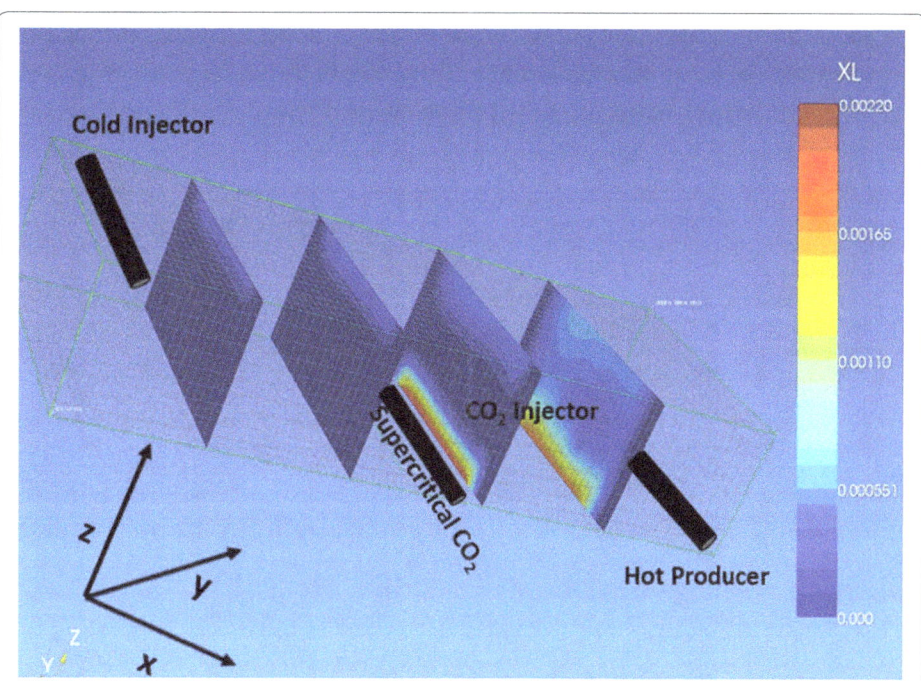

Fig. 14 Fraction of the supercritical CO_2 after 20 year of gas injection. The view cross-sectional planes from the side of the reservoir. From the scale it is evident that the plume has been dispersed and only small concentration of the supercritical gas reaches the upper portion of the reservoir (visualized with PetraSim)

allows to compare meaningfully alternative production designs. Statistical analysis of the simulation results confirmed the expectation that dip controlled the intensity of natural convection and aided forced convection at moderate production and injection rates. The juxtaposition between simulation suites with and without CO_2 injection revealed that injection of the supercritical greenhouse gas had a positive impact on the ultimate thermal energy recovery.

This study suggests several interesting directions for future research.

1. CO_2 sequestration at higher gas injection rates with simultaneous heat harvesting and dynamic control over the gas plume might further increase revenue. Because both CO2 sequestration and geothermal aquifer development are marginally profitable, this approach might make the combination project more commercially attractive.

2. A comparative study of alternative software tools might provide calibration of the obtained results and resolution for problems involving heat fluxes. In this study, the analysis of heat fluxes to the wellbore or in and out of the reservoir was used sparingly and qualitatively. The reason for this is limitations imposed by output from TOUGH2 software that does not separate conduction, convection, and radiation. It would be particularly helpful to have such capability for estimation of heat fluxes from bounding layers.

3. One can envision an investigation of effects of non-uniform salinity and heat sources due to the presence of salt domes on the natural convection pattern. Thermohaline convection is an important factor in heat transfer in the GOM coast environment that might have an impact on recovery of geothermal heat.

Author details
[1] Department of Petroleum Engineering, Texas A&M University, 907 Richardson Hall, College Station, TX 77843, USA.
[2] Department of Earth and Environmental Sciences, University of Tulane, Room 200 Blessey Hall, New Orleans, LA 70118, USA.

Acknowledgements
Professor Jeffrey Nunn and Tyler Gray (MS) of the Department of Geology and Geophysics, Louisiana State University, characterized the Camerina A reservoir, providing valuable data and guidance. Professor Jeffrey Hanor of the Department of Geology and Geophysics, Louisiana State University, advised on thermohaline flows in the Gulf of Mexico. Financial support for research assistant stipends, travel, training, and computing hardware was provided by the Chevron Distinguished Professorship in the College of Engineering, Louisiana State University. Additional support for purchase of software was provided by the Coast to Cosmos Focus Area at the Center for Computation and Technology, Louisiana State University.

References
Adams B, Kuehn T, Bielicki J, Randolph J, Saar M. On the importance of the thermosiphon effect in CPG (CO$_2$ plume geothermal) power systems. Energy. 2014;69:409–18.
Anchliya A. Aquifer management for CO2 sequestration. Master's thesis. Texas A&M University. 2009.
Curtice R, Dalrymple E. Just the cost of doing business. World Oil. 2004;225(10):77–8.
Esposito A, Augustine C. Geopressured geothermal resource and recoverable energy estimate for the Wilcox and Frio formations, Texas. GRC Transactions. 2011;35:1563–71.
Ewing T, Light P, Tyler N. Thermal and Diagenetic History of the Pleasant Bayou—Chocolate Bayou Area, Brazoria County Texas. Gulf Coast Association of Geological Societies Transactions. 1984;34:341–8.
Farajzadeh R, Salimi H, Zitha P, Bruining J. Numerical simulation of density-driven natural convection in porous media with application for CO$_2$ injection projects. Int J Heat Mass Transf. 2007;50(25–26):5054–64.

Freifeld B, Zakim S, Pan L, Cutright B, Sheu M, Doughty C, Held T. Geothermal energy production coupled with CCS: a field demonstration at the SECARB Cranfield Site, Cranfield, Mississippi, USA. Energy Procedia. 2013;37:6595–603.

Ganjdanesh R, Pope G, Sepehmoori K. Production of energy from saline aquifers: a Method to offset the energy cost of carbon capture and storage. Int J Greenhouse Gas Control. 2015;34:97–105.

Gray T. Geothermal resource assessment of the Gueydan salt dome and the adjacent Southeast Guyedan Field, Vermilion Parish, Louisiana. MS Thesis, Louisiana State University. 2010.

Goodman A, Hakala A, Bromhal G, Deel D, Rodosta T, Frailey S, Small M, Allen D, Romanov V, Fazio J, Huerta N, McIntyre D, Kutchko B, Guthrie G. U.S. DOE methodology for the development of geologic storage potential for carbon dioxide at the national and regional scale. Int J Greenhouse Gas Control. 2011;5:952–65.

Gustavson T, Kreitler C. 1979. An Environmental overview of geopressured-geothermal development: Texas Gulf Coast. Technical report, Bureau of economic geology, The University of Texas at Austin. DOE No. 7949703. http://igor.beg.utexas.edu/readingroom/fulltext.aspx?ID=74077. Accessed 11 Nov 2015.

Hanor J. Kilometer-scale thermohaline overturn of pore fluid in the Louisiana Gulf Coast. Nature. 1987;327:502–3.

Horne R. Transient effects in geothermal convective systems. Dissertation, University of Auckland, New Zealand. 1975.

Islam A, Sharif M, Carlson E. Numerical investigation of double diffusive natural convection of CO_2 in a brine saturated geothermal reservoir. Geothermics. 2013;48:101–11.

Karimnezhad M, Jalalifar H, Kamari M. Investigation of caprock integrity for CO_2 sequestration in an oil reservoir using numerical method. J Natural Gas Sci Eng. 2014;21:1127–37.

Kneafsey T, Pruess K. Laboratory flow experiments for visualizing carbon dioxide-induced, density-driven brine convection. Transp Porous Media. 2010;82:123–39.

MIT. The future of geothermal energy: impact of enhanced geothermal systems (EGS) on the United States in the 21st century. Technical Report, Renewable Energy and Power Department, Idaho National Laboratory. 2006. https://mitei.mit.edu/system/files/geothermal-energy-full.pdf. Accessed 13 July 2015.

NETL. Carbon Sequestration Atlas of the United States and Canada. Technical. 2008.

Nicot J. Evaluation of large-scale CO_2 storage on fresh-water sections of aquifers: an example from the Texas Gulf Coast Basin. Int J Greenhouse Gas Control. 2008;2:582–93.

Nield D, Bejan A. Convection in Porous Media. New York: Springer Science + Business Media Inc; 2006.

Pool M, Carrera J, Vilarrasa V, Silva O, Ayora C. Dynamics and design of systems of injecting dissolved CO_2. Adv Water Resour. 2013;62:533–42.

Plaksina T. Modeling effects of coupled convection and CO_2 injection in stimulating geopressured geothermal reservoirs. MS thesis, Louisiana State University. 2011.

Pruess K, Oldenburg C, Moridis G. TOUGH2 User's Guide, Version 2.0, Report LBNL-43134, Lawrence Berkeley National Laboratory, Berkeley, Calif. (Superseded by Pruess et al. 2012). 1999.

R Team. Development core. R: a language and environment for statistical computing. Vienna, Austria: R Foundation for Statistical Computing. 2013. http://web.mit.edu/r_v3.0.1/fullrefman.pdf. Accessed 13 July 2015.

Randolph J, Saar M. Coupling carbon dioxide sequestration with geothermal energy capture in naturally permeable, porous geologic formations: implications for CO_2 sequestration. Energy Procedia. 2011;4:2206–13.

Randolph J, Saar M, Bielicki J. Geothermal energy production at geologic CO_2 sequestration sites: impact of thermal drawdown on reservoir pressure. Energy Procedia. 2013;37:6625–35.

Salimi H, Wolf K. Integration of heat-energy recovery and carbon sequestration. Int J Greenhouse Gas Control. 2012;6:56–68.

Shukla R, Ranjith P, Haque A, Choi X. A review of studies on CO_2 Sequestration and caprock integrity. Fuel. 2010;89:2651–64.

Singh A, Goerke U, Kolditz O. Numerical simulation of non-isothermal compositional gas flow: application to carbon dioxide injection into gas reservoirs. Energy. 2011;36:3446–58.

Smith D, Reeve F. Salt piercement in shallow Gulf Coast salt structures. AAPG Bull. 1970;54(7):1271–89.

Szalkowski S, Hanor J. Spatial variations in the salinity of produced waters from Southwestern Louisiana. GCAGS/GCSSEPM Transactions. 2003;53:798–806.

Vilarrasa V, Silva O, Carrera J, Olivella S. Liquid CO_2 injection for geological storage in deep saline aquifers. Int J Greenhouse Gas Control. 2013;14:84–96.

Wang J, Ju Y, Gao F, Liu J. A simple approach for the estimation of CO_2 penetration depth into a caprock layer. J Rock Mech Geotechnical Eng. 2015;. doi:10.1016/j.jrmge.2015.10.002.

Warwick P, Verma M, Freeman P, Corum M, Hickman S. US Geological survey carbon sequestration—geologic research and assessment. Energy Procedia. 2014;63:5305–9.

White D, Williams D. Assessment of geothermal resource of the United States. Technical Report, US Geological Survey (USGS) Circular 726. 1975. http://pubs.usgs.gov/circ/1975/0726/report.pdf. Accessed 13 July 2015.

Zhang K, Moridis G, Pruess K. TOUGH + CO_2: a multiphase fluid-flow simulator for CO_2 geologic sequestration in saline aquifers. Comput Geosci. 2011;37:714–23.

Zhang L, Ezekiel J, Li D, Pei J, Ren S. Potential assessment of CO_2 injection for heat mining and geological storage in geothermal reservoirs of China. Appl Energy. 2014;122:237–46.

2

Geothermal exploration in a sedimentary basin: new continuous temperature data and physical rock properties from northern Oman

Felina Schütz[1][*] [iD], Gerd Winterleitner[1,2] and Ernst Huenges[1]

*Correspondence:
fschuetz@gfz-potsdam.de
[1] Section 6.2 Geothermal
Energy Systems, Helmholtz
Centre Potsdam-GFZ
German Research
Centre for Geoscience,
Telegrafenberg,
14473 Potsdam, Germany
Full list of author information
is available at the end of the
article

Abstract

The lateral and vertical temperature distribution in Oman is so far only poorly understood, particularly in the area between Muscat and the Batinah coast, which is the area of this study and which is composed of Cenozoic sediments developed as part of a foreland basin of the Makran Thrust Zone. Temperature logs (T-logs) were run and physical rock properties of the sediments were analyzed to understand the temperature distribution, thermal and hydraulic properties, and heat-transport processes within the sedimentary cover of northern Oman. An advective component is evident in the otherwise conduction-dominated geothermal play system, and is caused by both topography and density driven flow. Calculated temperature gradients (T-gradients) in two wells that represent conductive conditions are 18.7 and 19.5 °C km^{-1}, corresponding to about 70–90 °C at 2000–3000 m depth. This indicates a geothermal potential that can be used for energy intensive applications like cooling or water desalinization. Sedimentation in the foreland basin was initiated after the obduction of the Semail Ophiolite in the late Campanian, and reflects the complex history of alternating periods of transgressive and regressive sequences with erosion of the Oman Mountains. Thermal and hydraulic parameters were analyzed of the basin's heterogeneous clastic and carbonate sedimentary sequence. Surface heat-flow values of 46.4 and 47.9 mW m^{-2} were calculated from the T-logs and calculated thermal conductivity values in two wells. The results of this study serve as a starting point for assessing different geothermal applications that may be suitable for northern Oman.

Keywords: Continuous temperature logging, Physical rock properties, Sedimentary basin, Geothermal applications in Oman

Introduction

Background

Geothermal systems currently under exploitation are found in a number of geological environments, where temperatures and depths of the reservoirs vary accordingly. Many high-temperature (> 180 °C) hydrothermal systems are associated with recent volcanic activity and are found near plate tectonic boundaries (e.g., subduction zones, rift systems, oceanic spreading centers or transform margins), or at crustal and mantle hot spot anomalies. Such systems cannot be expected in Oman, but do occur in other parts of the Arabian Peninsula. In Saudi Arabia and Yemen several active volcanos are associated with the Afar Plume and the recent opening of the Red Sea (Chang and Van der Lee 2011). Intermediate- (100–180 °C) and low-temperature (< 100 °C) systems in continental settings are either linked to above-normal heat production and increased terrestrial heat flow through radioactive isotope decay, or to aquifers charged by hot water originating from circulation along deep (crustal-scale) fault zones (Huenges 2010).

Our study focused on the sedimentary infill of the foreland basin north of the Oman Mountains, which consists of Cenozoic sediments. The main aim of this study is to understand the temperature distribution in the subsurface and the physical rock properties of the carbonate and clastic sediments that accumulated after the obduction of the Semail Ophiolite complex during the late Cretaceous time. We evaluated temperature logs (T-logs) in terms of advective and conductive components to determine a geothermal gradient representative for the area. We further derived heat flow values for two well locations and used these data to calculate the temperature at depth as addition to the well logs.

Status quo of geothermal use in Oman and its surroundings

Geothermal energy is currently not represented in the national energy mix of Oman and only plays a very minor role in the Arabian Peninsula in general. In the whole region a large percentage of the energy consumption is used for cooling. In Oman 50% of the residential power supply is only used for residential cooling (Sweetnam et al. 2014) in Saudi Arabia even 80% and the average for the whole Gulf Cooperation Council (GCC), consisting of Bahrain, Kuwait, Oman, Qatar, Saudi Arabia, and the United Arab Emirates, is 40–50% (Lashin et al. 2015). The electricity demand will rise in the GCC by 7–8% per year on average; in the smallest and fastest-growing economies, demand will grow even faster (GCC 2010). The electricity sector in Oman is primarily based on natural gas (97.5%) and diesel (2.5%, Authority for Electricity Regulation, Oman). The same applies to the other GCC countries, which all fall in the top 25 countries of carbon dioxide emissions per capita (Reiche 2010). Geothermal energy could significantly contribute to a greener energy mix, especially in countries in possession of high-temperature and intermediate temperature hydrothermal systems, like Saudi Arabia and the Yemen. The usage of thermally driven cooling could significantly help to lower the energy demand in this sector. Thermally driven cooling requires temperatures of 70–100 °C. The energy for sustainable cooling supply can be developed from solar and geothermal sources. Solar heat supply is fluctuating, whereas geothermal heat can provide base load heat supply. The results of this study show that the target temperatures can be reached in around 2000 m depth. Fault structures which exist in the contact zone between the Tertiary sedimentary sequence and the Semail Ophiolite, correlated with hot springs, constitute a potential setting for drilling.

Research collaboration and implication of the study

The present study is part of a research collaboration between The Research Council (TRC) of the Sultanate of Oman and the German Research Centre for Geosciences (GFZ), which seeks to develop a cooling system driven by renewables. Both geothermal and solar energy are considered as potential options to deliver the hot water required to drive an absorption chiller. A prototype of this cooling system will be installed at the study site 40 km west of Muscat to cool one institute building of the TRC (Fig. 1). The temperature distribution with depth is an important parameter for evaluating the potential of geothermal energy as driving heat for an absorption chiller and is also required for the planning of other shallow geothermal installations. One example of a shallow geothermal system is underground storage where energy, in the form of hot or cold water, is temporarily stored in the subsurface and recovered when needed. These systems are increasingly used in Europe, and are also of interest for countries in the sun-belt, especially in combination with solar, wind or any other existing fluctuating energy source to utilize excess and surplus energies as auxiliary energy back-up systems during peak demand times. A further challenge of the region is the rejection of process waste heat during hot summer months. The ambient air temperature of Oman is too high to efficiently reject the heat to the surrounding environment. The conventional method in a climate like Muscat would be to use a cooling tower, but they become inefficient when the ambient temperature and humidity rise up in summer. A more energy-efficient approach is to inject the process waste heat to the subsurface, where conditions are cooler and more stable. Over the course of the year, the ground temperature only varies

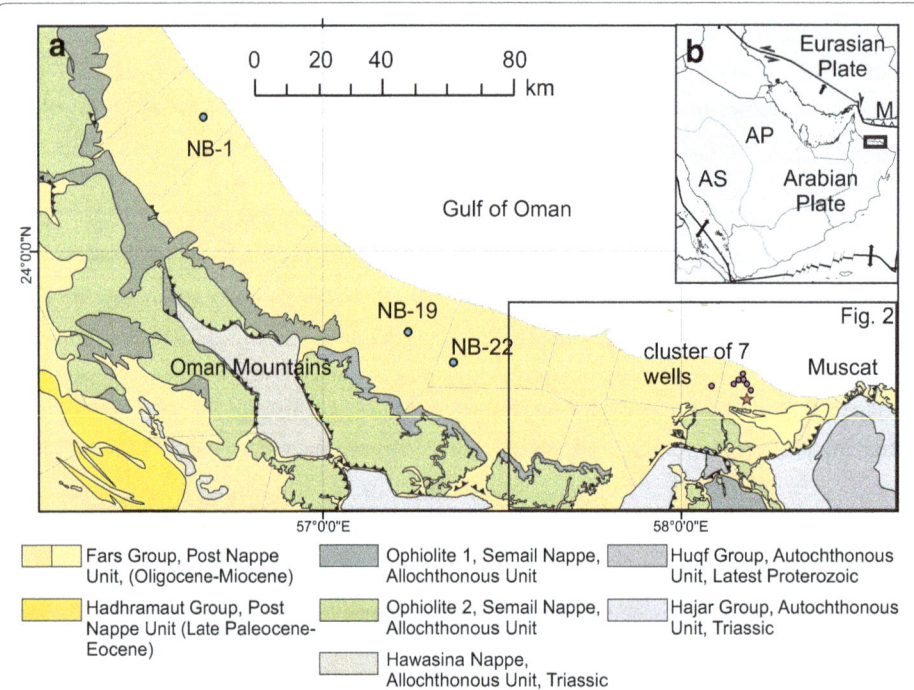

Fig. 1 Inset map is a general map of the Arabian Plate. *AS* Arabian Shield, *AP* Arabian Platform, *M* Makran Thrust Zone [modified from Mahmoud et al. (2005), Stern and Johnson (2010), Johnson et al. (2011)]. Main map is a geological map of the study area, with the major geological units (modified from NF40-03, Seeb sheet, 1992, Sultanate of Oman, Ministry of Petroleum and Minerals). Blue dots—wells, of which T-logs were provided by the MRMWR. Purple dots—wells in which new T-logs were recorded. Red star—site, where a thermally driven cooling system will be installed. Black triangle—major thrust fault [modified from NF40-03, Seeb sheet, 1992, Sultanate of Oman, Ministry of Petroleum and Minerals, Fournier et al. (2006)]. Black rectangle gives the outline of Fig. 2

slightly at a fixed depth and during the hot summer months when temperature in Muscat can rise to 47 °C (Directorate General of Meteorology, Oman, PACA) the subsurface temperature is significantly lower. It is therefore possible to reject the process waste heat into intermediate aquifers, where minimum temperatures were measured. To plan and simulate such an application the temperature gradient (T-gradient) and the physical rock properties of the underground are required. The data generated in this study serve as background for further analysis of an intelligent utilization of the subsurface in combination with a thermally driven cooling system at the study site, 40 km west of Muscat.

Geological setting

Oman is located at the south-eastern margin of the Arabian Plate, which can be divided into the Arabian Shield and the Arabian Platform. The Arabian Shield comprises juvenile continental crust of mainly Neoproterozoic age and is partly covered by younger sediments (Stern and Johnson 2010). The Arabian Platform, in contrast, is mostly covered by Phanerozoic sediments, which dominate the western part of the Arabian Plate (see inset of Fig. 1). The Phanerozoic cover is up to 10 km thick and defines the large stable platform of eastern Arabia. During late Cretaceous times a subduction zone and island arc complex developed in the Tethyan Ocean, whose remnants now constitute the Gulf of Oman. The Arabian Plate moved progressively northwards and early phases of collision between the Arabian continent and the Iranian block (Eurasian Plate) started. During the subduction process a thrust slice of the Tethys Ocean floor was obducted on the NE margin of the Arabian Platform, to form the Semail Ophiolite, which nowadays represent the Oman Mountains (Alsharhan 2014).

Five tectonic sequences are recognized in Oman. They are comprised of the following units: (1) Pre-Permian (Huqf Group), (2) the Hajar Supergroup (Permian to Cretaceous sediments), (3) the Allochthonous Sequence (Semail Nappe and Hawasina Nappe), (4) the Hadhramaut Group, and (5) the Fars Group. The latter two are part of the neoautochthonous sedimentary cover (Al-Lazki and Barazangi 2002).

The ophiolite sequence is divided into a unit of cumulate layered gabbro (Ophiolite 1), and a unit composed of harzburgite with minor lherzolite and dunite (Ophiolite 2). The Hadhramaut Group and the Fars Group are an assemblage of late Maastrichtian and Tertiary sediments, unconformably deposited on top of the allochthonous units (Stern and Johnson 2010) and forming the neoautochthonous sedimentary cover (termed post nappe unit in Fig. 1). Our study area is located in the Al Khawd Fan aquifer system (see Fig. 2), which is part of the neoautochthonous sedimentary cover between Muscat and the Batinah coast and thus part of the convergence zone between the Arabian and Eurasian Plates. In the following paragraphs the stratigraphic and structural characteristics of the area are described.

Post-obduction sequences

Development of the foreland basin started with the obduction of the Semail Ophiolite complex onto the Arabian Platform during the late Campanian. The first sediments deposited in the evolving basin were siliciclastic sediments originating from the unroofing of the emergent proto-Oman Mountains (Al Khawd Formation). They unconformably rest on the ophiolitic basement and are composed of fluviatile,

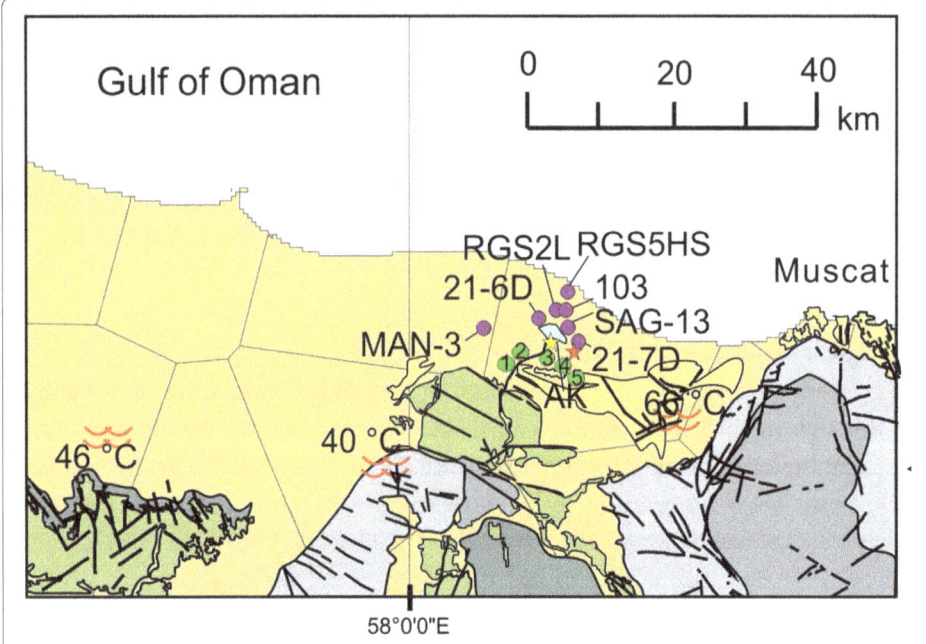

Fig. 2 Purple dots—location of the wells, where new continuous T-logs were recorded. Red waves—location of hot springs and temperature (data provided by the MRMWR). Green dots—location of outcrop sampling of the Barzaman Fm., Seeb Fm. and Oligocene carbonates (Ma'am reefs). AK—Al Khawd dam. Red star—site, where a thermally driven cooling system will be installed. Yellow star—well field with bottom-hole temperature (BHT) data (cf. Fig. 4). The geological units are described in Fig. 1. Black lines—major faults [modified from NF40-03, Seeb sheet, 1992, Sultanate of Oman, Ministry of Petroleum and Minerals; Kusky et al. (2005), Fournier et al. (2006)]

and localised beach to fan delta deposits. The boundary to the overlying Simsima (shallow shelf carbonate) Formation is described for the area as conformable and elsewhere a non-sequence is present (Nolan et al. 1990). The late Palaeocene to early-Eocene shallow-marine carbonates of the Jafnayn Formation were deposited on top of a major regional unconformity, separating Paleogene and Cretaceous strata in the central Oman Mountains (Özcan et al. 2015; Tomás et al. 2016). Subsequently, the shale and fine-grained limestones of the Rusayl Formation (early-Eocene age) were deposited in a fluvial and lagoonal environment (Dill et al. 2007). A marine transgression during the middle Eocene resulted in the development of a carbonate platform which is represented in the stratigraphic succession as the Seeb Formation (Fig. 3). Sediments were deposited on a storm-influenced carbonate ramp with well-defined inner, mid to outer ramp facies belts, and a general fining-upward trend. Deposition of the Seeb carbonates terminated at the end of the Eocene with the emersion of the Arabian Platform due to the opening of the Red Sea. Marine carbonate-dominated sedimentation resumed during the Oligocene (Ma'am platform reefs) followed by the mixed carbonate siliciclastic Barzaman Formation which developed as an alluvial fan system during Miocene to Pliocene times (Fig. 3). The latter started with the emergence and subsequent erosion of the Oman Mountains (Al-Lazki and Barazangi 2002). Nolan et al. (1990) described the general stratigraphy and depositional systems of the post-obduction sediments accumulated in the Batinah coast area in detail.

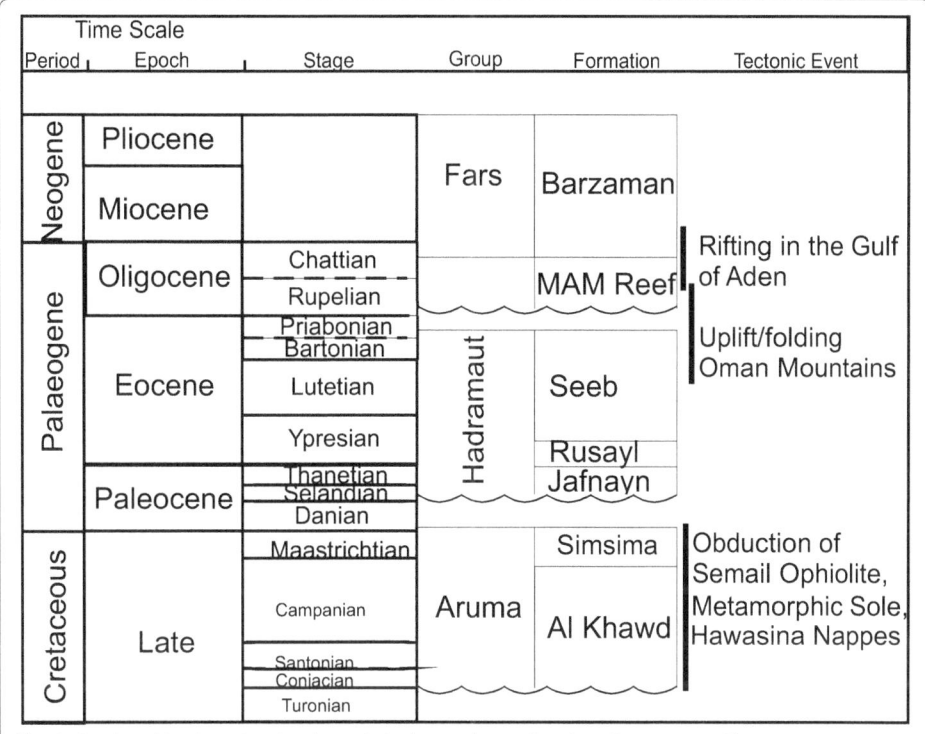

Fig. 3 Stratigraphic chart showing the units in the study area from late Cretaceous to Pliocene time [modified from Al-Lazki et al. (2002), Breton et al. (2004), Al-Husseini (2008)]

Structural characteristics

Faults, fracture zones and karst zones in carbonates are of special interest for geothermal exploitation. In the Batinah coastal plain, compression resumed after a late Cretaceous to Paleogene extensional phase that followed obduction of the ophiolite. The extensional phase caused large-scale normal faulting in the northeast Oman margin. Extensional tectonics were recognized from the observation of normal faults displacing the post-nappes sedimentary deposits against the autochthonous series or the ophiolites; for example at the Batinah coast and the Rusayl Embayment (Fournier et al. 2006). Convergence between the north Oman margin and the Zagros–Makran appears to have been re-established by the Eocene (Coleman 1981; Mann et al. 1990). The autochthonous sedimentary cover, allochthonous nappe complex, and the neoautochthonous sedimentary cover were further affected in the axial zone and the foreland basin of the Oman Mountains by large-scale folding, short-distance thrusting, and uplift (Searle et al. 1983; Michard et al. 1984; Searle 1985; Poupeau et al. 1998; Mount et al. 1998; Al-Lazki and Barazangi 2002; Fournier et al. 2006). Their broad structural style indicates that shortening was minor. Hence, the onset of compression in the Oman Mountains was either synchronous with late Oligocene to early Miocene rifting in the Gulf of Aden, or started immediately thereafter. Compressional deformation continued until the Pliocene and is recorded in the Mio-Pliocene deposits of the Barzaman Formation (Fournier et al. 2006).

Methods

Physical rock properties

Due to the lack of suitable core material we collected outcrop samples of the sedimentary succession in a field campaign (see Fig. 2 for location details). The outcrop samples might be affected by some weathering or diagenetic overprinting and therefore it cannot be guaranteed that they represent conditions of the subsurface rocks. The majority of outcrops are located close to the Al Khawd dam and only two samples of the Barzaman Formation were taken at the southern foothills of the Oman Mountains (Lat. 22.3041702; Long. 57.8264319), where the former alluvial stream deposits form complex systems of 'raised' or upstanding, sinuous, superimposed linear ridges and broad gravel sheets (Maizels 1987). To correlate T-logs with physical rock properties, it was necessary to obtain rock samples from the same stratigraphic intervals. Therefore, our sampling strategy focused on the sediments encompassing the uppermost 700 m of the succession. In total we took 25 samples of the Barzaman Formation, the Ma'am reef sequence and the Seeb Formation. The thermal conductivity, effective porosity, permeability, and matrix density of each sample was determined. The rock thermal conductivity was measured with a thermal conductivity scanner (TCS, optical scanning technique). The underlying technique of the TCS has been described in detail by Popov et al. (1999). Samples were dried in an oven at 80 °C for 24 h before the thermal conductivity measurements were performed. A black acryl lacquer line was drawn on the polished and plain sample surface to avoid different mineral colors affecting the measurement. Additionally, we measured the thermal conductivity of water-saturated samples. For this, samples were saturated under vacuum in a desiccator for 3 days. The optical scanning technique is a transient method, which yields a continuous profile of thermal conductivity along the core axis of the sample. The mean thermal conductivity is determined as the arithmetic mean of all conductivities measured along the scanning line. To determine anisotropy measurements were made parallel and perpendicular to bedding planes, where developed.

Effective porosity and density of the rock samples were analyzed using the Archimedes method, which is based on the principle of mass displacement. Additionally, plugs of 25 mm diameter and 50 mm length were drilled for bulk permeability measurements. This was done using a conventional gas permeameter, and the results were adjusted with the Klinkenberg-correction method (Milsch et al. 2011), a procedure described in detail by Tanikawa and Shimamoto (2009). The sampled rocks comprise a range of different lithologies including: conglomerate, sandstone, limestone, mudstone, and siltstone and thus cover the major lithotypes comprising the shallow sedimentary cover of the Batinah coast.

Temperature data

We analyzed temperature measurements of three wells (NB-1; NB-19; NB-22) located along the Batinah coast, to get a first approximation of the temperature pattern of the study area. The data were kindly provided by the Ministry of Regional Municipalities and Water Resources (MRMWR). These exploration wells are between 70 and 316 m

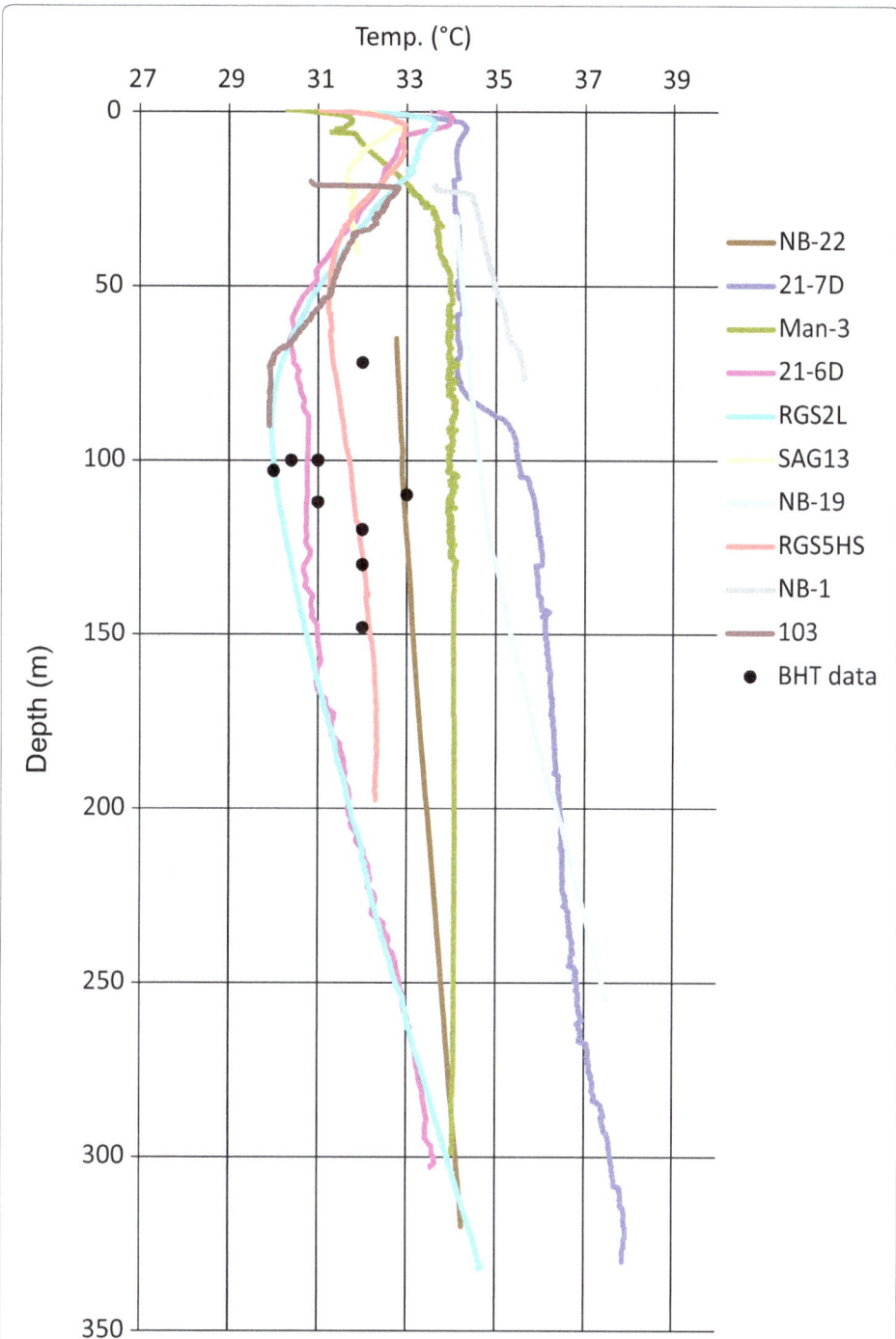

Fig. 4 Continuous T-logs measured with a conventional temperature probe in seven wells during this study. Additionally three continuous T-logs (NB-1; NB-19; NB-22) and bottom-hole temperatures (BHT), provided by the MRMWR, measured in a cluster of nine wells close to the Al Khawd dam (for location see Figs. 1, 2)

deep (cf. Fig. 4) and were drilled between 2005 and 2006 to evaluate the thickness, yield potential and water quality of the coastal aquifer. Additionally, litho-stratigraphic and gamma ray logs were available for these three wells.

Our field campaign aimed to acquire additional high precision temperature data of existing water wells. The MRMWR provided a database with more than 300 boreholes, which were screened and evaluated as potential wells for temperature logging. The wells spread over the entire study area, but the deepest wells cluster around the Al Khawd dam, a recharge dam built in 1982 (see Fig. 2). The wells are mainly shallow (< 100 m), only nine wells are deeper than 300 m and 35 wells are between 100 and 300 m. The wells exceeding 300 m in total depth were the main target for the temperature logging field campaign. Finally, seven wells were measured, five of which had a depth between 193 and 330 m, and two of them were only accessible to a depth of 90 and 45 m, respectively (Table 1). To identify the intervals in the T-logs that were measured above and below the water table it was important to know in advance in which depth the water level can be expected. For several wells the monitoring data of the water table were provided by the MRMWR. The data indicated that the water table around the Al Khawd dam varies between 20 and 30 m below ground level. The logging equipment consisted of a cable (500 m length), a conventional temperature probe and a mobile winch. The temperature probe was lowered to the borehole with the winch and lifted with a constant velocity of 3 m min^{-1}. Thereby it measured a temperature point every 10 cm, resulting in a spatial continuous T-log. The slow speed guaranteed that turbulences affecting the measurements were kept as low as possible. Subsequently, the T-logs were smoothed with a 10 m running average (arithmetic mean). The results are plotted for all wells (temperature vs. depth) in Fig. 4 in combination with the data provided by the MRMWR. All ten wells are either exploration or monitoring wells, and are not used to produce water.

Temperature gradient and surface heat flow

For six wells litho-stratigraphic logs were available (NB-1, NB-19, NB-22, 21-6D, 21-7D, RGS2-L), and based on the T-logs we calculated the T-gradient for these six wells (cf. Fig. 5). A mean T-gradient is given for each of the well. For the calculation the negative T-gradient values of the shallow intervals were disregarded, as they do not represent steady state conditions. The T-gradient is plotted next to the litho-stratigraphic log to correlate changes in the lithological composition with changes in the T-gradient. If available gamma ray logs were also plotted. The T-logs from the two deepest wells, which show no significant advective component (NB-19, RGS2-L), were chosen for analysis of

Table 1 Coordinates, completion date and total depth of wells, which were used for T-logging and wells of which T-logs were provided by the MRMWR (NB-1, NB-22, NB-19)

ID	Coord. Y WGS	Coord. X WGS	Completion date	Total depth (m)
21-7D	23.6249	58.1887	1995-04-10	326
RGS-2L	23.6487	58.1659	1984-09-15	330
MAN-3	23.6306	58.0889	2012-01-12	300
21-6D	23.6385	58.1523	1995-04-10	300
RGS-5HS	23.6608	58.1747	1985-02-01	193
103	23.5804	58.1716	Unknown	90
SAG-13	23.6338	58.1804	1993-12-14	45
NB-1	24.7227	56.4159	2005-10-01	70
NB-22	23.6969	57.3677	2006-06-03	316
NB-19	23.7814	57.2416	2006-05-03	256

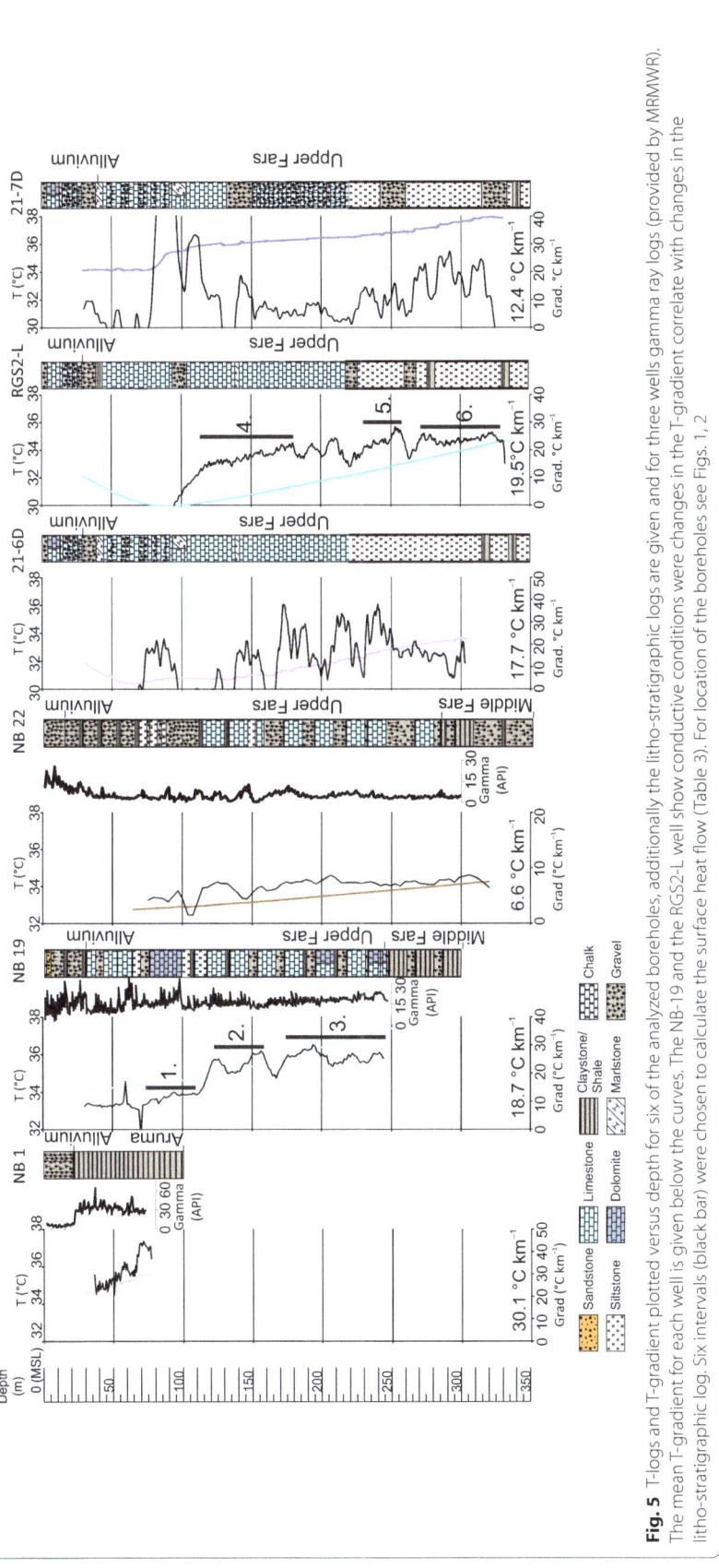

Fig. 5 T-logs and T-gradient plotted versus depth for six of the analyzed boreholes, additionally the litho-stratigraphic logs are given and for three wells gamma ray logs (provided by MRMWR). The mean T-gradient for each well is given below the curves. The NB-19 and the RGS2-L well show conductive conditions were changes in the T-gradient correlate with changes in the litho-stratigraphic log. Six intervals (black bar) were chosen to calculate the surface heat flow (Table 3). For location of the boreholes see Figs. 1, 2

the surface heat flow. This facilitates the overall understanding of the conductive temperature pattern of the region. For this reason, the interval method was applied. This method is described by Powell et al. (1988) and has been applied successfully in recent heat flow studies (Förster et al. 2007; Schütz et al. 2014). The underlying principle is to identify intervals where changes in the T-gradient are clearly related to changes in the lithology, and hence changes in the thermal conductivity of the interval. The heat flow is defined by the Fourier equation:

$$q = -\lambda \cdot \Gamma, \tag{1}$$

where λ is the thermal conductivity (W m^{-1} K^{-1}) and Γ is the T-gradient (K km^{-1}) for the interval. For the two well locations we calculated mean heat flow values from the interval heat flow values.

Results

Physical rock properties

The thermal conductivity of the water-saturated rock samples ranges from 1.7 to 2.9 W m^{-1} K^{-1}. The Barzaman conglomerates show the lowest values (1.7–1.9 W m^{-1} K^{-1}), and fossiliferous limestone the highest ones (2.4–2.7 W m^{-1} K^{-1}). No effect of anisotropy was determined, therefore only the results of the measurements perpendicular to the bedding are shown in Table 2. The difference between dry and water-saturated samples is relatively small. The thermal conductivity of dry rock samples (λ_{dry}) ranges from 0.9 to 2.7 W m^{-1} K^{-1} and the ratio of thermal conductivity measured on dry versus saturated samples (λ_{dry} vs. λ_{sat}) ranges from 0.4 to 1.0 (mean value of 0.8) and is directly linked to the effective rock porosity. The carbonates reveal a relatively small difference between dry and water-saturated thermal conductivity values due to the little effective rock porosity. Very dense grainstones of the Ma'am reef sequence show slightly higher thermal conductivity values (mean λ_{sat} 2.8 W m^{-1} K^{-1}) than the less dense, mixed carbonate siliciclastic rocks of the Barzaman Formation (mean λ_{sat} 2.3 W m^{-1} K^{-1}). Effective porosity ranges between 1.1 and 36.3%, and is highest in the sandstones and sandy limestone samples. The sediments of the Seeb Formation deposited in a carbonate ramp system show a wide range of effective porosity values (2.2–33.8%, mean 16.8%). The main lithologies in the Seeb Formation are *Nummulites* grainstones/packstones at the base and wackestone to mudstones in the upper half of the sequence. The highest effective porosity can be found in the *Nummulite* grainstones/packstones and sandy limestones, which is in accordance with observations made in other regions. The *Nummulite limestone* of the Seeb Formation are excellent hydrocarbon reservoirs (e.g., Tunisia and Libya, Beavington-Penney et al. 2008) due to the high effective porosity and permeability. The lowest effective porosity values are observed in the Ma'am reef sequence (range 1.1–4.0%, mean 2.6%). The investigated section represents the main reef-top, which generally has a high effective porosity but was here probably affected by later diagenesis. The coarse-grained clastics of the alluvial fan system (Barzaman Formation) show a smaller range in effective porosity but a higher mean value (8.9–36.3%, mean 17.1%) than the Seeb Formation. Bulk permeability ranges between 0.01 and 7.5 mD, with highest values in the sandstones of the Seeb Formation (Table 2).

Table 2 Overview of the sampled rocks with the lithology, geological formation, measured porosity (φ, %), density (ρ, 10^3 kg m^{-3}), permeability (k, D), the thermal conductivity measured on dry (λmean$_{dry}$) and water-saturated (λmean$_{sat}$, W m^{-1} K^{-1}) samples

Sample ID	Sample point	Lithology	Geological formation	φ (%)	ρ (g cm^{-3})	k [D]	λmean$_{dry}$ (W m^{-1} K^{-1})	λmean$_{sat}$ (W m^{-1} K^{-1})
B1	x	Conglomerate, diagenetic alteration	Barzaman	17.0	2.2		1.79	1.90
B2	x	Conglomerate, diagenetic alteration	Barzaman	15.0	2.1		2.13	2.38
B3	3	Brownish pebbly sandstone	Barzaman	21.3	2.1	2.00E−04	1.56	2.28
B4	3	Sandy pebbly conglomerate	Barzaman	12.4	2.3	1.00E−05	2.24	2.50
B5	3	Coarse sized sandstone	Barzaman			3.50E−03	1.11	
B6	3	V.C sandstone	Barzaman	36.3	1.7	7.50E−03	0.74	1.93
B7	3	Limestone with pebble size clasts	Barzaman	21.2	2.1	1.00E−04	1.65	2.14
B8	3	Limestone (reef limestone)	Barzaman	14.6	2.3	6.00E−04	2.42	2.89
B9	3	Limestone breccia	Barzaman	12.8	2.3	8.00E−06	2.04	2.26
B10	3	Siltstone, partially cong	Barzaman	18.6	2.1		1.65	1.97
B11	3	Mudstone, beige	Barzaman				0.93	
B12	3	Polymictic conglomerate	Barzaman	18.6	2.1	6.00E−04	1.36	1.73
B13	3	Fossiliferous limestone	Barzaman	8.9	2.5	–	2.36	2.49
B14	3	Grainstone	Barzaman	10.0	2.4		2.27	2.49
B15	2	Conglomerate, diagenetic alteration	Barzaman	15.0	2.2	2.00E−04	2.27	2.72
MAM1	4	Limestone	MAM reef	2.5	2.6		2.71	2.80
MAM2	4	Fossiliferous-rich grainstone	MAM reef	1.1	2.7		2.64	2.71
MAM3	4	Fossiliferous-rich grainstone	MAM reef	4.0	2.6	–	2.63	2.76

Table 2 (continued)

Sample ID	Sample point	Lithology	Geological formation	φ (%)	ρ (g cm^{-3})	k [D]	λmean$_{dry}$ (W m^{-1}K^{-1})	λmean$_{sat}$ (W m^{-1}K^{-1})
S1	5	Foraminiferal packstone–wackstone	Seeb	32.3	1.8	2.00E−04	1.09	1.83
S2	5	Foraminiferal grainstone	Seeb	8.8	2.5		2.15	2.42
S3	4	Foraminiferal packstone–wackstone	Seeb	33.8	1.8	3.80E−03	0.94	1.87
S4	1	Sandstone (sandy limestone)	Seeb	2.2	2.6		2.59	2.65
S5	4	Foraminiferal limestone	Seeb	7.2	2.5		2.35	2.73

Sampling points can be found in Fig. 2, B1 and B2 are not on the map

Temperature data

The T-log of the RGS2L well shows decrease in temperature from 10 m (33.3 °C) to 95 m (30.0 °C) (Fig. 4). This is followed by a relatively smooth and homogeneous increase to the total depth (332 m, 34.6 °C). The temperature in the 21-6 D well shows a similar overall pattern. Between 83 m and 138 m the temperature remains relatively stable, followed by an increase from 30.7 to 33.7 °C and a final temperature of 33.5 °C at the total depth. However, at some depths a high noise is evident. Bodri and Cermak (2011) suggest that sharp discontinuities in a temperature-depth profile (so called "spike" anomalies) may reflect the in- and/or outflow of fluid and its movement within the borehole as well as groundwater flow along a fault or narrow horizontal layer. Transient signals from drilling and mud circulation process can be excluded as the well was completed long in the past (Table 1). The temperature curve in the RGS5HS well generally shows a similar trend as the previously described logs, but with overall slightly higher values and a minor temperature increase from 31.3 °C at 50 m to 32.3 °C at 197.5 m. The T-logs of the wells close to the shoreline show a strong decrease in the temperature in the first tens of meters. The RGS5HS well is located 2.65 km away from the shoreline and the temperature decreases from 32.9 °C at 13 m depth to 31.2 °C at 60 m depth. The subsequent temperature rise is almost negligible (~ 1.1 °C down to the final well depth of 198 m). The data of the 21-7D well show anomalous temperatures. After an increase in the first meters the temperature remains static to a depth of 80 m and then increases first abruptly and afterwards more gradually to 37.8 °C at the total depth of 330 m. As in the well 21-6D a certain noise is observed over the whole length of the log, the source is unknown. The wells were measured at the first (21-7D) and the fourth day (21-6D) of the field campaign with smooth and undisturbed measurements in between (RGS2L). However, a problem with the power unit cannot be ruled out. Background noise might have been generated by a dysfunctional wavelet generator. The temperature pattern of the MAN-3 T-log is also disturbed. After an increase until a depth of 30 m no further temperature increase is observed to the total depth (300 m, 34.0 °C). This might indicate that the temperature probe was blocked by an obstacle. However, determination of such an outage at the surface is problematic. To avoid jamming of the probe, we attached a weight of 5 kg to the cable to ensure continuous downhole logging. This technique was also applied during this measurement, but gave no clear result at the bottom of the hole. A similar observation was made for the T-log of the SAG-13 well. The T-log of the well 103 is plotted only between 20 and 90 m, the shallower part was not considered due to extremely low temperatures. At 90 m the well was blocked. The temperature curve shows a very similar trend like the one of RGS2-L well. In the NB-22 well the water table is at 60 m depth, the temperature data are only given for the interval below the water table. The temperature curve shows a very smooth and minor temperature increase from 32.8 °C at 65 m depth to 34.3 °C at 320 m depth. The water table in the NB-19 well is recorded in 30 m depth, also here the T-logs starts below the water table. The temperature increase with depth is more significant in this borehole. The temperature rises from 34.1 °C at 30 m depth to 37.5 at 255 m depth. The temperature pattern shows no disturbance by advective flow. The temperature curve of the NB-1 well shows first an abrupt and then a moderate increase from 33.6 to 35.6 °C.

There is a notable temperature minimum between 50 and 90 m at the four well locations that are located closest to the shoreline (21-6D, RGS2L, RGS5HS, 103). The bottom-hole temperatures that were measured in a cluster of nine water wells close to the Al Khwad dam show a range of 30.0–33.0 °C at depth between 72 and 148 m. The results are plotted for all wells (temperature vs. depth) in Fig. 4 in combination with the data provided by the MRMWR.

Temperature gradient

For six well locations the T-log and the calculated T-gradient are plotted next to the litho-stratigraphic log and gamma ray logs (Fig. 5). The calculated mean T-gradient for each of the six wells show a relatively large spread (6.6–30.1 °C km^{-1}). T-gradients down to 30 m were disregarded as these are influenced by annual climate effects. Four wells (NB-19, 21-6D, 103, RGS5HS) show disturbances also in deeper parts or even over the whole logging interval. The T-gradient of the NB-22 well is anomalously low (mean 6.6 °C km^{-1}) without any correlation to changes in the lithology. The T-gradient of the 21-6D well fluctuates strongly within small intervals, which is an indication for advective flow. The same observations can be made in the 21-7D well, which shows significant temperature perturbations between 80 and 110 m. In the conglomeratic limestone interval of the Upper Fars Group (155 and 200 m) the T-gradient is very low ($\sim$8 °C km^{-1}). In the lowermost interval which consists of siltstone and claystone the T-gradient shows strong fluctuations around a mean of 18 °C km^{-1}. The T-gradient of the NB-1 well is comparatively high with a mean of 30.1 °C km^{-1}. This could be a result of the low thermal conductivity of the claystone which dominates this interval. However, the T-log covers only 50 m and hence it is not possible to achieve reliable results. The T-gradient of wells NB-19 and RGS2L seem to reflect a thermal equilibrium as changes can be correlated to the lithology. Therefore, these two sites were chosen for a surface heat flow calculation.

Surface heat flow

The T-gradient of the NB-19 well shows two peaks at 60 and 70 m depth which cannot be explained by changes in the lithology. Below these peaks the gradient seems undisturbed and three intervals were chosen for a surface heat flow calculation (see black bars in Fig. 5). The first interval is mainly composed of dolomite, with a slightly higher thermal conductivity (2.6 W m^{-1} K^{-1}) value and a mean T-gradient of 11.0 °C km^{-1} (Table 3). The second interval is composed of limestone, conglomerate and some shale and the T-gradient is significantly higher (23.0 °C km^{-1}). The thermal conductivity was calculated based on the percentages of different lithotypes making up the depth interval and the corresponding thermal conductivity values from Table 2. The third interval consists of conglomerates limestone and minor chalk, shale and dolomitic limestone. This results in a mean T-gradient of 24.0 °C km^{-1} and a mean thermal conductivity of 2.4 W m^{-1} K^{-1}. The interval surface heat flow values range between 28.6 and 57.6 mW m^{-2} with a mean of 47.9 mW m^{-2}. In the RGS2-L well an undisturbed T-gradient can be identified below 100 m. Three interval T-gradients corresponding to lithological changes are marked by a black bar (Fig. 5). In between these intervals minor disturbances occur, which are not reflected in the litho-stratigraphic log. The first interval is dominated by limestone (mean T-gradient 18 °C km^{-1}, mean thermal conductivity 2.5 W m^{-1} K^{-1}) and the interval surface heat flow is 45.0 mW m^{-2}. The second interval is mainly dominated by siltstone, reflected in a relatively low mean thermal conductivity (2.2 W m^{-1} K^{-1}) and a slightly higher T-gradient compared to the shallower interval (23.0 °C km^{-1}). The lowermost interval is comparable to the one described before. The litho-stratigraphic log indicates shale in addition to the siltstone, which results in slightly lower thermal conductivity values but a constant T-gradient. The interval surface heat flow values range from 44.0 to 50.1 mW m^{-2} the mean value is 46.4 mW m^{-2}.

Implication for temperature predictions

The determined surface heat flow can be used to calculate the temperature at depth beyond the logging depth. Knowledge of the sedimentary infill of the basin is required for such an analysis to be able to estimate thermal conductivity data for the different

Table 3 Surface heat flow calculation with the interval method for two well locations

Well	Interval		Γ (°C km^{-1})	Lith.	λ (W m^{-1} K^{-1})	$q_{interval}$ (mW m^{-2})	q_{mean} (mW m^{-2})
	Nr.	(m)					
NB-19	1	74–109	11.0	Dol.	2.6	28.6	47.9
	2	122–158	23.0	Ls. Cong.	2.5	57.5	
	3	174–245	24.0	Ls. Cong. Dol.	2.4	57.6	
RGS2-L	4	112–180	18.0	Ls	2.5	45.0	46.4
	5	230–267	23.0	Si.	2.2	50.1	
	6	271–329	22.0	Si.	2.0	44.0	

The intervals are marked with a black bar in Fig. 5. For location of the wells see Fig. 2. The interval gradient (Γ) is given in °C km^{-1}, major lithological units of the interval are given (lith.), interval thermal conductivity (λ) is given in W m^{-1} K^{-1}. Interval surface heat flow ($q_{interval}$) and mean surface heat flow (q_{mean}) are given in mW m^{-2}

lithological layers. For this purpose we used litho-stratigraphic information from Nolan et al. (1990). We estimated the thermal conductivity of stratigraphic formations by taking the proportion of different rock types into account. We used mean thermal conductivity data determined in this study, for the shallow sedimentary formations (Alluvial, Barzaman Fm., Ma'am reef sequence, Seeb Fm.). For rock types, not measured in this study, we used literature data (Norden and Förster 2006; Fuchs and Förster 2010). The Al Khawd Formation is composed of conglomerates and sandstones and a thermal conductivity of 3.2 is assumed (Beach et al. 1986; Norden and Förster 2006). The Simsima Formation is composed of bioclastic limestone; therefore, we assigned a thermal conductivity of 2.7 W m^{-1} K^{-1}, typical for limestone (Schütz et al. 2012). Below follows the Semail Ophiolite complex consisting of basalt and gabbro with a thermal conductivity of 2.2 W m^{-1} K^{-1} (Norden and Förster 2006). The thickness of the stratigraphic formations derives from Nolan et al. (1990), however, the depth of the transition from the post-obduction sediments to the ophiolites is not yet determined with certainty. The temperature at depth is calculated (based on the Fourier equation) by summing the results of the incremental temperature calculations across individual layers of uniform conductivity:

$$T_z = T_s + \sum_{i}^{n} \left[z_i \frac{q_s}{\lambda_i} \right] \tag{2}$$

where T_s (K) and q_s (mW m^{-2}) are the temperature and heat flow at the surface, z_i is the thickness and λ_i (W m^{-1} K^{-1}) is the thermal conductivity of unit i. An annual average surface temperature of 29 °C and two different values of surface heat flow (46 and 48 mW m^{-2}) are used. Pressure correction was performed of the thermal conductivity to reflect the in situ pressure conditions with the following equation (Fuchs and Förster 2013):

$$\lambda_{p,cor} = (1.095 \cdot \lambda_{lab} - 0.172) \cdot p^{(0.0088 \cdot \lambda_{lab} - 0.0067)}, \tag{3}$$

where λ_{lab} (W m^{-1} K^{-1}) is the thermal conductivity determined in the lab with zero pressure and p (MPa) is the assumed in situ pressure.

To correct the thermal conductivity to the expected in situ temperature the approach of Somerton (1992) was used:

$$\lambda_{T,cor} = \lambda_{20} - 10^{-3}(T - 293) \cdot (\lambda_{20} - 1.38) \cdot \left[\lambda_{20}(1.8 \cdot 10^{-3} T)^{-0.25\lambda_{20}} + 1.28 \right] \lambda_{20}^{-0.67}, \tag{4}$$

where λ_{20} (W m^{-1} K^{-1}) is the thermal conductivity at 20 °C and T (K) is the expected in situ temperature.

The results of the temperature at depth calculation are plotted in Fig. 6.

Discussion

Physical rock properties

The rock properties determined in this study give first insights in the thermal characteristics of the sedimentary rocks in the foreland basin of northern Oman. The thermal conductivities of the rocks are moderate without any significant high values.

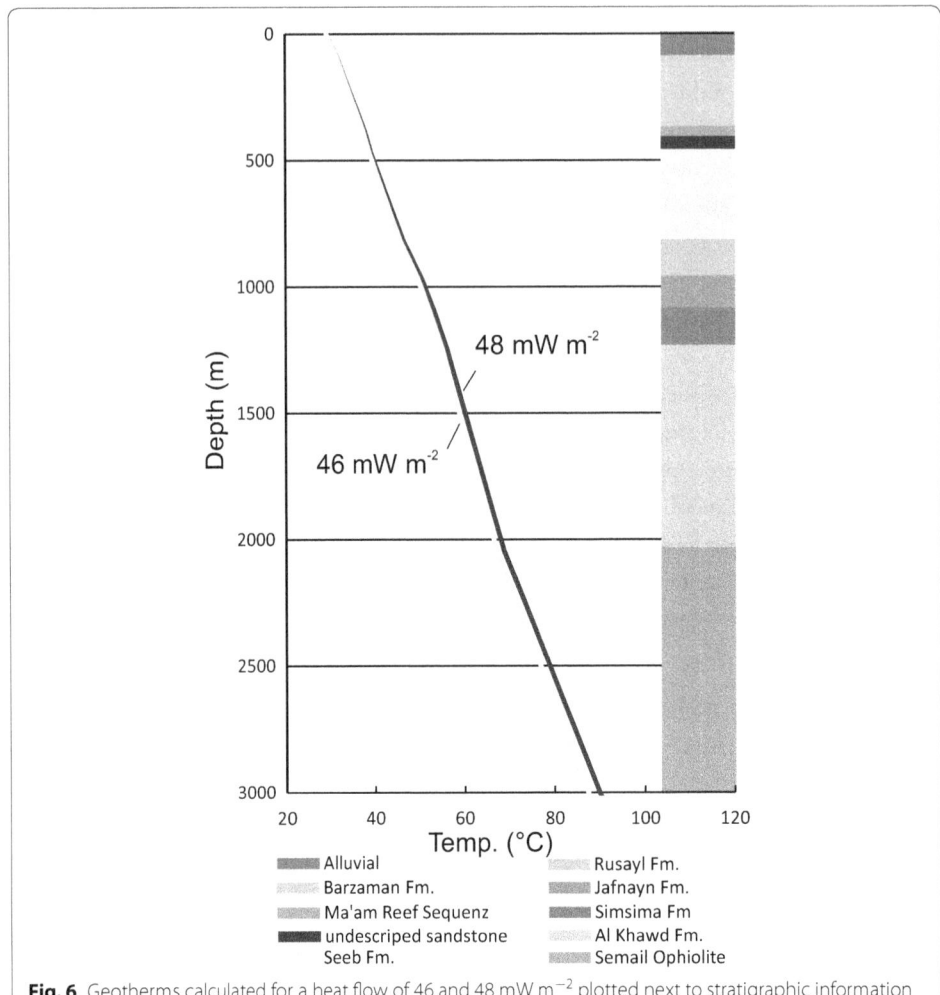

Fig. 6 Geotherms calculated for a heat flow of 46 and 48 mW m^{-2} plotted next to stratigraphic information determined from Nolan et al. (1990)

Only a mudstone from the Barzaman Formation shows noticeable low values (mean 0.9 W m^{-1} K^{-1}). The effective porosities and permeabilities are the controlling parameters for the type of geothermal energy usage. Sedimentary layers with high effective porosity and permeability values in principle allow hydrothermal exploitation of the system, whereas very tight sedimentary rocks require additional stimulation to enhance hydraulic performance (Enhanced Geothermal Systems, EGS). Although some sandstone samples from the Seeb Formation and the Barzaman Formation show a high effective porosity (32.3–36.3%) the corresponding permeability values are low (2.0 × 10^{-4}–7.5 × 10^{-3} mD). This indicates that the measured porosity is not only reflecting effective porosity, but also ineffective porosity, which is in contradiction to values that have been determined on rock samples from the equivalent sedimentary formation from Tunisia (Beavington-Penney et al. 2008). We determined the physical rock properties that only represent the shallow sedimentary cover (< 700 m, depth of the Seeb Formation). These data are important for the planning of shallow geothermal systems like aquifer thermal energy storage or heat sinks to dispense process waste heat to the underground, especially if no appraisal wells exist. The thermal conductivity also cannot be determined through downhole geophysics. More accurate data could only be determined if measurements were carried out on core samples. Another option is to determine the thermal conductivity indirectly by use of different petrophysical logs (Fuchs et al. 2015). Winterleitner et al. 2018 used the database presented here as the basis for a study on the impact of subsurface heterogeneities on high-temperature aquifer thermal energy storage systems for northern Oman. To understand the deeper parts of the foreland basin and to calculate the temperature at depth beyond logging data the thermal conductivities were interpolated to depth based on litho-stratigraphic information from Nolan et al. 1990 (cf. "Implication for temperature predictions" section). Temperature and pressure impacts on the thermal conductivity were considered. The effect of pressure seems to be slightly higher, which results in an overall higher temperature of 143.4 °C in 4631 m depth compared to 139.4 °C when no pressure correction is applied (for the case of 48 mW m^{-2}). However, the two effects basically cancel each other out.

Subsurface temperature distribution in the study area

In general, the determined temperature pattern of the ten wells is relatively consistent (Fig. 4). In all cases, the temperature curves show higher temperatures down to ~ 30 m. This reflects a response of soil and bedrock to warming from the atmosphere. The first tens of meters of the temperature profiles are clearly not representing steady state conditions, but are disturbed by advective flow of heat. A reconstruction of the undisturbed geothermal gradient would be possible if satisfactory hydrological information of the area were available, especially the groundwater flow. Further, the southern Batinah coast is affected by saltwater intrusion into the coastal aquifer system due to excessive groundwater withdrawal for irrigated agriculture (Walther et al. 2012; Grundmann et al. 2014; Chitrakar and Sana 2016). This might be reflected in negative temperature gradients close to the sea in the first tens of meters after the climate related temperature peak. However, no significant decrease in groundwater elevation has been reported around the Al Khawd dam (MRMW, cf. Fig. 3) and therefore the negative temperature gradients might also reflect advective fluid flow related to topography.

The temperature gradient is comparable low along the Batinah coast (Fig. 5) with a mean temperature gradient of around 17.5 °C km^{-1} for the six wells plotted in Fig. 5. However, only two of the six wells with litho-stratigraphic log correlation show undisturbed conditions and seem to be under a thermal equilibrium (NB-19, RGS2-L) with a mean temperature gradient of 19.1 °C km^{-1}. It is likely that a relatively low temperature gradient is representative for the Batinah coast, even if advective disturbance is present in the shallow intervals of the wells. This is in accordance with T-gradients determined by Rolandone et al. (2013) in 13 water wells (6.6–26.1 °C km^{-1}) in south and north Oman. The closest water wells that were analyzed by the authors are in the northern Batinah coast around 230 km to the northwest of our study area. In two water wells the authors determined T-gradients of 6.6 and 11.6 °C km^{-1}, but only the higher value is considered to be reliable.

The target temperature for geothermal exploitation is 70–100 °C, if the heat is to be used to drive an absorption chiller or for water desalinization. When the annual average surface temperature of Muscat (29 °C) is applied the target temperatures can be calculated using the determined surface heat flow and the thermal conductivity of stratigraphic formations as model input. This results in a target depth of about 2000–3000 m (Fig. 6). The use of geothermal energy in the sedimentary section would be possible, but only in the deeper parts of the foreland basin.

Besides suitable hydraulic properties the rejection of process waste heat to the underground requires stable and low temperatures. The temperature logs give an idea which depth levels might be considered for such an application. Varying between 29.9 and 34.4 °C the temperature logs show a minimum between 50 and 100 m depth. These values are only slightly above the average annual surface temperature of Muscat (29 °C), and significantly lower than peak temperatures during summer time (47 °C) and therefore constitute a suitable target for rejecting process waste heat.

Surface heat flow of the Arabian Platform

Rolandone et al. (2013) determined a uniformly low (45 mW m^{-2}) surface heat flow in the entire region of the eastern Arabian Platform, which is in accordance with the results determined from the newly measured temperature profiles of this study. Determination of the surface heat flow of two selected boreholes resulted in values of 46.4 and 47.9 mW m^{-2} with a mean of 47.2 mW m^{-2}. However, if such low values are representative for all Oman is a question which has to be solved by a broader heat flow study. The determination of heat flow values for the Batinah coast allows the generation of 2D or 3D thermal models of the sedimentary basin. The basis for such a model would be a geological model incorporating all the major geological units. The required input parameters that need to be assigned to each geological unit are the thermal conductivity, the radiogenic heat production and the density. The heat flow then constitutes the boundary condition, together with the annual surface temperature. Such a model would allow the generation of temperature-depth maps. The radiogenic heat production of the sedimentary cover can contribute significantly to the surface heat flow when clastic sediments are dominant. In case of the neoautochthonous sedimentary cover the amount of radiogenic heat production is probably minor due to the prevalence of carbonates. This is also reflected in several gamma ray logs provided by the MRMWR. The logs are recorded in API units and among six well logs the values are below 50 API with a mean of 15 API (cf. Fig. 5), which is characteristic for carbonatic dominated composition (Vila et al. 2010).

Conclusions

Newly measured temperature logs of six well locations and the three T-logs provided by the MRMW were used as a basis to calculate the geothermal gradient of the Batinah coast. The calculation indicates a relatively low geothermal gradient (~19.1 °C km^{-1}), and a temperature window of 70–90 °C reached at 2000–3000 m depth (Fig. 6). The results show that geothermal energy as heat source for a thermally driven absorption chiller is an option even in areas with low geothermal gradients. The required temperatures are reached in a drillable depth within the sedimentary basin. Geothermal energy can significantly contribute to cover the base load of such a cooling system that is not available through solar thermal energy. With geothermal energy as heat source for a thermally driven cooling system a storage system would become less relevant. Whereas, it is vital when a fluctuating heat source, like the sun, is used. To further evaluate the geothermal potential of the study area it is crucial to carry out structural analysis. Especially the hot springs in the area indicate a potential geothermal target. At one spring a temperature of 66 °C was measured (data provided by the MRMWR); it is located where the post-nappes sedimentary deposits have been faulted against the autochthonous series. Fournier et al. 2006 reconstructed the local stress tensor and identified two phases of extension followed by a compressional phase. The authors analyzed fractures on outcrop scale which could be used to understand the hydraulic behavior of the faults. Therefore, a coupled heat and fluid transport analysis is required to better understand the potential of fault/fracture controlled geothermal activity in the study area.

The surface heat flow calculated for two sites show relatively low values (47.9 and 46.4 mW m^{-2}). Such low heat flow values have been determined at different sites in Oman (Rolandone et al. 2013) but are nevertheless atypical for the crustal setting. Heat-flow values in the middle and upper 50 mW m^{-2} are determined in other regions for provinces of Neoproterozoic age (Nyblade and Pollack 1993; Rudnick et al. 1998). A crustal transect generated from seismic data (Al-Lazki and Barazangi 2002) indicates a Moho depth of about 40 km under the passive continental margin of north-eastern Oman. The metamorphic basement is around 25 km thick, and the average thickness of the sedimentary cover is around 15 km. In this study a transition from the upper to the lower crust (Conrad discontinuity) is not indicated. The upper crust consists of the sedimentary cover, metamorphic rocks and granites/granodiorites, whereas the lower crust is defined by a more mafic composition with plagioclase-rich and pyroxenite-rich granulite. Assuming a normal heat flow at the Moho (qm = 25 mW m^{-2}, Artemieva and Mooney 2001) and moderate radiogenic heat production in each of the crustal layers a surface heat flow of around 58 mW m^{-2} would be expected. Thus, the surface heat flow determined in the present and previous study (Rolandone et al. 2013) are around 10 mW m^{-2} lower, than the crustal setting would imply. Further studies on the crustal heat flow of the Arabian Peninsula in general, and Oman in particular are essential to better understand the temperature pattern of the broader region.

The temperature gradient, surface heat flow and thermal properties determined in this study serve as input parameter for simulations of the underground to plan potential shallow geothermal applications. Knowledge of the temperature distribution in the subsurface and the physical rock properties is fundamental in the development of underground storage system. The rejection of process waste heat to the underground is a relevant topic for the whole Arabian Peninsula, where temperatures during summer can reach more than 47 °C (in Muscat, PACA) causing all conventional technologies to be either inefficient or expensive. A stable temperature below 35 °C is required to efficiently reject process waste heat of an absorption chiller. The temperature logs of this study suggest temperatures that would be sufficient to reject the heat into the ground. Minimum temperatures (~ 30 °C) were measured between 50 and 100 m at the four well locations located closest to the shoreline (21-6D, RGS2L, RGS5HS, 103) and these temperatures are significantly lower than the maximum temperatures that can be reached during the hottest month. Due to the low temperature gradient deeper parts of the basin might also be suitable. Only three of the ten T-logs show temperatures > 35 °C (21-7D, NB-19, NB-1).

Authors' contributions

FS carried out the field work and developed the results. FS also drafted the manuscript. The second and third authors, GW and EH, supervised the research and guided the interpretation of results. GW and EH also considerably edited and improved the drafts. GW advised the geological field work and gave technical support during the data acquisition process. All authors read and approved the final manuscript.

Author details

[1] Section 6.2 Geothermal Energy Systems, Helmholtz Centre Potsdam-GFZ German Research Centre for Geoscience, Telegrafenberg, 14473 Potsdam, Germany. [2] Department for Earth and Environmental Sciences, University of Potsdam, Karl-Liebknecht Straße 24-25, Golm, Germany.

Acknowledgements

We particularly wish to thank the Ministry of Regional Municipalities and Water Resources for granting permission for the use of wells, data and their library. We would like to thank Hilal Al Julandani and Musaab Al Sarmi (SQU) for logistic support and assistance in the field work as well as Ronny Giese and Tanja Ballerstedt (GFZ). Christian Wenzlaff (GFZ) is thanked for his support in the lab. The manuscript profited from valuable remarks by two anonymous referees and Christina von Nicolai.

Competing interests

The authors declare that they have no competing interests.

Funding

The study is funded through Institute of Advanced Technology Integration (IATI)/The Research Council of the Sultanate of Oman and part of the GeoSolCool-project (https://www.gfz-potsdam.de/en/section/geothermal-energy-systems/projects/geosolcool/).

References

Al-Husseini M. Launch of the Middle East geologic time scale. GeoArabia. 2008;13(4):11, 185–8.

Al-Lazki AI, Seber D, Sandvol E, Barazangi M. A crustal transect across the Oman Mountains on the eastern margin of Arabia. GeoArabia. 2002;7:47–78.

Alsharhan AS. Petroleum systems in the Middle East. In: Rollinson HR, Searle MP, Abbasi IA, Al-Lazki A, Al Kindi MH, editors. Tectonic evolution of the Oman Mountains, vol. 392. London: Geological Society, Special Publications; 2014. p. 361–408.

Artemieva IM, Mooney WD. The thermal thickness and evolution of pre-cambrian lithosphere: a global study. J Geophys Res B. 2001;106:16387–414.

Beach RDW, Jones FW, Majorowicz JA. Heat flow and heat generation estimates for the Churchill basement of the Western Canadian Basin in Alberta, Canada. Geothermics. 1986;16(1):1–16.

Beavington-Penney SJ, Nadin P, Wright VP, Clarke E, McQuilken J, Bailey HW. Reservoir quality variation on an Eocene carbonate ramp, El Garia formation, offshore Tunisia: structural control of burial corrosion and dolomitisation. Sediment Geol. 2008;209(1–4):42–57.

Bodri L, Cermak V. Borehole climatology: a new method how to reconstruct climate. Amsterdam: Elsevier; 2011. p. 352.

Breton J-P, Béchennec F, Le Métour J, Moen-Maurel L, Razin P. Eoalpine (Cretaceous) evolution of the Oman Tethyan continental margin: insights from a structural field study in Jabal Akhdar (Oman Mountains). GeoArabia. 2004;9(2):2004.

Chang S-J, Van der Lee S. Mantle plumes and associated flow beneath Arabia and East Africa. Earth Planet Sci Lett. 2011;302:448–54.

Chitrakar P, Sana A. Groundwater flow and solute transport simulation in Eastern Al Batinah Coastal Plain, Oman: case study. J Hydrol Eng. 2016;21(2):05015020.

Coleman RG. Tectonic setting for ophiolite obduction in Oman. J Geophys Res. 1981;6(B4):2497–508.

Dill HG, Wehner H, Kus J, Botz R, Berner Z, Stüben D, Al-Sayigh A. The Eocene Rusayl Formation, Oman, carbonaceous rocks in calcareous shelf sediments: environment of deposition, alteration and hydrocarbon potential. Int J Coal Geol. 2007;72(2):89–123.

Förster A, Förster H-J, Masarweh R, Masri A, Tarawneh K. The surface heat flow of the Arabian Shield in Jordan. J Asian Earth Sci. 2007;30:271–84.

Fournier M, Lepvrier C, Razin P, Jolivet L. Late cretaceous to paleogene post-obduction extension and subsequent Neogene compression in Oman Mountains. GeoArabia. 2006;11(4):17–40.

Fuchs S, Förster A. Rock thermal conductivity of mesozoic geothermal aquifers in the Northeast German Basin. Chem Erde. 2010;70(Supplement 3):13–22.

Fuchs S, Förster A. Well-log based prediction of thermal conductivity of sedimentary successions: a case study from the North German Basin. Geophys J Int. 2013;196:291–311.

Fuchs S, Balling N, Förster A. Calculation of thermal conductivity, thermal diffusivity and specific heat capacity of sedimentary rocks using petrophysical well logs. Geophys J Int. 2015;203(3):1977–2000.

GCC. The GCC in 2020: resources for the future. The Economist Intelligence Unit Limited; 2010.

Grundmann J, Schütze N, Heck V. Optimal integrated management of groundwater resources and irrigated agriculture in arid coastal regions. Evolving water resources systems: understanding, predicting and managing water–society interactions. In: Proceedings of ICWRS2014, Bologna, Italy, June 2014 (IAHS Publ. 364, 2014). 2014.

Huenges E, editor. Geothermal energy systems: exploration, development and utilization. Weinheim: Wiley-VCH; 2010.

Johnson PR, Andresen A, Collins AS, Fowler AR, Fritz H, Ghebreab W, Kusky T, Stern RJ. Late Cryogenian-Ediacaran history of the Arabian–Nubian Shield: a review of depositional, plutonic, structural, and tectonic events in the closing stages of the northern East African Orogen. J Afr Earth Sci. 2011;61:167–232.

Kusky T, Robinson C, El-Baz F. Tertiary-quaternary faulting and uplift in the northern Oman Hajar mountains. J Geol Soc. 2005;162(5):871–88.

Lashin A, Al-Arifi N, Chandrasekharam D, Al Bassam A, Rehman S, Pipan M. Geothermal energy resources of Saudi Arabia: country update. In: Proceedings World Geothermal Congress, Melbourne, Australia. 2015.

Mahmoud S, Reilinger R, McClusky S, Vernant P, Tealeb A. GPS evidence for northward motion of the Sinai Block: implications for E, Mediterranean tectonics. Earth Planet Sci Lett. 2005;238:217–24.

Maizels JK. Plio-Pleistocene raised channel systems of the western Sharqiya (Wahiba), Oman. In: Frostick L, Reid I, editors. Desert sediments: ancient and modern, vol. 35. Geological Society, Special Publication: London; 1987. p. 31–50.

Mann A, Hanna SS, Nolan SC, Mann A, Hanna SS. The post-campanian tectonic evolution of the Central Oman Mountains: tertiary extension of the Eastern Arabian Margin, vol. 49, no. 1. London: Geological Society, Special Publications; 1990. p. 549–63.

Michard A, Bouchez JL, Ourzzani-Touhami M. Obduction related planar and linear fabrics in Oman. J Struct Geol. 1984;6:39–50.

Milsch H, Priegnitz M, Blöcher G. Permeability of gypsum samples dehydrated in air. Geophys Res Lett. 2011;38:L18304.

Mount VS, Crawford RIS, Bergman SC. Regional structural style of the central and southern Oman mountains: Jebel Akhdar, Saih Hatat, and the northern Ghaba Basin. GeoArabia. 1998;3:475–90.

Nolan SC, Skelton PW, Clissold BP, Smewing JD. Maastrichtian to early tertiary stratigraphy and palaeogeography of the Central and Northern Oman Mountains, vol. 49, no. 1. London: Geological Society, Special Publications; 1990. p. 495–519.

Norden B, Förster A. Thermal conductivity and radiogenic heat production of sedimentary and magmatic rocks in the Northeast German Basin. AAPG Bull. 2006;90(6):939–62.

Nyblade AA, Pollack HN. A global analysis of heat flow from Precambrian terrains: implications for the thermal structure of Archean and proterozoic litho-sphere. J Geophys Res B. 1993;98:12207–18.

Özcan E, Abbasi IA, Drobne K, Govindan A, Jovane L, Boukhalfa K. Early Eocene orthophragminids and alveolinids from the Jafnayn Formation, N Oman: significance of Nemkovella stockari Less & Özcan, 2007 in Tethys. Geodin Acta. 2015;28(3):160–84.

Popov YA, Pribnow DFC, Sass JH, Williams CF, Burkhardt H. Characterization of rock thermal conductivity by high-resolution optical scanning. Geothermics. 1999;28:253–76.

Poupeau G, Saddiqi O, Michard A, Goffé B, Oberhänsli R. Late thermal evolution of the Oman Mountains subophiolitic windows: apatite fission-track thermometry. Geology. 1998;26(12):1139–42.

Powell WG, Chapman DS, Balling N, Beck AE. Continental heat flow density. In: Haenel R, Rybach L, Stegena L, editors. Handbook of terrestrial heat-flow density determination. Dordrecht: Kluwer; 1988. p. 167–222.

Reiche D. Energy Policies of Gulf Cooperation Council (GCC) countries—possibilities and limitations of ecological modernization in rentier states. Energy Policy. 2010;38:2395–403.

Rolandone F, Lucazeau F, Leroy S, Mareschal J-C, Jorand R, Goutorbe B, Bouquerel H. New heat flow measurements in Oman and the thermal state of the Arabian Shield and Platform. Tectonophysics. 2013;589:77–89.

Rudnick RL, McDonough WF, O'Connell RJ. Thermal structure, thickness and composition of continental lithosphere. Chem Geol. 1998;145:395–411.

Schütz F, Norden B, Förster A; DESIRE Group. Thermal properties of sediments in southern Israel: a comprehensive data set for heat flow and geothermal energy studies. Basin Res. 2012;24(3):357–76.

Schütz F, Förster H-J, Förster A. Thermal conditions of the central Sinai Microplate inferred from new surface heat-flow values and continuous borehole temperature logging in central and southern Israel. J Geodyn. 2014;76:8–24.

Searle MP. Sequence of thrusting and origin of culminations in the northern and central Oman Mountains. J Struct Geol. 1985;7:129–43.

Searle MP, James NP, Calton TJ, Smewing JD. Sedimentological and structural evolution of the Arabian continental margin in the Musandam Mountains and Dibba Zone, United Arab Emirates. Bull Geol Soc Am. 1983;94:1381–400.

Somerton WH. Thermal properties and temperature-related behavior of rock/fluid systems. Amsterdam: Elsevier Science Publishers B.V; 1992. p. 257.

Stern RJ, Johnson P. Continental lithosphere of the Arabian Plate: a geologic, petrologic, and geophysical synthesis. Earth Sci Rev. 2010;101:29–67.

Sweetnam T, Al Ghaithi H, Almaskari B, Calder C, Patterson J, Mohaghedi S, Oreszczyn T, Rasla R. Residential energy use in Oman: a scoping study. Project Report. 2014.

Tanikawa W, Shimamoto T. Comparison of Klinkenberg-corrected gas permeability and water permeability in sedimentary rocks. Int J Rock Mech Min. 2009;46(2):229–38.

Tomás S, Frijia G, Bömelburg E, Zamagni J, Perrin C, Mutti M. Evidence for seagrass meadows and their response to paleoenvironmental changes in the early Eocene (Jafnayn Formation, Wadi Bani Khalid, N Oman). Sediment Geol. 2016;341:189–202.

Vila M, Fernández M, Jiménez-Munt I. Radiogenic heat production variability of some common lithological groups and its significance to lithospheric thermal modelling. Tectonophysics. 2010;490:152–64.

Walther M, Delfs J-O, Grundmann J, Kolditz O, Liedl R. Saltwater intrusion modeling: verification and application to an agricultural coastal arid region in Oman. J Comput Appl Math. 2012;236(18):4798–809.

Winterleitner G, Schütz F, Wenzlaff C, Huenges E. The impact of subsurface heterogeneities on high-temperature aquifer thermal energy storage systems. A case study from Northern Oman. Geothermics. 2018. (Accepted).

3

Optimization of operating parameters of earth air tunnel heat exchanger for space cooling: Taguchi method approach

Kamal Kumar Agrawal[1*], Mayank Bhardwaj[2], Rohit Misra[3], Ghanshyam Das Agrawal[1] and Vikas Bansal[3]

*Correspondence:
kamal.rightway@gmail.com
[1] Mechanical Engineering
Department, Malaviya
National Institute
of Technology, Jaipur 302017,
India
Full list of author information
is available at the end of the
article

Abstract

In the present study, CFD-based parametric analysis is carried out to optimise the parameters affecting the temperature drop and heat transfer rate achieved from earth air tunnel heat exchanger (EATHE) system. ANSYS FLUENT 15.0 is used for CFD analysis, and k-ε model and energy equation were considered to define the turbulence and heat transfer phenomena. For a straight EATHE system configuration, four design and operating parameters, i.e., diameter of the pipe (A), length of pipe (B), inlet air velocity (C), and inlet air temperature (D), are considered at four different levels in Taguchi method. The Taguchi method is used to obtain maximum air temperature drop and heat transfer rate. The best combination of parameters for achieving a maximum drop in air temperature is $A_1B_4C_1D_4$ and that for obtaining maximum total heat transfer rate is $A_4B_4C_4D_4$. Statistical analysis reveals the percentage contribution of different factors for air temperature drop in the following order: inlet air temperature (57.80%), diameter of pipe (20.66%), length of pipe (12.03%), and air velocity (9.51%), while, for heat transfer rate, pipe diameter (53.28%), inlet air temperature (30.87%), air velocity (9.40%), and length of pipe (6.45%).

Keywords: Earth air tunnel heat exchanger, Design and operating parameters, Taguchi optimization, ANOVA, Air temperature drop, Heat transfer rate

Background

The significance of renewable energy is increasing due to depletion of fossil fuels and rise in fossil fuels prices. In last 2 decades, the energy demands in buildings have raised significantly due to increasing living standards and population. Space cooling and heating utilized about 33% of total energy consumption world over (Nejat et al. 2015; Omer 2008). The conventional cooling and heating systems are energy intensive. Therefore, many countries are adopting passive and low-grade energy systems for cooling and heating of buildings. Earth air tunnel heat exchanger (EATHE) system uses earth as a heat source/sink to transfer heat to/from fluid flowing through the buried pipes. At a depth of 3–4 m, soil temperature remains constant round the year, and when the air is passed through buried EATHE pipe, it produces a heating effect in winter and cooling effect in summer (Bansal et al. 1983; Bansal and Sodha 1986; Bharadwaj and Bansal 1981; Wang et al. 2009). The working principle of EATHE system is shown in Fig. 1.

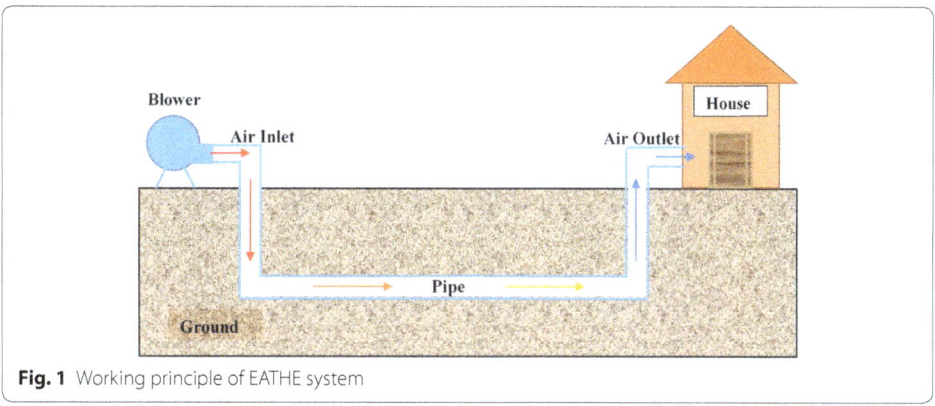

Fig. 1 Working principle of EATHE system

For the efficient working of the EATHE system during the summer, the estimation of maximum air temperature drop and heat transfer rate that can be produced by EATHE system are two key performance indicators. Large temperature drops and heat transfer rate also improve the economic viability of the EATHE system. Hence, there should be a methodology to investigate the maximum air temperature drop obtainable from EATHE system for space cooling applications.

Mihalakakou (2003) calculated the heating potential of EATHE using a dynamic and deterministic numerical model. The estimated values of soil temperature were compared with measured soil temperature values and observed that the neural network could efficiently simulate the outlet air temperature of EATHE system. It was found that the ground temperature is the most critical parameter to estimate the outlet air temperature. Kumar et al. (2006) developed a deterministic model and an intelligent model by using the artificial neural network (ANN). The intelligent model calculates outlet air temperature of EAHE with an accuracy of ±2.6%, whereas the accuracy of the deterministic model was ±5.3%. It was found that the cooling/heating potential of the 80 m-long EAHE was 7.49 kW in winter and 12.25 kW in summer. Zhang and Haghighat (2009, 2010) developed a method to estimate the convective heat transfer in an EAHE system of the large rectangular cross-sectional area. Parametric studies were carried out using CFD simulations, and ANN models were trained using simulation results. A mathematical relationship was developed between six design parameters and area-weighted local average Nusselt numbers. It was observed that average Nusselt number was not affected by the turbulence intensity of air at the inlet, the surface temperature variation, nor the size of the outlet section of the duct. However, the heat convection was influenced by the duct length, width, and height; the size of the inlet; the temperature difference between inlet air and surface; air velocity; mode of operation. Diaz et al. (2013) applied a fuzzy logic controller to optimise the power consumption in an EATHE system. It was observed that an EATHE consumes less energy when the fuzzy logic controller is implemented instead of an on–off controller.

Kumar et al. (2008) used the concept of the goal-oriented genetic algorithm (GA) for evaluation and optimisation of different aspects of EAHE system. Four parameters, viz., air humidity, ambient air temperature, ground temperature at burial depth, and ground surface temperature, were considered, and through sensitivity analysis, it was found that outlet air temperature was significantly affected by the ground temperature at burial depth and ambient air temperature. Kaushal et al. (2015) predicted the thermal performance of an EATHE-coupled solar air heater by applying finite-volume method. Response surface methodology (RSM) was used to optimise the process parameters. Five independent parameters were considered, viz., soil thermal conductivity, inlet air velocity, inlet air temperature, depth of solar air heater channel, and solar radiation intensity, and two output responses, viz., temperatures difference between outlet to inlet air for simple EATHE and hybrid EATHE, were taken. It was observed that the soil thermal conductivity is the most important factor followed by the depth of the solar air heater channel and the intensity of solar radiation.

It has been seen that, with an increase in pipe diameter, total temperature drop/rise in summer/winter falls and overall heating/cooling capacity enhances (Ahmed et al. 2016; Ghosal and Tiwari 2006; Krarti and Kreider 1996; Santamouris et al. 1995). Sodha et al. (1994) compared the single air-pipe EATHE system with the multi-air-pipe EATHE system at the same air mass flow rate. It was observed that, for a given mass flow rate, the heating potential (HP) and cooling potential (CP) increase with increasing the number of pipes of smaller diameter, because effective heat transfer area increases with increase in number of pipes. Mihalakakou et al. (1994a) found that, by reducing pipe diameter from 0.5 to 0.25 m, air temperature drop increases by 1.5–2.5 °C at the pipe's outlet in summer cooling operation. Mihalakakou et al. (1996a) observed that the convective heat transfer coefficient reduces with increase in the radius of the buried pipe; this leads to lower air temperature at the outlet of pipe in winter and thus reduces the heating capacity of the system. Typical diameters of pipe are 10–30 cm but may be as large as 1 m for commercial applications.

By increasing the pipe length of EATHE, the temperature difference between inlet and outlet air increases, but the rate of change of temperature decreases (Derbel and Kanoun 2010; Kabashnikov et al. 2002). It was observed that, after a certain length, heat transfer does not increase by increasing the pipe length (Benhammou and Draoui 2015), and it is termed as saturation length (L_{sat}). The saturation length increases with increase in the air flow rate. According to Lee and Strand (2008), there are no significant advantages in using pipes over 70 m length. In a parametric study, Ahmed et al. (2016) found that the pipe length is a dominating parameter over other parameters (air velocity, pipe diameter, pipe material, and pipe depth) which affects thermal performance of EATHE system. In an EATHE system, pipe length is the critical factor which affects the performance of EATHE. In many studies, it was identified that, by increasing pipe length, the drop/rise in air temperature is increased (Ghosal and Tiwari 2006; Mihalakakou et al. 1994c; Santamouris et al. 1995), but, after a certain extent, the effect of pipe length on temperature drop/rise decreases.

The velocity of flowing air in the buried pipe also significantly influences the performance of EATHE system. It was observed that, by increasing air velocity, the total temperature difference between inlet and outlet air temperature decreases (Kabashnikov et al. 2002; Mihalakakou et al. 1994a, b, 1996a, b). Niu et al. (2015) considered five flow velocities, viz., 0.5, 1.0, 1.5, 2, and 2.5 m/s, for cooling operation, and found that the air temperature drop rate was the highest at 0.5 m/s, because low velocity provides more contact time between air and pipe. Bansal et al. (2009) considered four flow velocities (2, 3, 4, and 5 m/s) for winter heating application, and observed that maximum air temperature rise was obtained at 2 m/s, while the maximum hourly heat gain was found at air velocity 5 m/s. In another study for summer cooling, it was noted that maximum air temperature drop was observed at 2 m/s and maximum hourly cooling was observed at air velocity 5 m/s (Bansal et al. 2010). Similar results were also obtained by Wu et al. (2007) for cooling operation, by increasing air velocity (from 1 to 4 m/s); the outlet air temperature jumps to higher values, but the heat transfer rate increases because of increase in mass flow rate. Dubey et al. (2013) observed that, by increasing air velocity from 4.1 to 11.6 m/s, the air temperature drop reduced from 8.6 to 4.18 °C, and COP also decreased from 6.4 to 3.6. In an experimental study, Bisoniya et al. (2014) considered three different velocities (2, 3.5, and 5 m/s) for an EATHE pipe of 0.1 m diameter and 19.2 m length. The observed maximum and minimum drops in air temperature were 12.9 and 11.3 °C, at air flow velocities of 2 and 5 m/s, respectively.

Yusof et al. (2018) simulated different input parameters (inlet air temperature varied between 31 and 35 °C, ground temperature from 23 to 25 °C, and air flow rate between 0.03 and 0.07 kg/s) of an EAHE system under laboratory conditions. The highest temperature drop (9.62 °C) was obtained at the air flow rate of 0.03 kg/s and ground temperature of 23 °C, while the maximum heat transfer rate (558.3 W) was achieved at air flow rate of 0.07 kg/s and ground temperature of 23 °C.

Air temperature at the inlet of EATHE pipe plays a significant role, because the rate of heat transfer between air and soil is governed by the temperature difference between air and ground. Kumar et al. (2006) studied the impact of ambient air temperature on outlet air temperature. It was noticed that, with the increase in inlet air temperature, the outlet temperature of air also increases, but the amplitude decreases significantly. When the inlet temperature varied from 27 to 45.9 °C, outlet air temperature varied from 23.8 to 27.9 °C for an 80 m-long earth air tunnel. Elminshawy et al. (2017) used a small laboratory scale EAPHE experimental set-up under controlled conditions of operating parameters such as soil bulk temperature, air flow rate, and induced air temperature at three different compaction levels. It was noticed that, for a particular compaction level (at highest density), the cooling capacity of EAPHE system increases by 227% when the induced air inlet temperature is increased from 40 to 55 °C. Niu et al. (2015) established a 1-D steady-state control volume model and observed the impact of inlet air temperature on the performance of EAHE. It was found that, when the inlet air temperature was high, the decline rate of air temperature in EAHE pipe was higher. For the inlet air temperatures of 34, 32, 30, 28, and 26 °C, the highest and lowest decline rate of air temperature was found for 34 and 26 °C, respectively.

Vidhi et al. (2014a, b) presented an application of EAHE system for condenser cooling of a supercritical Rankine cycle (SRC) power generation. A 2-D model was developed in MATLAB to analyze the effect of various parameters on cooling of air in the EAHE and the efficiency of thermodynamic cycle. For the parametric study, pipe length, diameter, and installation depth were kept as 25–75 m, 25–50 cm, and 1–4 m, respectively. It was noticed that, by increasing the depth and length of pipe, the efficiency of the SRC increases; however, after a certain limit, the rate of improvement in SRC efficiency is very small.

Constructal Design has been widely used to seek for the optimal geometries, i.e., which leads to the best performances. Rodrigues et al. (2015) performed a numerical investigation on different geometrical configurations of an EAHE based on Constructal Design for achieving the highest thermal potential. The results revealed that the thermal performance was improved up to 115 and 73% for heating and cooling, respectively, by increasing the number of buried pipes for the same occupied area and fixed mass flow rate of air.

A detailed literature survey indicates that the performance of EATHE system depends on various parameters such as inlet air temperature and humidity ratio, soil temperature, pipe diameter and length, burial pipe depth, soil thermal conductivity, air flow velocity, etc. These parameters have been optimised using different techniques by various researchers, but limited research outcomes have been reported in which the key parameters affecting the performance of EATHE were simultaneously optimised and the contribution of each parameter has been discussed. Therefore, in the present study, four performance-affecting parameters, i.e., pipe diameter, pipe length, air velocity, and inlet air temperature, have been considered to determine the contribution of each parameter and the best combination of these parameters for achieving maximum temperature drop and heat transfer rate for space cooling application. CFD simulations are performed for all 16 cases which are obtained by the Taguchi method for the considered parameters at different levels as CFD analysis proved as an efficient method of calculation with reduced operating time.

Simulation model description

A three-dimensional simulation model of the EATHE system has been prepared in ANSYS FLUENT package (15.0) for analysis. This CFD software package uses finite-volume method to change the complex governing equations into the numerically solvable algebraic equations. The control volume of EATHE system was defined by creating a cylindrical volume of soil around the pipe, as shown in Fig. 2a. The developed physical model of EATHE system was discretised using 3D hybrid (hexahedral and tetrahedral) meshing (Fig. 2b) with minimum and maximum element size of 0.0015 and 2.99 m, respectively, with a growth rate of 1.2 in ANSYS Workbench Meshing.

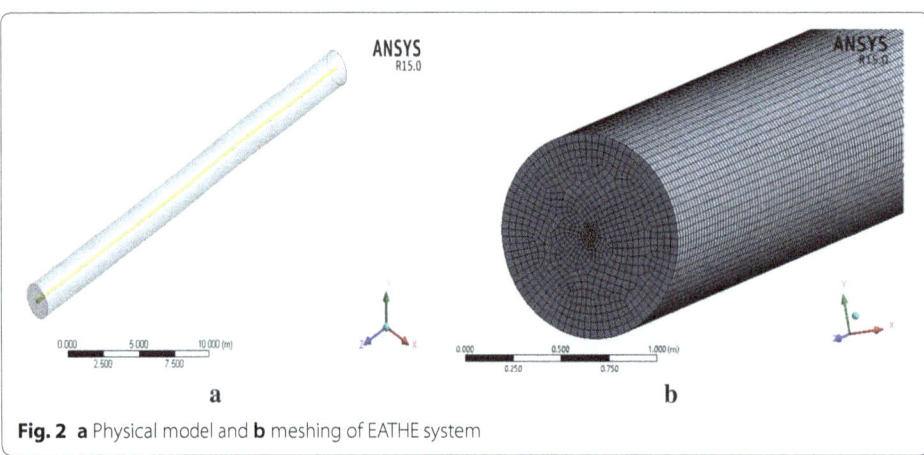

Fig. 2 a Physical model and **b** meshing of EATHE system

Governing equations

The following set of governing equations is used to perform simulation in FLUENT software to describe the heat and mass transfer and flow of fluid within any systems (ANSYS Inc. 2013).

- Law of mass conservation: The equation for mass conservation law or continuity equation is written as

$$\frac{\partial u}{\partial x} + \frac{\partial v}{\partial y} + \frac{\partial w}{\partial z} = 0 \tag{1}$$

- Law of energy conservation: the first law of thermodynamics or law of energy conservation stated as neither the energy can be created nor destroyed, it only changes its form in nature. The equation can be written as follows:

$$u\frac{\partial T}{\partial x} + v\frac{\partial T}{\partial y} + w\frac{\partial T}{\partial z} = \alpha\left[\frac{\partial^2 T}{\partial x^2} + \frac{\partial^2 T}{\partial y^2} + \frac{\partial^2 T}{\partial z^2}\right] \tag{2}$$

- Law of momentum conservation (Navier–Stokes equation, also known as Newton's second law): the equation for momentum conservation is as follows:

X-momentum equation:

$$u\frac{\partial u}{\partial x} + v\frac{\partial u}{\partial y} + w\frac{\partial u}{\partial z} = -\frac{1}{\rho}\frac{\partial p}{\partial x} + \vartheta\left[\frac{\partial^2 u}{\partial x^2} + \frac{\partial^2 u}{\partial y^2} + \frac{\partial^2 u}{\partial z^2}\right] \tag{3a}$$

Y-momentum equation:

$$u\frac{\partial v}{\partial x} + v\frac{\partial v}{\partial y} + w\frac{\partial v}{\partial z} = -\frac{1}{\rho}\frac{\partial p}{\partial y} + \vartheta\left[\frac{\partial^2 v}{\partial x^2} + \frac{\partial^2 v}{\partial y^2} + \frac{\partial^2 v}{\partial z^2}\right] \tag{3b}$$

Z-momentum equation:

$$u\frac{\partial w}{\partial x} + v\frac{\partial w}{\partial y} + w\frac{\partial w}{\partial z} = -\frac{1}{\rho}\frac{\partial p}{\partial z} + \vartheta\left[\frac{\partial^2 w}{\partial x^2} + \frac{\partial^2 w}{\partial y^2} + \frac{\partial^2 w}{\partial z^2}\right] \tag{3c}$$

In the above Eqs. (1–3), u, v, and w are the velocity components in x-, y-, and z-directions, and T and p are the temperature and pressure of the flowing air, respectively.

Material properties

Different thermo-physical properties of air, soil, and PVC pipe used in CFD simulation are presented in Table 1.

Boundary conditions

The following conditions are taken at different sections during the simulation in CFD software.

Table 1 Material properties used in simulation

Material	Density (kg/m³)	Specific heat (J/kg-K)	Thermal conductivity (W/m-K)
Air	1.22	1006	0.02
Soil	2050	1840	0.52
PVC pipe	1380	900	0.16

- At pipe inlet: uniform velocity is provided as input in the normal direction to the pipe; heat flux is taken as zero.
- At soil-pipe interface: no-slip condition is considered at pipe wall and soil interface.
- At pipe outlet: pressure outlet; heat flux is zero.

Turbulence model description

In the simulation, the pressure-based Navier–Stokes algorithm has been adopted and SIMPLE scheme was selected for solving pressure–velocity coupling. The pressure gradients are solved by second order and LSCB (least square cell-based), respectively. Second-order upwind for the kinetic energy of turbulence and second-order upwind for turbulent momentum were taken. The step size of 60 s is used for calculation. During simulation, the far-field boundaries were treated as an adiabatic wall, and EATHE pipe wall and surrounding soil temperatures were initialised at 27 °C as the average sub-soil temperature (at 3–4 m depth) remains 27 °C throughout the year in Ajmer, India. Air is treated as an incompressible ideal gas and the viscous k-epsilon (k-ε) realizable turbulence model with standard wall function is applied. The constants in the viscous model are as follows: $C_{1\epsilon} = 1.44$; $C_2 = 1.9$; $\sigma_k = 1.0$; $\sigma_\epsilon = 1.2$, and $\mathrm{Pr}_{\mathrm{wall}}$ and $\mathrm{Pr}_{\mathrm{energy}}$ are 0.85. The viscosity of air was kept constant as 1.78 e−05 kg/m-sec.

Grid independence test

Grid independence tests were conducted in FLUENT 15.0 to assess the quality of the developed CFD model. In the present analysis, CFD simulations have been performed using 3D hybrid (hexahedral and tetrahedral) meshing. A grid-independency test was carried out to check the effect of mesh size on the accuracy of the solution. Mesh size varies from 0.0015 and 2.99 m from pipe surface to soil outer layer. The independent grid size was determined by successive refinements, increasing the number of elements from 681,320 (Coarse mesh) to 948,231 (Fine mesh). A medium size mesh having 820,830 elements was also checked for suitability.

Figure 3 shows the simulated air temperatures along the length of EATHE pipe for coarse, medium, and fine mesh. Simulated air temperatures for medium and fine meshes have very close approximation with each other, and a maximum difference of 0.15 °C was observed between the temperatures for two. Hence, medium mesh used in the simulation meets the simulation requirement to produce mesh-independence results, and therefore, to save computation effort and time, medium mesh has been chosen for further parametric study.

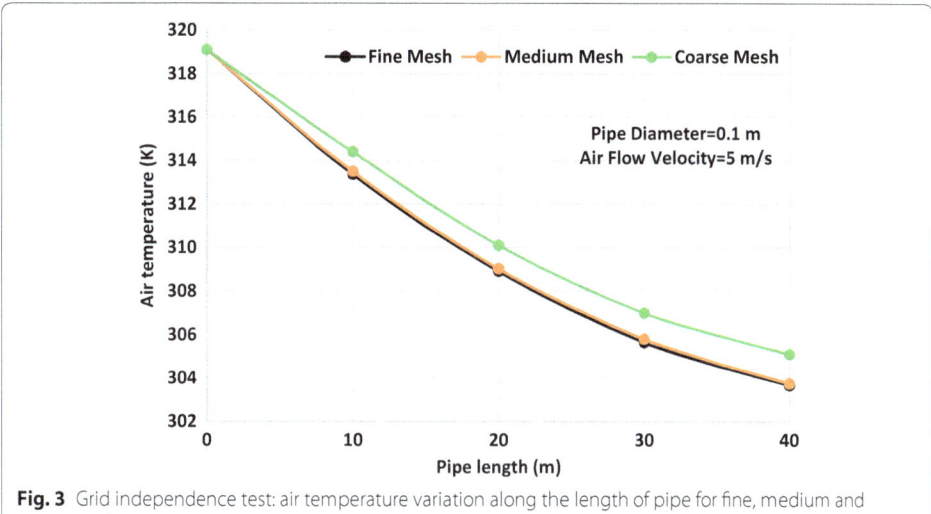

Fig. 3 Grid independence test: air temperature variation along the length of pipe for fine, medium and coarse mesh

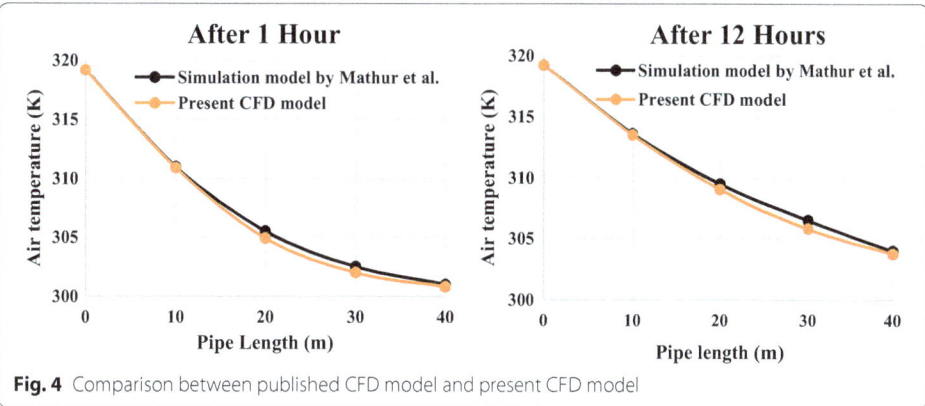

Fig. 4 Comparison between published CFD model and present CFD model

Validation of simulation model

The developed CFD model has been validated against numerical results obtained by Mathur et al. (2015). It is observed from the Fig. 4 that the maximum deviation in the temperature profile achieved by present simulation model and numerical model of Mathur et al. (2015) is 0.7 °C along the pipe length. There is a good agreement between the previously published and present simulation model, and hence, the proposed CFD model is suitable for further analysis.

Methodology

Taguchi technique

The Taguchi method is a statistical technique of laying out the conditions of experiments involving multiple factors to optimise the parameters and improve the performance of the system. The method is popularly known as the factorial design of experiments. A full factorial design will identify all possible combinations for a given set of factors and results in a large number of experiments. To reduce the number of experiments, Taguchi proposed an orthogonal array-based method to study the entire parameter space (matrix of experiments) with a smaller number of experiments. With the help of this matrix, one can get maximum information from a minimum number of trials and also the best level of each parameter can be obtained for an objective function. For the analysis of data, signal-to-noise (S/N) ratios are used to calculate the response of the experimental trials. There are three types of analysis of the response results: lower is better, nominal is better, higher is better, and these can be expressed by the following equations (Inc 2015, 2016).

The lower is better characteristics are expressed by the following equation:

$$\text{SNR} = -10 \log_{10} \left(\frac{1}{n} \sum h^2 \right) \tag{4}$$

The nominal is better characteristics are expressed by the following equations:

$$\text{SNR} = -10 \log_{10} \left(s^2 \right) \tag{5}$$

or

$$\text{SNR} = 10 \log_{10} \left(\overline{Y}^2 \big/ s^2 \right) \tag{6}$$

The higher is better characteristics which are expressed by the following equation:

$$\text{SNR} = -10 \log_{10} \left(\frac{1}{n} \sum \frac{1}{h^2} \right) \tag{7}$$

In the above equations, n = number of cases/test runs; h^2 = experimental results/data; where Y = mean of responses for a combination of selected factor level; s = standard deviation of the responses for given factor-level combination.

Taguchi design of experiments

For performing the Taguchi optimisation, four parameters at four levels have been considered, as shown in Table 2. The minimum number of experimental trials to be conducted can be preset using the following relation:

$$N_{\text{Taguchi}} = 1 + \text{nv} \, (L - 1) \tag{8}$$

where N_{Taguchi} is the minimum number of experiments or cases required to be conducted, nv is the number of variables or control parameters, and L is the number of levels selected. For the present study, nv = 4 and L = 4. Therefore, a minimum of thirteen computational trials have to be conducted, and the nearest orthogonal array is L16, and

Table 2 EATHE factors considered at different levels in Taguchi method

Factor	Parameter	Level			
		L1	L2	L3	L4
A	Diameter of pipe (m)	0.10	0.15	0.20	0.25
B	Length of pipe (m)	30	40	50	60
C	Inlet air velocity (m/s)	2	3	4	5
D	Inlet air temperature (K)	307.35	311.35	315.35	319.35

Table 3 Taguchi L16 orthogonal array

Experiment no.	Factor A	Factor B	Factor C	Factor D
1	0.10	30	2	307.35
2	0.10	40	3	311.35
3	0.10	50	4	315.35
4	0.10	60	5	319.35
5	0.15	30	3	315.35
6	0.15	40	2	319.35
7	0.15	50	5	307.35
8	0.15	60	4	311.35
9	0.20	30	4	319.35
10	0.20	40	5	315.35
11	0.20	50	2	311.35
12	0.20	60	3	307.35
13	0.25	30	5	311.35
14	0.25	40	4	307.35
15	0.25	50	3	319.35
16	0.25	60	2	315.35

hence, L16 is selected for the experimental trials (however, based on the factorial design, $4^4 = 256$ cases are required to be conducted). Each computational trial is carried out according to L16 array combinations, as shown in Table 3 in FLUENT. Higher is better concept has been applied for calculation of S/N ratios (SNR), because maximum air temperature drop and heat transfer rate are the objective of present study.

Results and discussion

The primary aim of this study is to find out the best combination of considered parameters to maximize the air temperature drop and heat transfer rate for achieving the maximum cooling effect from EATHE.

Taguchi method—analysis of S/N ratio

Table 4 shows the standard experimental design of L16 orthogonal array with computed temperature drop in air temperature and total heat transfer rate in EATHE system for cooling as per experimental trials. The S/N ratio values of the temperature difference and heat transfer rate are calculated using higher is better concept. The average responses for S/N ratios for each level of four parameters for temperature difference and heat transfer rate are presented in Tables 5 and 6.

To solve the orthogonal array (Table 3), the steps are given in the following :

- Calculate the higher is better response characteristic for every factor-level combination.
- For every factor, the average response characteristic at each level is calculated using Minitab software.
- For every factor, calculate the delta value.
- Finally, calculate the rank of the factor.

Table 4 Taguchi L16 experimental plan with corresponding temperature drop in air and *S/N* ratios

Experimental no.	Factor A	Factor B	Factor C	Factor D	Outlet air temperature (K)	Temperature drop (response)	SNR	Heat transfer rate (Watt) (response)	SNR
1	0.10	30	2	307.35	300.53	6.82	16.67	129.13	42.22
2	0.10	40	3	311.35	300.70	10.65	20.54	302.47	49.61
3	0.10	50	4	315.35	301.08	14.27	23.08	540.38	54.65
4	0.10	60	5	319.35	301.15	18.20	25.20	861.51	58.70
5	0.15	30	3	315.35	304.84	10.51	20.43	671.62	56.54
6	0.15	40	2	319.35	302.90	16.45	24.32	700.80	56.91
7	0.15	50	5	307.35	301.91	5.43	14.69	579.39	55.25
8	0.15	60	4	311.35	301.58	9.77	19.79	832.44	58.40
9	0.20	30	4	319.35	310.72	8.62	18.71	1307.22	62.32
10	0.20	40	5	315.35	307.43	7.92	17.97	1505.27	63.55
11	0.20	50	2	311.35	301.98	9.37	19.43	715.71	57.09
12	0.20	60	3	307.35	301.52	5.83	15.31	664.59	56.45
13	0.25	30	5	311.35	307.60	3.75	11.48	1109.43	60.90
14	0.25	40	4	307.35	303.98	3.37	10.55	797.60	58.03
15	0.25	50	3	319.35	306.94	12.44	21.89	2202.88	66.86
16	0.25	60	2	315.35	302.87	12.48	21.92	1476.87	63.38

Table 5 Taguchi response table for air temperature drop in EATHE system

Level	Factor A	Factor B	Factor C	Factor D
1	21.38	16.82	20.59	14.31
2	19.81	18.35	19.55	17.82
3	17.86	19.78	18.04	20.85
4	16.46	20.56	17.34	22.53
Delta	4.91	3.73	3.25	8.22
Rank	2	3	4	1

Table 6 Taguchi response table for heat transfer rate in EATHE system

Level	Factor A	Factor B	Factor C	Factor D
1	51.30	55.50	54.90	52.99
2	56.78	57.03	57.37	56.50
3	59.86	58.47	58.36	59.53
4	62.30	59.24	59.60	61.20
Delta	11.00	3.74	4.70	8.21
Rank	1	4	3	2

Figures 5, 6 show the average *S/N* ratio values plots for all four levels of four parameters for the EATHE temperature drop and heat transfer rate, respectively. From Tables 5 and 6, the optimum set of parameters for obtaining maximum air temperature drop and heat transfer rate can be determined by selecting the highest value of *S/N* ratio for each factor. Hence, the optimum control parameter level is A_1 (factor A at level 1), B_4 (factor B at level 4), C_1 (factor C at level 1), and D_4 (factor D at level 4), for maximum air

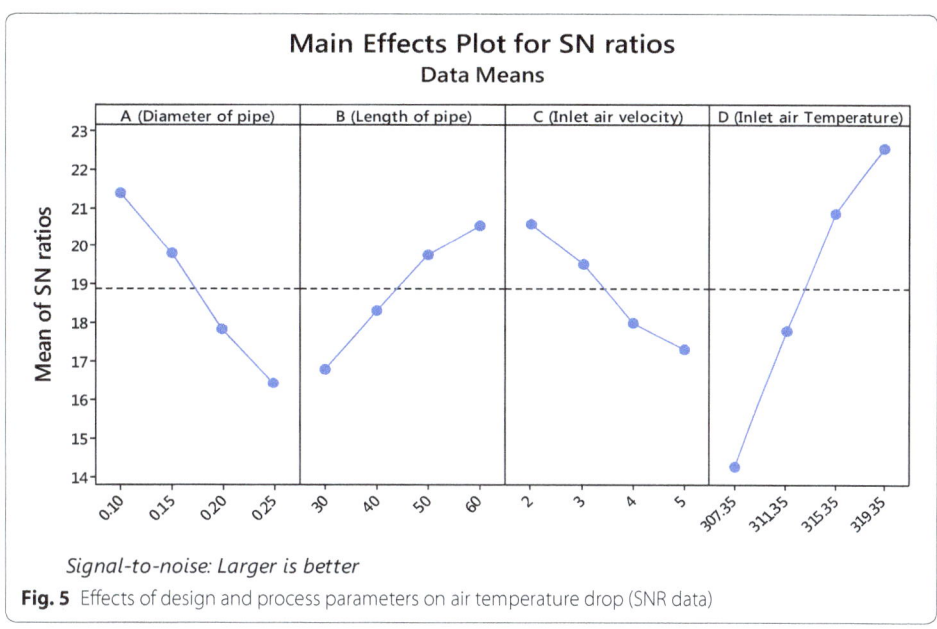

Fig. 5 Effects of design and process parameters on air temperature drop (SNR data)

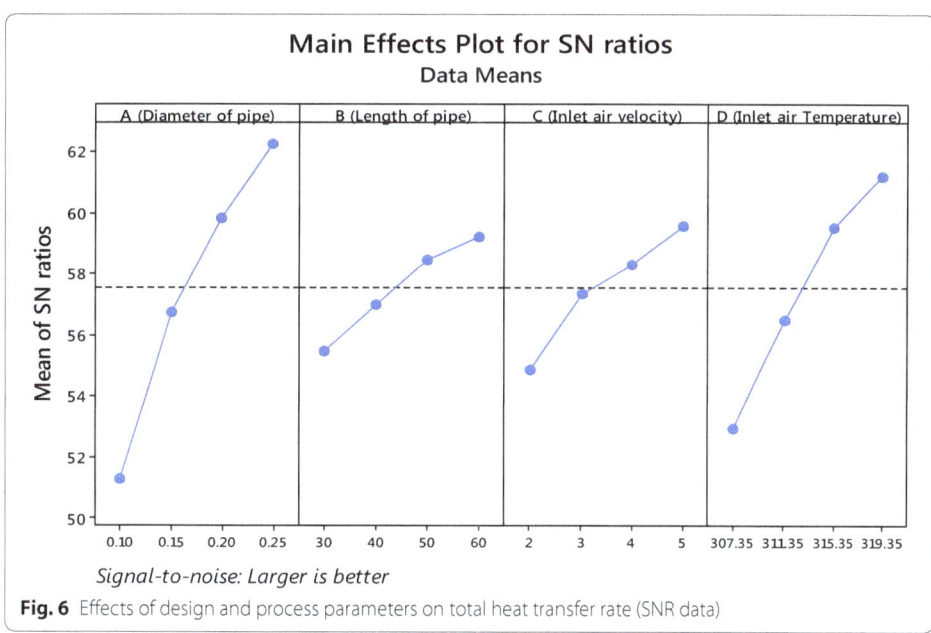

Fig. 6 Effects of design and process parameters on total heat transfer rate (SNR data)

temperature drop. However, optimum control parameter level for maximum heat transfer rate is A_4 (factor A at level 4), B_4 (factor B at level 4), C_4 (factor C at level 4), and D_4 (factor D at level 4).

ANOVA analysis

Analysis of variance (ANOVA) is used to estimate the relative importance of the control parameters by computing the percentage contribution of each parameter in overall response. The sum of squares (SS) and percentage of contribution are included in the ANOVA table (Tables 7 and 8). The parameter with the highest percentage of

Table 7 Contribution of each parameter for maximum air temperature drop in EATHE

Parameter	Factor	$\overline{SNR}$	SS_i	SS	Contribution, %
Diameter of pipe	A	18.87	14.01	7.83	20.66
Length of pipe	B		8.15		12.03
Inlet air velocity	C		6.45		9.51
Inlet air temperature	D		39.21		57.80

Table 8 Contribution of each parameter for maximum heat transfer rate in EATHE

Parameter	Factor	$\overline{SNR}$	SS_i	SS	Contribution, %
Diameter of pipe	A	57.55	67.55	126.78	53.28
Length of pipe	B		8.17		6.45
Inlet air velocity	C		11.91		9.40
Inlet air temperature	D		39.13		30.87

contribution is ranked highest in terms of relative importance among all the control parameters and also has a major contribution to the overall response.

Percentage contribution of each parameter

ANOVA provides the contribution of the individual parameter. The following steps are being used for calculating the percentage contribution of each parameter on different SNR responses.

Step I   Calculation for mean of signal-to-noise response $\overline{(SNR)}$ for all the 16 cases, i.e.,

$$\overline{(SNR)} = \frac{1}{16} \sum_{i=1}^{16} (SNR)_i \tag{9}$$

Step II   Calculating the sum of squares (SS_i) for each parameters based on their SNR response, i.e.,

$$SS_i = \sum_{i=1}^{16} \left((SNR)_i - \overline{(SNR)} \right)^2 \tag{10}$$

Step III   Calculating the sum of squares (SS) for each parameter, i.e.,

$$SS = \sum_{j=1}^{4} \left((SNR)_i - \overline{(SNR)} \right)^2 \tag{11}$$

Step IV   At last, the contribution of each parameter in percentage is determined by the following equation:

$$\text{Contribution, \%} = \frac{\text{Sum of squares (SS) of each parameter}}{\sum_{i=1}^{n} \text{SS}_i} \times 100 \qquad (12)$$

Here, $i =$ for individual values of the sum of squares; $n =$ num of experimental cases in the array.

Table 7 shows that the inlet air temperature has got the maximum influence on air temperature drop obtainable having a percentage contribution of 57.80%. The contribution of other parameters such as diameter of pipe, length of pipe, and inlet air velocity is found to be 20.66, 12.03, and 9.51%, respectively. The air temperature drop produced by EATHE system highly depends on inlet air temperature, because when the difference between inlet air temperature and soil temperature increases, more drop in temperature of air flowing through the buried pipes is achieved.

Table 8 shows the percentage contribution of each parameter for heat transfer rate in an EATHE system. It is seen that the diameter of pipe has a maximum contribution, i.e., 53.28% of the total. The contribution inlet air temperature, inlet air velocity, and length of pipe are found to be 30.87, 9.40, and 6.45%, respectively. The contribution of pipe diameter is found to be highest on heat transfer rate, because the heat transfer rate is directly proportional to the mass flow rate of air which is proportional to the square of the pipe diameter.

In the present study, the percentage contribution of the pipe length is least for heat transfer rate. It is noticed from the present study that the maximum heat transfer occurs in the initial 30 m length of pipe in EATHE system. Therefore, pipe length in the range of 30–60 m does not produce a significant effect on heat transfer rate in EATHE system. Similarly, the contribution of air flow velocity is least for air temperature drop in EATHE, because maximum air temperature drop is achieved at an air velocity of 2 m/s and by increasing air velocity from 2 to 5 m/s; the drop in air temperature is not significantly affected.

Confirmation test—Taguchi method

The optimum combination of parameters is investigated using CFD and compared with the 16 cases of Taguchi method for maximum temperature difference and heat transfer rate. It is found that the optimum case for achieving maximum temperature difference, i.e., $A_1B_4C_1D_4$ gave temperature difference of 19.05 and the optimum case for having maximum heat transfer rate, i.e., $A_4B_4C_4D_4$ provide heat transfer rate of 3085 W which are higher than all the previous experimental trials taken in Taguchi method.

Conclusions

In this study, a methodology is proposed to maximize the air temperature drop and heat transfer rate in EATHE system for cooling applications using the Taguchi method. For this purpose, four operating parameters at four levels of operation of the EATHE system were considered. A total of 16 trials were run using a validated simulation model in the ANSYS FLUENT software. ANOVA is employed to optimise the operating parameters. The major findings and observations of the study are the following:

- Taguchi optimisation analysis inferred that, for the EATHE system, inlet air temperature is the most influencing (57.80%) control parameter in cooling mode for achieving maximum air temperature drops and the diameter of pipe has a maximum contribution (53.28%) in maximizing the heat transfer rate in EATHE system.
- The least influencing parameter for a temperature drop of air is air velocity while that for the heat transfer rate is pipe length.
- The air temperature drop and heat transfer rate characteristics are very much influenced by the geometric and flow parameters (control factors), viz., diameter of the pipe, length of pipe, air flow velocity, and inlet air temperature. The contribution ratio of each of these parameters on air temperature drop is 20.66, 12.03, 9.51, and 57.80% respectively, and that of heat transfer rate is 53.28, 6.45, 9.40, and 30.87% respectively.
- The optimum combination of parameters for achieving a maximum drop in air temperature is $A_1B_4C_1D_4$ and that for obtaining maximum total heat transfer rate is $A_4B_4C_4D_4$.

Highlights

- Performance of earth-air-tunnel heat exchanger (EATHE) system has been investigated using CFD simulation.
- Effect of geometric and flow parameters on the performance of EATHE has been analysed.
- Taguchi method is applied for maximization of air temperature drop and heat transfer rate in EATHE system.
- Analysis of variance (ANOVA) and signal to noise (S/N) ratio is used for evaluation of simulation results.
- Contribution ratio of each control factor has been evaluated.

Abbreviations
C_p: specific heat of air (J/kg K); $\dot{m}$: mass flow rate of air (kg/s); Q: heat transfer rate (W); T_i: inlet air temperature (K); T_o: outlet air temperature (K); ΔT: temperature difference $(T_i - T_o)$; ANN: artificial neural network; ANOVA: analysis of variance; CFD: computational fluid dynamics; EAPHE: earth air-pipe heat exchanger; EATHE: earth air tunnel heat exchanger; HEAHE: hybrid earth to air heat exchanger; GA: genetic algorithm; PVC: poly-vinyl chloride; RSM: response surface methodology; SNR: signal-to-noise (S/N) ratios.

Authors' contributions
All authors contributed equally towards preparing this manuscript. All authors read and approved the final manuscript.

Author details
[1] Mechanical Engineering Department, Malaviya National Institute of Technology, Jaipur 302017, India. [2] Department of Renewable Energy, Rajasthan Technical University, Kota 304010, India. [3] Mechanical Engineering Department, Government Engineering College, Ajmer, Ajmer 305001, India.

Acknowledgements
The first author is thankful to the Ministry of Human Resources and Development, Government of India, for providing the fellowship for pursuing Ph.D. at Malaviya National Institute of Technology Jaipur, Jaipur, India.

Competing interests
The authors declare that they have no competing interests.

Funding
Not applicable.

References

A friendly guide to Minitab. Acad. Support Cent. 2015. https://www.rit.edu/studentaffairs/asc/sites/.../s7_minta bguide_BP_3_17_2015.pdf. Accessed 13 Oct 2018.

Ahmed SF, Amanullah MTO, Khan MMK, Rasul MG, Hassan NMS. Parametric study on thermal performance of horizontal earth pipe cooling system in summer. Energy Convers Manag. 2016;114:324–37. https://doi.org/10.1016/j.enconman.2016.01.061.

ANSYS Inc. ANSYS FLUENT User's Guide ANSYS. 2013. http://cdlab2.fluid.tuwien.ac.at/LEHRE/TURB/Fluent.Inc/v140/flu_ug.pdf. Accessed 12 Oct 2017.

Bansal NK, Sodha MS. An earth-air tunnel system for cooling buildings. Tunn Undergr Space Technol. 1986;1(2):177–82.

Bansal NK, Sodha MS, Bharadwaj SS. Performance of earth air tunnels. Energy Res. 1983;7:333–45.

Bansal V, Misra R, Agrawal GD, Mathur J. Performance analysis of earth pipe air heat exchanger for winter heating. Energy Build. 2009;41:1151–4.

Bansal V, Misra R, Das Agrawal G, Mathur J. Performance analysis of earth pipe air heat exchanger for summer cooling. Energy Build. 2010;42:645–8.

Benhammou M, Draoui B. Parametric study on thermal performance of earth-to-air heat exchanger used for cooling of buildings. Renew Sustain Energy Rev. 2015;44:348–55. https://doi.org/10.1016/j.rser.2014.12.030.

Bharadwaj SS, Bansal NK. Temperature distribution inside ground for various surface conditions. Build Environ. 1981;16(3):183–92.

Bisoniya TS, Kumar A, Baredar P. Cooling potential evaluation of earth-air heat exchanger system for summer season. Int J Eng Tech Res. 2014;2(4):309–16.

Derbel HBJ, Kanoun O. Investigation of the ground thermal potential in tunisia focused towards heating and cooling applications. Appl Therm Eng. 2010;30:1091–100.

Diaz SE, Sierra JMT, Herrera JA. The use of earth—air heat exchanger and fuzzy logic control can reduce energy consumption and environmental concerns even more. Energy Build. 2013;65:458–63.

Dubey MK, Bhagoria J, Atullanjewar. Earth air heat exchanger in parallel connection. Int J Eng Trends Technol. 2013;4(6):2463–7.

Elminshawy NAS, Siddiqui FR, Farooq QU, Addas MF. Experimental investigation on the performance of earth-air pipe heat exchanger for different soil compaction levels. Appl Therm Eng. 2017;124:1319–27. https://doi.org/10.1016/j.applthermaleng.2017.06.119.

Ghosal MK, Tiwari GN. Modeling and parametric studies for thermal performance of an earth to air heat exchanger integrated with a greenhouse. Energy Convers Manag. 2006;47:1779–98.

Inc. M. Getting Started with Minitab 17, 2016. https://www.minitab.com/uploadedFiles/Documents/getting-started/Minitab17_GettingStarted-en.pdf. Accessed 13 Oct 2017.

Kabashnikov VP, Danilevskii LN, Nekrasov VP, Vityaz IP. Analytical and numerical investigation of the characteristics of a soil heat exchanger for ventilation systems. Int J Heat Mass Transf. 2002;45:2407–18.

Kaushal M, Dhiman P, Singh S, Patel H. Finite volume and response surface methodology based performance prediction and optimization of a hybrid earth to air tunnel heat exchanger. Energy Build. 2015;104:25–35. https://doi.org/10.1016/j.enbuild.2015.07.014.

Krarti M, Kreider JF. Analytical model for heat transfer in an underground air tunnel. Energy Convers Manag. 1996;37(10):1561–74.

Kumar R, Kaushik SC, Garg SN. Heating and cooling potential of an earth-to-air heat exchanger using artificial neural network. Renew Energy. 2006;31(8):1139–55.

Kumar R, Sinha AR, Singh BK, Modhukalya U. A design optimization tool of earth-to-air heat exchanger using a genetic algorithm. Renew Energy. 2008;33:2282–8.

Lee KH, Strand RK. The cooling and heating potential of an earth tube system in buildings. Energy Build. 2008;40:486–94.

Mathur A, Srivastava A, Agrawal GD, Mathur S, Mathur J. CFD analysis of EATHE system under transient conditions for intermittent operation. Energy Build. 2015;87:37–44. https://doi.org/10.1016/j.enbuild.2014.11.022.

Mihalakakou G. On the heating potential of a single buried pipe using deterministic and intelligent techniques. Renew Energy. 2003;28:917–27.

Mihalakakou G, Santamouris M, Asimakopoulos D. On the cooling potential of earth to air heat exchangers. Energy Convers Manag. 1994a;35(5):395–402.

Mihalakakou G, Santamouris M, Asimakopoulos D. Modeling the thermal performance of earth-to-air heat exchangers. Sol Energy. 1994b;53(3):301–5.

Mihalakakou G, Santamouris M, Asimakopoulos D, Papanikolaou N. Impact of ground cover on the efficiencies of earth-to-air heat exchangers. Appl Energy. 1994c;48:19–32.

Mihalakakou G, Lewis JO, Santamouris M. The influence of different ground covers on the heating potential on earth to air heat exchangers. Renew Energy. 1996a;7(1):33–46.

Mihalakakou G, Lewis JO, Santamouris M. On the heating potential of buried pipes techniques—application in Ireland. Energy Build. 1996b;24:19–25.

Nejat P, Jomehzadeh F, Taheri MM, Gohari M, Abd Majid MZ. A global review of energy consumption, CO_2 emissions and policy in the residential sector (with an overview of the top ten CO_2 emitting countries). Renew Sustain Energy Rev. 2015;43:843–62. https://doi.org/10.1016/j.rser.2014.11.066.

Niu F, Yu Y, Yu D, Li H. Heat and mass transfer performance analysis and cooling capacity prediction of earth to air heat exchanger. Appl Energy. 2015;137:211–21. https://doi.org/10.1016/j.apenergy.2014.

Omer AM. Energy, environment and sustainable development. Renew Sustain Energy Rev. 2008;12:2265–300.

Rodrigues MK, da Silva Brum R, Vaz J, dos Rocha ED, Santos LAO, Isoldi LA. Numerical investigation about the improvement of the thermal potential of an earth-air heat exchanger (EAHE) employing the constructal design method. Renew Energy. 2015;80:538–51. https://doi.org/10.1016/j.renene.2015.02.041.

Santamouris M, Mihalalalou G, Balaras CA, Argiriou A, Asimakopoulos D, Vallindras M. Use of buried pipes for energy conversion in cooling of agricultural greenhouses. Sol Energy. 1995;55(2):111–24.

Sodha MS, Mahajan U, Sawhney RL. Thermal performance of a parallel earth air-pipes system. Int J Energy Res. 1994;18:437–47.

Vidhi R, Goswami DY, Stefanakos E. Supercritical Rankine cycle coupled with ground cooling for low temperature power generation. Energy Procedia. 2014a;57:524–32. https://doi.org/10.1016/j.egypro.2014.10.206.

Vidhi R, Goswami DY, Stefanakos EK. Parametric study of supercritical Rankine cycle and earth-air-heat- exchanger for low temperature power generation. Energy Procedia. 2014b;49:1228–37. https://doi.org/10.1016/j.egypro.2014.03.132.

Wang H, Qi C, Wang E, Zhao J. A case study of underground thermal storage in a solar-ground coupled heat pump system for residential buildings. Renew Energy. 2009;34(1):307–14.

Wu H, Wang S, Zhu D. Modelling and evaluation of cooling capacity of earth—air—pipe systems. Energy Convers Manag. 2007;48:1462–71.

Yusof TM, Ibrahim H, Azmi WH, Rejab MRM. Thermal analysis of earth-to-air heat exchanger using laboratory simulator. Appl Therm Eng. 2018;134:130–40. https://doi.org/10.1016/j.applthermaleng.2018.01.124.

Zhang J, Haghighat F. Convective heat transfer prediction in large rectangular cross-sectional area earth-to-air heat exchangers. Build Environ. 2009;44(9):1892–8. https://doi.org/10.1016/j.buildenv.2009.01.011.

Zhang J, Haghighat F. Development of artificial neural network based heat convection algorithm for thermal simulation of large rectangular cross-sectional area earth-to-air heat exchangers. Energy Build. 2010;42(4):435–40. https://doi.org/10.1016/j.enbuild.2009.10.011.

4

Borehole damaging under thermo-mechanical loading in the RN-15/IDDP-2 deep well: towards validation of numerical modeling using logging images

M. Peter-Borie[1]*[iD], A. Loschetter[1], I. A. Merciu[2], G. Kampfer[2] and O. Sigurdsson[3]

*Correspondence:
m.peter-borie@brgm.fr
[1] BRGM, 3 av. C. Guillemin,
BP36009, 45060 Orléans
Cedex 2, France
Full list of author information
is available at the end of the
article

Abstract

A wider exploitation of deep geothermal reservoir requires the development of Enhanced Geothermal System technology. In this context, drilling and stimulation of high-enthalpy geothermal wells raise technical challenges. Understanding and predicting the rock behavior near a deep geothermal wellbore are decisive to implement stimulation strategies to reach the couple temperature/flowrate target. Numerical modeling can contribute to enhanced stimulation processes thanks to a better understanding of impact of stress release, pressure changes and rock cooling in the near-wellbore area. In this paper, we use Discrete Element Method (code PFC2D, © Itasca Consulting Group), and more specifically bonded-particle model to capture the thermo-mechanical processes at metric scale. The application case corresponds to the beginning of thermal stimulation at Reykjanes in well RN-15/IDDP-2 (Iceland, IDDP-2 project and H2020 project DEEPEGS). A cold fluid is injected at a depth of 4.5 km where the rock temperature is above 430 °C and the well pressure is around 34 MPa. Since we have site-specific data and logging images after drilling, we attempt to link the simulations with the reality. The numerical results are confronted with incipient interpretation of logging images and with analytical solution to go towards validation of the modeling approach. Numerical results show breakouts and thermally and/or mechanically induced fractures consistent with the analytical solutions. Moreover, the sensitivity analysis on uncertain parameters yields important clues regarding some logging features as, for example, asymmetric damaging or caving.

Keywords: EGS (Enhanced Geothermal System), Borehole, Thermal stimulation, Fracture initiation, DEM (Discrete Element Model), PFC2D, LWD, Borehole logging images

Background

EGS (enhanced/engineered geothermal system) constitutes a potential renewable energy technology to produce heat and electricity from geothermal reservoirs deficient in fluid or in permeability. In most cases, reaching an economically viable temperature target requires drilling down to several kilometers depth, where the permeability of the system is generally naturally low. The implementation of stimulation strategies is then necessary to increase the injectivity or the productivity of the wells (Tester et al. 2006). The deployment of such EGS method in a wide range of geological contexts is still a technical challenge, and the number of projects in operation is currently limited. According to EGEC (2017), concerning EGS, three electricity plants (Insheim and Landau in Germany, Soultz-sous-Forêts in France; the reservoir temperatures are, respectively, > 160 °C, 160 °C and > 180 °C; Lu 2018) and one heat plant (Rittershoffen in France, the reservoir temperature is 177 °C; Baujard et al. 2017) are now in operation, with a further ten plants under development.

In this context, the EU-funded H2020 DEEPEGS (Deployment of DEEP Enhanced Geothermal Systems for sustainable energy business) project aims at demonstrating the feasibility of EGS in high-enthalpy reservoirs (temperature up to 550 °C, with an Icelandic demonstrator) and in deep hydrothermal reservoirs (temperatures around 200 °C, with French demonstrators), to deliver new innovative solutions and models for wider deployments of EGS. The first demonstrator deployed in the frame of the DEEPEGS project is located in the Reykjanes geothermal system in southwest Iceland. The deepening of the wellbore RN-15 from 2500 m depth (RN-15/IDDP-2) began in August 2016 and the well is completed at a depth of 4659 m MD (measured depth, ~ 4.5 km vertical depth) in January 2017 (temperature around 500–530 °C) (Friðleifsson et al. 2017; Friðleifsson and Elders 2017a, b; Stefanson et al. 2017). RN-15/IDDP-2 provides information about the deep geology and the deep rock behavior in the Icelandic context. Notably, borehole images provide a unique view of the geological structure of the Icelandic crust. To improve the productivity of the well (estimated injectivity index around 1.7 L s^{-1} bar^{-1} at the end of drilling), stimulations in the form of cold-water injection (mainly thermal stimulations) have been performed to connect the wellbore to existing hydraulic pathways, i.e., pre-existing natural fracture network.

Understanding all the processes that lead to fracture initiation in the EGS near-wellbore remains challenging due to the high temperatures. In this context, numerical modeling contributes to improve our understanding and it allows for predictions in the future. The objective of this article is to propose a physical modeling approach contributing to the understanding of phenomena occurring in the wellbore vicinity during drilling and EGS operations. We focus on the thermo-mechanical processes induced by the rock cooling. The choice of the numerical method is based on assumptions drawn from onsite information. The results are compared with results from analytical equations and with observations made during drilling to critically discuss the numerical results and go towards validation of the chosen approach.

In this paper, after briefly describing the geological and geothermal context, we first present data and borehole images from well RN-15/IDDP-2. Then we describe the numerical modeling tool, based on the Discrete Element Method (DEM), and the chosen setup for the numerical simulations. Numerical results show breakouts and thermally

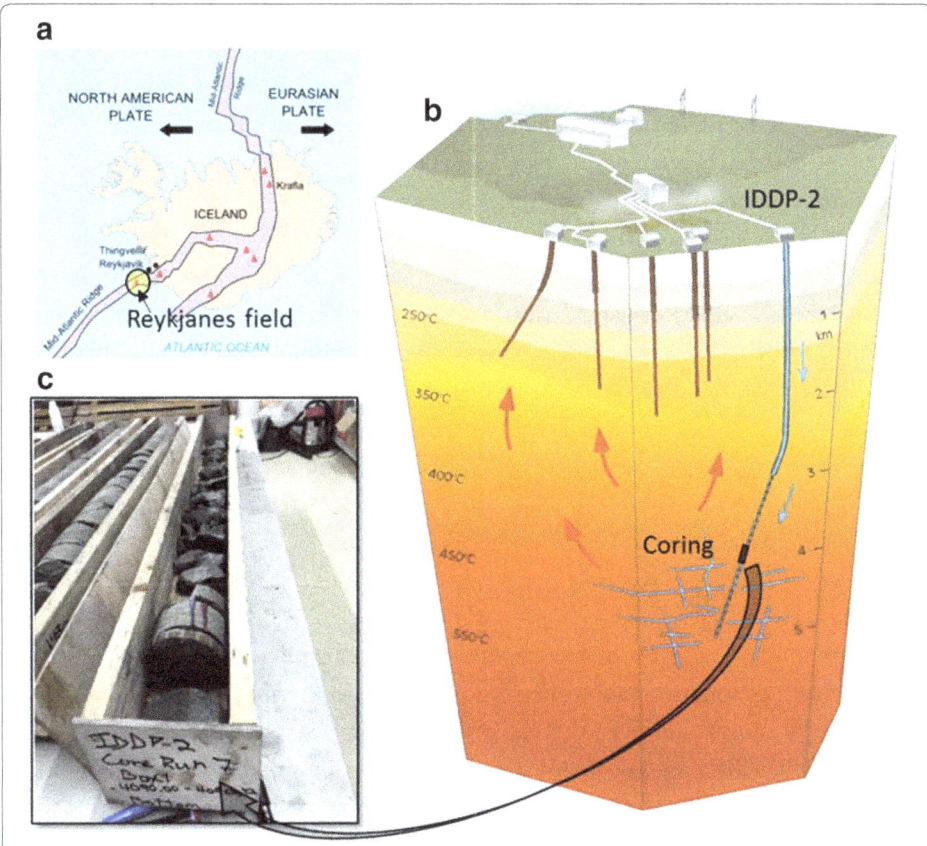

Fig. 1 **a** Map of the major deformation zones in Iceland and Reykjanes field location. The purple area corresponds to the Mid-Atlantic Ridge. Red triangles correspond to the locations of some Iceland's active volcanoes. **b** The RN-15/IDDP-2 wellbore configuration (modified from Friðleifsson et al. 2016). The blue arrows symbolize the injection of cold fluid while the red ones symbolize the moving of warmed fluid to the production wells. **c** Picture of the dolerite cored at 4-km depth (courtesy of HS Orka)

and/or mechanically induced fractures consistent with the analytical solutions and with observations made during drilling. We finish with words of conclusion and with discussion on the experienced limitations and perspectives.

Geological knowledge

Regional data

The in situ geological, mechanical and thermal conditions are little known in the deep part of the Reykjanes field. The deepest well in this area was shallower than 3 km before drilling the IDDP-2 well. Besides, geophysical methods are limited for investigations at several kilometers depth. We summarize below the information concerning the regional geology, the regional stress state and the rock behavior.

Geology

The Reykjanes geothermal system is located at the tip of the Reykjanes peninsula, SW Iceland (Fig. 1a) at the landward extension of the Reykjanes Ridge. From the surface to around 2.5 km depth, the lithology consists of sub-aerial basaltic lavas and to a lesser degree of hyaloclastites. Below, typical sheeted dyke complex of an ophiolite is assumed to take place, including a swarm of tectonic fractures and faults. These intrusive rocks are assumed to overlay a lower gabbroic crust (Pálmason 1970; Gudmundsson 2000; Foulger et al. 2003; Karson 2016; Stefanson et al. 2017; Friðleifsson and Elders 2017b).

Regional stress state

The in situ stress field is poorly characterized in the deep part of the Reykjanes field. The World Stress Map (Heidbach et al. 2008, 2016) indicates that the stress regime varies by short distances around Reykjanes. Most data (e.g., Ziegler et al. 2016 in the vicinity of wellbore RN-15/IDDP-2) consist of principal stress directions, with no indication of the stress magnitudes. It is not even certain that the vertical direction is a principal stress axis (Keiding and Lund 2009; Kristjánsdóttir 2013). Pieces of information concerning the orientation and magnitude of principal stresses were found in Keiding and Lund (2009), Batir et al. (2012), and Kristjánsdóttir (2013) but the characterization of in situ stress at such depth in this complex area remains very uncertain.

Rock behavior

Foulger et al. (2003) suggest that the brittle–ductile transition occurs deeper than the targeted depth considering gabbro-like rocks and the geothermal gradient. The analysis of earthquake swarms indicates that the brittle–ductile boundary is at 5.5–6 km depth under Reykjanes (Khodayar et al. 2017), thus below the considered depth. Observations from core retrieved between 4643 and 4652 m MD show fractures, which are supposed to be open and fluid-filled downhole, indicated by precipitations on the fracture surface. For that reason, we only assume brittle formation behavior in the presented study.

Site-specific data acquired during drilling operations

The drilling of IDDP-2 provides new knowledge concerning the rock composition, the rock properties and the in situ temperature at depth.

Rock composition

In-depth logging and coring lend credibility to the thesis of sheeted dyke complex. Cores retrieved from 4 km depth show mainly rocks with fine-grained igneous texture: micro-gabbro/dolerite to fine-grained basaltic intrusive (cf. Fig. 1c), with heterogeneous grain size (Friðleifsson et al. 2017). The mineral composition was assessed (see "Numerical settings and scenarios" section) and the porosity is found to be very low (matrix porosity between 3.6 and 0.1%—Claudia Kruber, Equinor internal report in progress).

Rock properties

Knowing the rock mineralogy and an estimate of the in situ temperature range, we can use results of Keshavarz (2009) to confirm the assumption of brittle rock behavior. His experimental results indeed show that the physical and mechanical properties of this gabbro remain on the same trend up to the critical temperature of 600 °C, thus sufficiently above the estimated formation temperature in the IDDP-2 well.

Rock temperature

At the end of drilling, the fluid temperature measured at 4560 m MD of IDDP-2 was 426 °C (Friðleifsson et al. 2017), after the deepest part of the well had the possibility to warm up for 6 days. It should be noted that this measurement is probably an underestimation of the in situ formation temperature since extensive cooling occurred during drilling the well. The in situ formation temperature was estimated in the range 536–549 °C, based on warm-up measurements and a Horner plot at 4565 m MD (Tulinius 2017).

Borehole response to drilling

During drilling, shear and tensile rock failures may threaten wellbore stability. We mainly distinguish between shear failure-induced breakouts and drilling-induced fractures (opening mode fractures) as the two main sets of mechanical instabilities when drilling with overly low and overly high mud weights, respectively. Breakouts are aligned with the minimum horizontal stress whereas drilling-induced fractures are aligned with the maximum horizontal stress in a vertical well. In the present case, severe mud losses were observed during drilling, leading to the conclusion that the pressure in the well exceeded the minimum compressive hoop stress around the wellbore, inducing a drilling-induced fracture or opening a pre-existing fracture. Since the volumes of mud loss are high, it is very likely due to leakage into a naturally existing fracture network (swarm of fractures of the sheeted dyke complex). Either the well directly crossed such a discontinuity, or induced damages connected the wellbore and natural discontinuities. Logging images (see next section) give insight into possible damages in the wellbore. It should be noted that as a consequence of the total mud loss, it was not possible to influence the well pressure. The pressure measured at the bottom of the well after completing the drilling operations was 34 MPa (thus below the hydrostatic pressure expected at such depths, around 45 MPa) and is supposed to represent an equilibrium between gain and loss from the formation along the whole open section which intersects several fracture zones of different productivity and injectivity, respectively. The low pressure, however, can also result from the change in density as the formation water is heated up to above-supercritical reservoir conditions.

Logging images

Borehole images recorded during drilling campaign of IDDP-2 well provide a unique view into the geological structure of the Icelandic crust.

For a selected IDDP-2 drilled interval from 2940 to 3410 m MD in a 21.6-cm (8.5 in.) hole, two sets of images are available for our exemplification: ultrasonic images and electrical microimages (Stefanson et al. 2017) with no caliper measurement available. Ultrasonic amplitude images are collected using wireline standard televiewer ABI 43 (ALT advanced logic technology 2018) from 9 5/8 in. (24.5 cm) casing shoe at 2940 m MD down to 3410 m MD. Electrical microimages are collected using Logging While Drilling SineWave™ Micro-Imager Tool (Weatherford International logging while drilling SinWave 2018) images from 9 5/8 in (24.5 cm) casing shoe at 2940 m MD to 4513 m MD (Friðleifsson et al. 2017; Friðleifsson and Elders 2017a, b; Stefanson et al. 2017).

Ultrasonic amplitude image data are affected by poor centralization and lack of measurement references at surface (Stefanson et al. 2017). The eccentralization of the sensor affects heavily the reflection coefficients that may be extracted from the amplitude envelope and post-processing artifacts are present on the images in the form of vertical shades. Further processing on ultrasonic amplitude images is limited due to challenging acquisition conditions.

Electrical microimages are affected by the high resistivity of the formation and the general raw values are accumulating towards 0 mA and exhibit sandy texture in dynamic normalization window which is further corrected by applying a median filter.

As a large data integration effort is ongoing at the time of this publication, we are selecting representative image examples for the scope of numerical simulation validation with intervals where both ultrasonic and electrical microimages are recorded and refraining from an in-depth evaluation of logging results.

Qualitative analysis of recorded images reveals a feature-rich borehole with clear evidence of vertical drilling-induced features and petals in both static and dynamic normalization window especially on ultrasonic amplitude images (Fig. 2) (Menger 1994; Deltombe and Schepers 2001; Holl and Barton 2015). The borehole breakouts manifest themselves largely in images with a clear 180° opposite directions (Fig. 2). Tensile fractures and other drilling-induced fractures manifest themselves at about 90° azimuth with respect to observable large breakouts forming rib-like structures which may emerge in larger petals—centerline features visible especially on the ultrasonic image towards the bottom of exemplified image, see Figs. 2 and 3 (Davatzes and Hickman 2005; Tingay et al. 2008; Rajabi et al. 2016). A closer look into lateral extension of the breakout manifestation in electrical microimage compared with ultrasonic image shows that the aperture extracted from the electrical microimage is three orders larger than the aperture observed on ultrasonic images. Furthermore, we chose an exemplification interval where the eccentralization of ultrasonic images is not very prominent and display the images side by side (3 times 360°) to eliminate visual obstruction at azimuths 0° and 360°. Images are displayed in Fig. 4 and reveal aggressive hole damages with petal features which can be observed along the well especially in ultrasonic images. These observations, corroborated with core sample analyses (Zierenberg et al. 2017) and inverse multigeophysical inversion (Hokstad and Tanavasuu-Milkeviciene 2017), lead to hypothesizing a mechanism of fracturing which is driven by temperature, low pressure and intersections with vertical sheeted dyke structure. The scope of the current article is to investigate further the initial fracture mechanism based on thermomechanical stress mechanism.

(See figure on next page.)

Fig. 2 Example of electrical microimage and ultrasonic image from RN-15/IDDP-2 wells with associated preliminary feature extractions (courtesy of Equinor and HS Orka). Observe large variation on azimuthal breakouts picking given by angular extension of the feature on the given images. Vertical white stripe on the ultrasonic image is an artifact from tool eccentralization, which adds to complexity of analysis

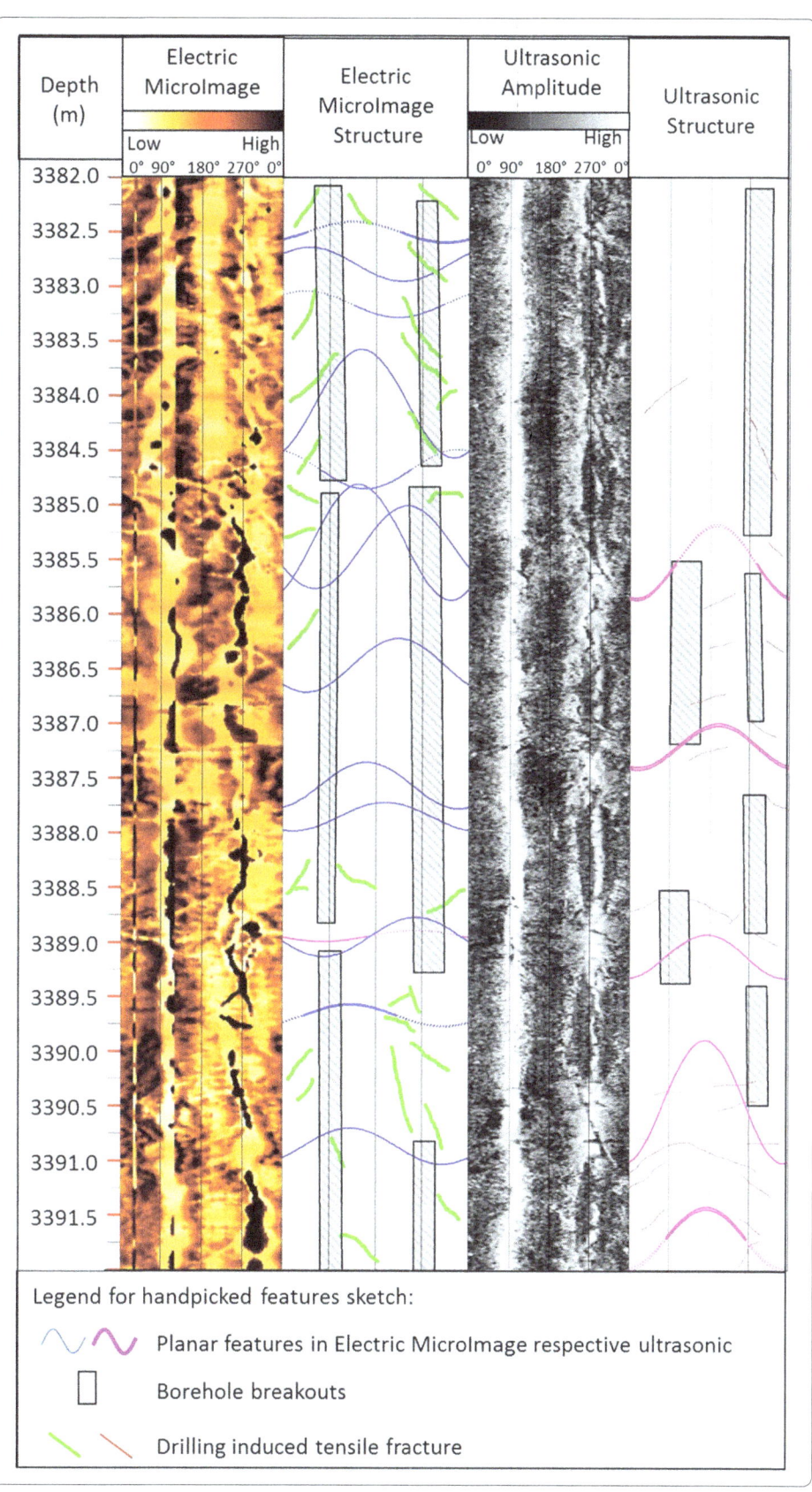

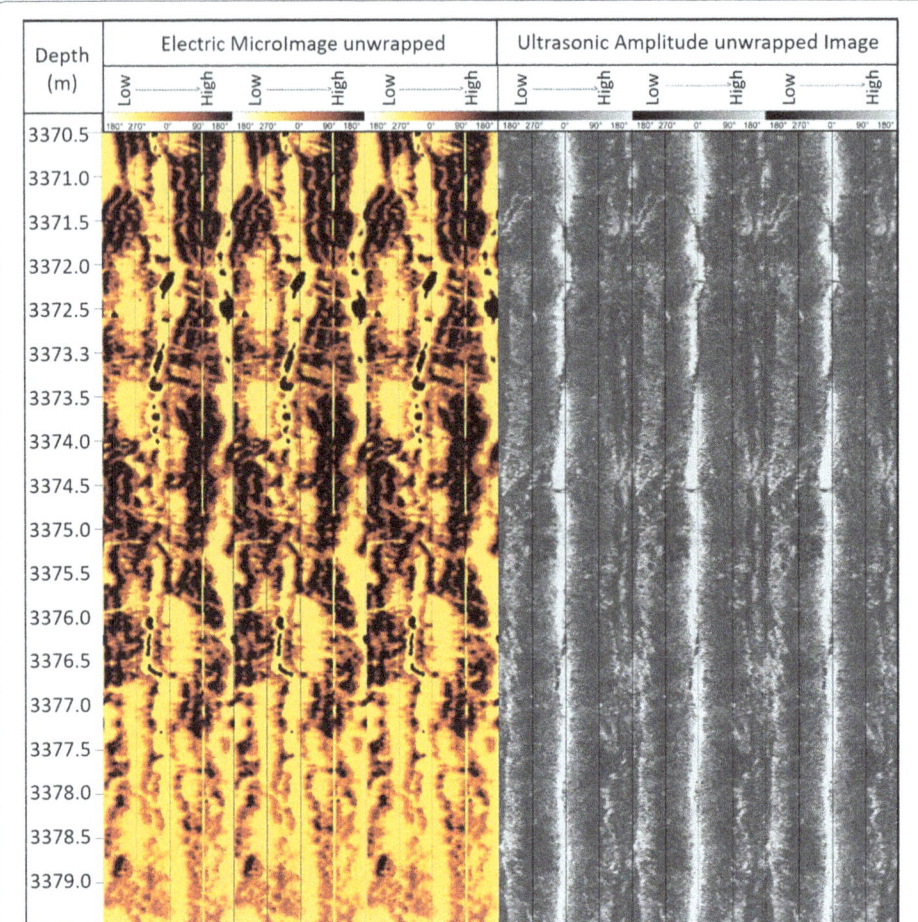

Fig. 3 Electrical microimage vs. ultrasonic image, side by side unwrapped three times. Vertical scale in meters. We can observe petal centerline structures extending from 3371.5 to 3374 m on both images (courtesy of Equinor and HS Orka)

Numerical simulations

Observation-driven modeling

Based on data and observations, the following main assumptions are held:

- The rock matrix has brittle elasto-plastic material properties.
- The porosity of the matrix is below 3%. As a consequence, it is considered reasonable to neglect the poroelastic effects (referring to the poroelasticity theory, this would mean assuming a zero Biot coefficient, which can be supported in such a situation, see Fjar et al. (2008), sections 1.3, 2.9 and 6.2; subsequently effective stresses are simplified and assumed as total stresses).
- The hole stability is ensured mainly by the rock matrix rigidity, with little influence from fluid pressure in the fractures of the surrounding rock.
- The rock has a fine-grained texture, with heterogeneous grain size and different mineral grain composition.

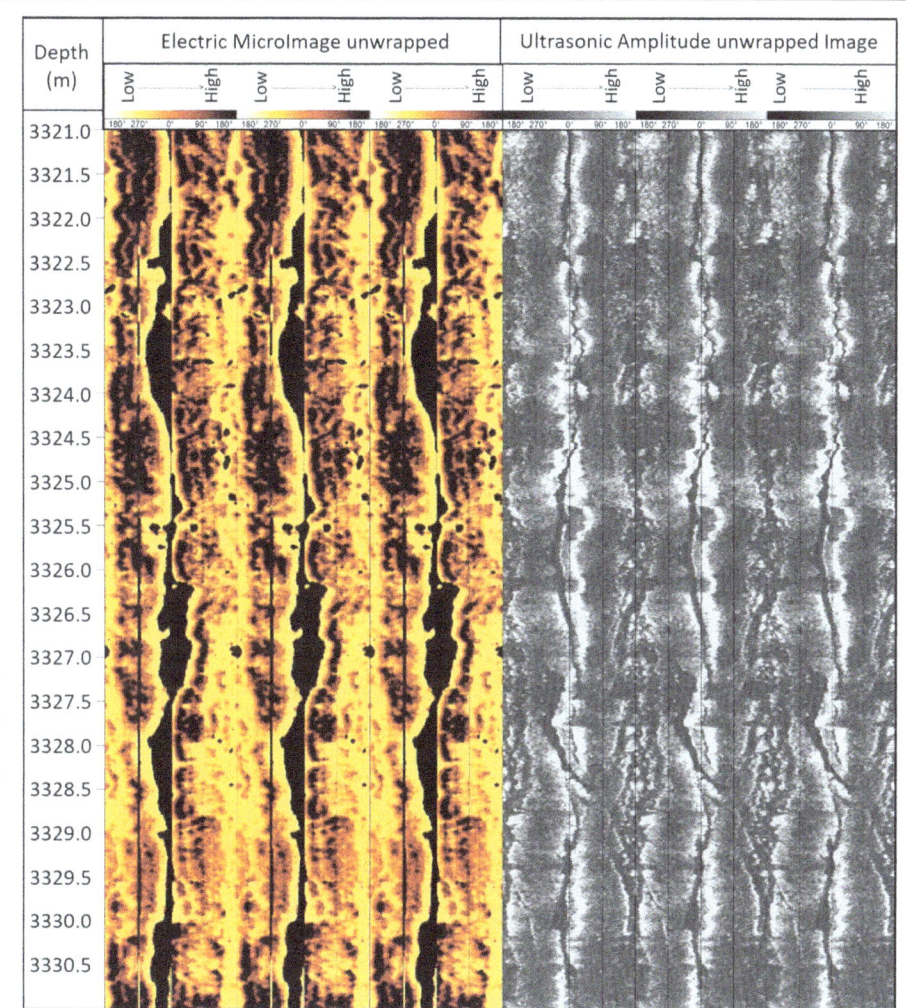

Fig. 4 Electrical microimage vs. ultrasonic image, side by side unwrapped three times. We can observe aggressive damage of the borehole in the ultrasonic image with variation of vertical drilling-induced breakout on both aperture and shape along the borehole. The electrical microimage presents larger aperture on breakout and inverted behavior on azimuthal opposite breakouts (courtesy of Equinor and HS Orka)

The stress state has a strong influence on the failure initiation and propagation, but we lack quantitative data to make solid assumptions. Hence, it was decided to work with a series of possible scenarios for the stress state (see "Numerical approach" section).

Logging images reveal numerous features, between other breakouts and induced fractures. From the drilling operation without any return of fluid to surface (for depths beyond around 3300 m) we can assume that the pressure in the wellbore during drilling was very close to the pressure in the fluid-filled fracture systems intersected by the well. With this limited hydraulic pressure in the well, it is not expected to observe drilling-induced fractures in conventional wells. A possible explanation for this observation is cooling-induced fracture (e.g., Yan et al. 2014). In such high-temperature environments, cooling of the rock necessarily occurs during drilling operations (even before dedicated thermal stimulation). In the following, we provide insight into the initial fracture mechanism based on thermomechanical stress mechanism.

The role of the thermal stimulation is often unclear, and determining which mechanisms lead to observed injectivity increase is still challenging (Flores et al. 2005; Grant et al. 2013; Héðinsdóttir 2014). Covell (2016) shows that thermal stimulation is driven by thermal contraction caused by the significant temperature difference between cold injection fluid and hot reservoir rock. The involved mechanisms lead to opening pre-existing discontinuities (contraction of discontinuity walls) or creating new ones. The thermal solicitation induces differential strains at the origin of thermo-mechanical stresses. When these stresses exceed the mechanical resistance of the rock, micro-cracks and failures could appear. Strains at the origin of this process can be mainly due to two causes: on the one hand, a thermal gradient in the rock mass, on the other hand, the heterogeneity of the grain contraction in the rock matrix. Because of this heterogeneity, two adjacent minerals can contract at different rates and this can generate uneven strains at the grain boundary (Wanne and Young 2008). In addition, petrographic characteristics (including grain size, grain shape, packing density, packing proximity, degree of interlocking, type of contacts and mineralogical composition) are known to affect mechanical properties (Ulusay et al. 1994). A critical review concerning DEM and its application to borehole stability was proposed by Kang et al. (2009). Santarelli et al. (1992) were among the first to study borehole stability using DEM. Yamamoto et al. (2002) used DEM to study the wellbore instability of laminated and fissured rocks. Karatela et al. (2016) studied the effect of in situ stress ratio and discontinuity orientation on borehole stability in heavily fractured rocks using DEM. DEM seems fairly adapted to take into account the physical phenomena at the granular phase level (micro scale), and to analyze their impact on the mechanical behavior of the near-wellbore zone (macro scale). We propose to implement this approach using the code Particle Flow Code—2 Dimensions (PFC2D) (Itasca Consulting Group Inc. 2008a, b), and to question the role of thermal loadings in the wellbore. Chemical interaction of the drilling fluid may play a role in the thermal stimulation, for instance, through dissolution or precipitation of the minerals, triggered by temperature change. The quantification of these chemical effects in the specific context of IDDP-2 remains a scientific challenge. Thus, in the absence of available data, indirect chemical effects of thermal stimulation are not considered in this study.

It is worth mentioning that the thermal stimulation by injecting cold water may not necessarily have a long-term effect because of the thermal expansion and closure of fractures during production. Only the naturally propped fractures keep some permeability and hence improve productivity.

Contrary to common analytical approaches (see Appendix), the proposed numerical approach enables quantifying the depth and shapes of damages. The results will be compared with the logging observations (breakouts, induced fractures, petal fractures), trying to identify what the simulation captures successfully and what it does not. Note as a limit of the method that the logging is performed several days after the drilling: throughout this time lapse, the well has been exposed to more mechanical and thermal stresses than simulated in the numerical approach.

Numerical approach

PFC2D calculates the movement and interaction of stressed assemblies of rigid circular particles using the DEM. As a discrete element code, it allows finite displacements and rotations of discrete bodies (including complete detachment), and recognizes new contacts automatically as the calculation progresses. The setup is composed of distinct particles that displace independently of one another, and interact only at contacts or interfaces between them. The calculations performed in the DEM alternate between the application of Newton's second law to the particles and a force–displacement law at the contacts, characterized by normal and tangential stiffnesses. Newton's second law is used to determine the motion of each particle arising from the contact and body forces acting upon it, while the force–displacement law is used to update the contact forces arising from the relative motion at each contact (Itasca Consulting Group Inc. 2008a).

For a plutonic rock, we choose bonding behavior for contacts (also called "parallel bond"—PB), which allows to reproduce the behavior of cohesive materials (Potyondy and Cundall 2004; Itasca Consulting Group Inc. 2008b). A rupture criterion based on the beam theory is used for PB; when the bond stress exceeds its yielding strength (in tension or in shear), the bond breaks.

The thermal option of PFC2D allows simulation of transient heat conduction and storage in particles and development of thermally induced displacements and forces. Each particle can be seen as a heat reservoir. The temperature (T_i, °C), the specific heat coefficient (C_v, J kg^{-1} °C^{-1}) and the linear thermal expansion coefficient (α, °C^{-1}) are initialized for each particle. The thermal power can be transmitted between two particles via a thermal pipe (contact between two particles). The parameters related to a thermal pipe are pipe length (L_p, m) and thermal resistance (R_{th}, °C W^{-1} m^{-1}). When particles (i and j) at different temperatures (T_i and T_j) are connected by a thermal pipe, a heat flux (Q_p, W) takes place in the thermal pipe (Itasca Consulting Group Inc. 2008b):

$$Q_p = \frac{T_i - T_j}{R_{th}L_p}. \tag{1}$$

The temperature increment (ΔT, °C) of the reservoir can be obtained by

$$\Delta T = \frac{Q_p}{mC_v}\Delta t_{th}, \tag{2}$$

where m (kg) is the mass of the reservoir and Δt_{th} (s) is the thermal time step. Note that by convention an increase of temperature is associated with a positive thermal power. Finally, the radius of the particle (R m) is changed as a consequence. We compute the radius increment (ΔR m) through

$$\Delta R = \alpha R\Delta T. \tag{3}$$

The integration of radii increment in the force–displacement law creates induced mechanical response of the system.

Table 1 Description of the four configurations considered to address uncertainties on the stress state

Name	"Andersonian" stress state (MPa)				Transformed stress state, aligned with a local coordinate system which is aligned with the wellbore axis (MPa)		
	Tectonics	σ_v	σ_H	σ_h	σ_{ss}	σ_{dd}	τ_{sd}
Case A	Intermediary normal/strike-slip fault ($\sigma_v = \sigma_H = \sigma_1$)	134	134	60	125	85	21
Case B	Normal fault—"low" horizontal isotropic stress ($\sigma_v = \sigma_1$)	134	60	60	60	79	0
Case C	Strike-slip fault ($\sigma_H = \sigma_1$ and $\sigma_v = \sigma_2$)	134	180	60	166	89	33
Case D	Normal fault—"high" horizontal isotropic stress ($\sigma_v = \sigma_1$)	134	80	80	80	94	0

The stress magnitudes are given in 3D in the vertical/horizontal system (here assumed to be the principal stress state) and their components in the 2D plane perpendicular to the wellbore axis (obtained by a linear transform, see Fig. 5). σ_1: major principal stress, σ_2: intermediate principal stress. σ_v: vertical stress, σ_H: maximum horizontal stress, σ_h: minimum horizontal stress, σ_{dd}: stress component in dip direction of the 2D plan, σ_{ss}: stress component in strike direction of the 2D plan, τ_{sd}: tangential shear stress component in the plane perpendicular to the well

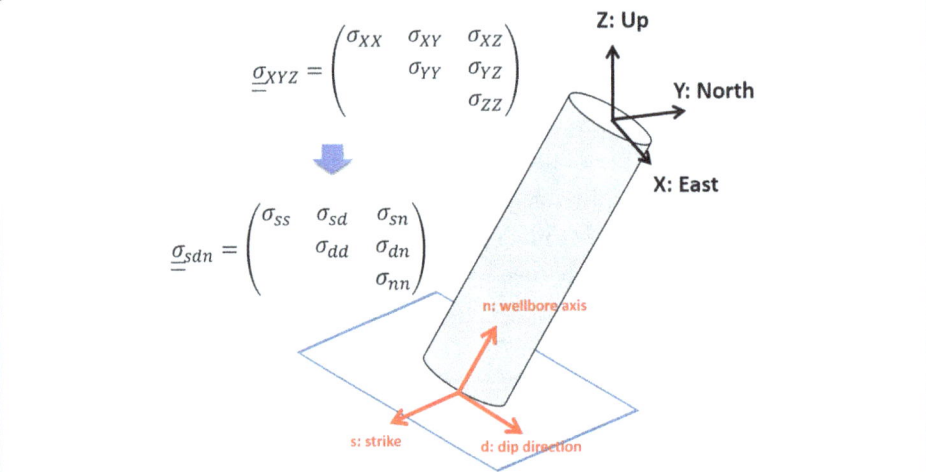

Fig. 5 Sketch of the stress state rotation. The transformed stress is obtained through (i) rotation of principal stresses from 30° along the vertical axis to be in the global coordinate system (Y south–north, X west–east, Z up–down); (ii) rotation of 40° along the vertical axis to align with the dip direction of the wellbore (Y_2 horizontal axis aligned with dip direction, X_2 horizontal axis perpendicular to dip direction, Z vertical axis); (iii) rotation of 30° along X_2 to be in the wellbore axis coordinate system (axis parallel to the dip direction of the plane perpendicular to the wellbore axis, noted as "d", axis parallel to the strike direction of the plane perpendicular to the wellbore axis, noted as "s")

Numerical settings and scenarios

The calculation setup consists of a two-dimensional cross section perpendicular to the well. The numerical simulations focus on the deepest part of the well. As far as possible, the conditions observed at 4560 m MD of IDDP-2 are used in the numerical simulations. The wellbore section is assumed to be 21.6 cm (8.5 in.). The temperature of the rock is assumed to be 426 °C (corresponding to the fluid temperature measured at the end of drilling). For thermal stimulation, we assume a temperature of 30 °C for the injected fluid (corresponding to the targeted temperature of the cooling fluid). Please note that temperatures recorded during logging operations are above 70 °C, thus using 30 °C

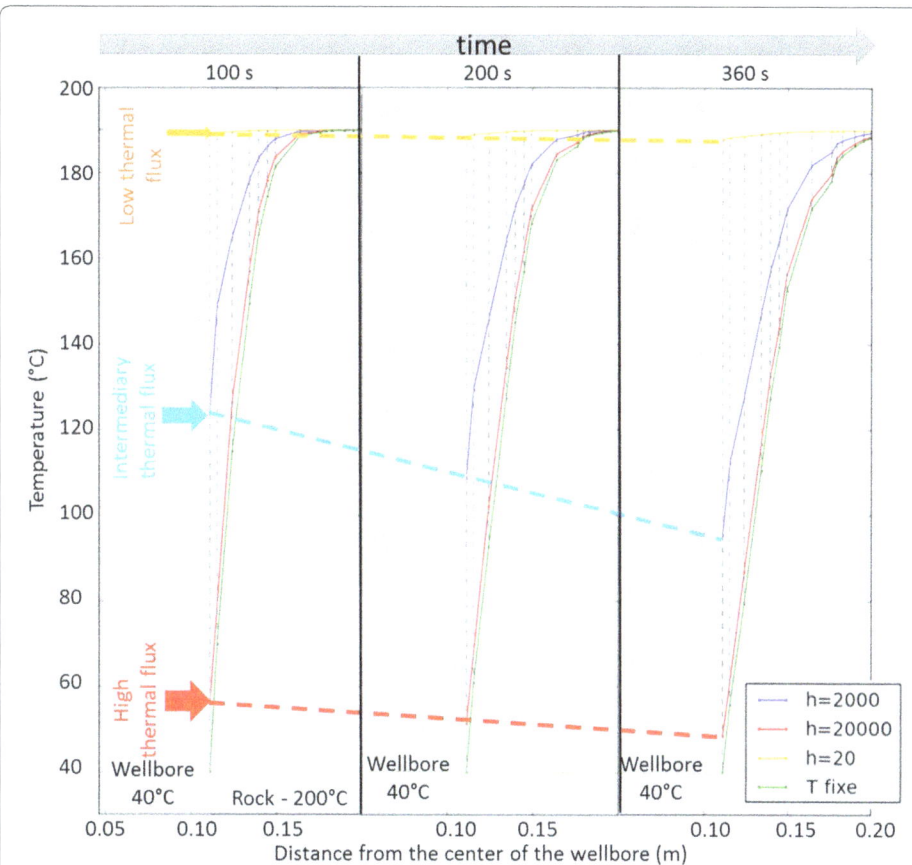

Fig. 6 Examples of the temperature gradient in a rock for different heat transfer coefficient (h, from 20 W m^{-2} K^{-1} to 20,000 W m^{-2} K^{-1}) inducing different heat flux, and evolution with the time. In this example (corresponding to a different numerical setting), the rock temperature is 200 °C and the wellbore fluid temperature is 40 °C. As a matter of comparison, the green curves (T fixed) present the thermal gradients obtained by setting directly the temperature of the particles of the wellbore at the target temperature (used in the analytical model, see Appendix)

overestimates the cooling during drilling operations (but may be appropriate for the subsequent thermal stimulation). The direction of the wellbore dip direction is N220°E and the deviation from the vertical is approximated at 30°. The modeled 2D cross section is thus oriented N130°E–60°NE.

We focus our study on the behavior of the matrix of the fine-grained-textured rock. The well-detailed description of the dolerite (weekly report IDDP-2, 2016) cored at 4 km depth is used as a reference for the numerical rock model. Deeper coring shows that similar rocks exist in the deeper part of the wellbore.

The four scenarios proposed to scan the range of possible stress states are described in Table 1 for classical coordinate system defined by Andersonian faulting theory, and the corresponding stress state in the 2D cross section normal to the wellbore axis (Fig. 5). In all cases, we consider the vertical direction as the principal stress axis, and we estimate its magnitude at 134 MPa at 4560 m MD depth. The expected horizontal stress magnitudes are within the range of data extrapolated from Batir et al. (2012), i.e., between 60 and 180 MPa.

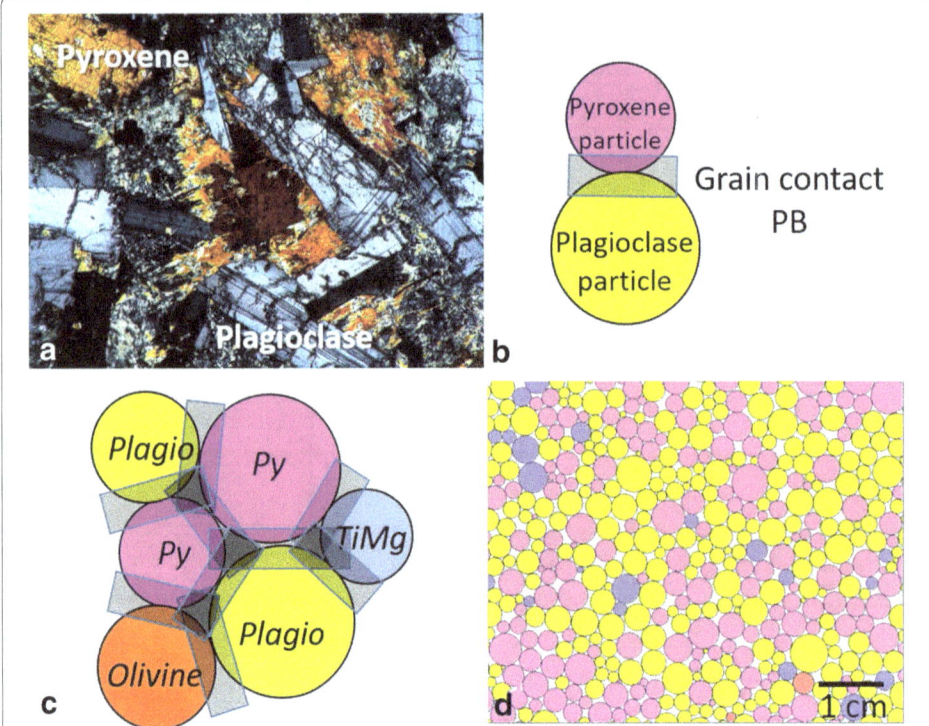

Fig. 7 Elaboration of the numerical rock model: conceptualization and geometric setup. **a** Thin section of the dolerite from RN-15/IDDP-2 (courtesy of HS Orka). **b** Conceptualisation: each crystal grain is modeled by a numerical particle; contacts between grains are modeled by parallel bonds (PB). **c** Schematic view of a detail of the numerical rock model: bonded particles aggregate (*Py* pyroxene, *Plagio* plagioclases, *TiMg* titanomagnetite). **d** View of the particles of the numerical dolerite model

The impact of the thermal flux at the wellbore boundary and of the fluid pressure in the wellbore are evaluated through a parametric study. Investigated fluid pressures in the wellbore are from 34 MPa (measured pressure in the well) to 104 MPa. The heat flux is a linear function of the temperature differential between the rock and the fluid in the wellbore and of the heat transfer coefficient. This coefficient is a quantity that empirically translates the heat exchanges between the circulating fluid and the solid. For the sake of clarification, Fig. 6 illustrates the impact of the choice of heat transfer coefficients on temperature fields. With low values, the heat transfer is slower and the particles of the wellbore need more time to reach the target temperature (Fig. 6). The heat transfer coefficient depends notably on the fluid velocity in the wellbore and on the fluid properties. Due to insufficient data, we chose two extreme values for the heat transfer coefficient:

- a low value (1000 W m^{-2} K^{-1}) simulating a slow cooling of the rock mass on the boundary of the wellbore;
- a very high value (10,000 W m^{-2} K^{-1}) simulating an instantaneous cooling of the rock mass on the boundary of the wellbore.

Table 2 Mechanical, thermal and thermo-mechanical properties of particles after calibration

	Plagioclase	Pyroxene	Olivine	Titanomagnetite
Young's modulus (mean value, MPa)	84	162	178	230
Ratio normal stiffness/shear stiffness	2.5	2.2	2.6	2.6
Friction coefficient	0.9	0.9	0.9	0.9
Tensile strength (MPa)	30	75	27	45
Cohesion (MPa)	140	350	126	210
Thermal conductivity (W m^{-1} K^{-1})	1.98	4.52	4.48	2.10
Specific heat (J kg^{-1} K^{-1})	1112	800	800	910
Linear thermal expansion coefficient (K^{-1})	6.81×10^{-6}	1.00×10^{-5}	3.85×10^{-6}	3.40×10^{-5}

Numerical rock setup

The elaboration of the numerical rock model setup is a preliminary essential task, often difficult due to insufficient data and thus high level of uncertainties. Before any numerical modeling, the real rock (here a dolerite) is conceptualized according to the role of each mineral phases in the behavior of the rock (more details in Peter-Borie et al. 2011, 2015). The dolerites are typical shallow intrusive bodies; they are micrograined, composed entirely of 1–5-mm-wide crystallized minerals without glassy matter. Grains or crystals are interlocked as the growth of each crystal has stopped on other crystals (MacKenzie and Adams 1999). For the numerical dolerite model, each numerical particle represents a mineral grain or crystal (Fig. 7). Parallel bonds (PB) model the contacts between grains (Fig. 7). Note that in grained rock, failure can occur between the grains as well as inside a grain (following then the weakest paths like twins leading to cleavage plans—Kranz 1983). In the numerical rock model, failure can occur only between particles. To fit with the reality, mechanical properties of the PB are the mean properties of the surrounded particles, a failure between two particles can then be interpreted as an intergrain or intragrain failure.

Following the conceptualization step, the properties of the numerical particles and bond are assigned. In this regard, quantified data from studied or analogue rock are required. For the present application case, the coring of fine-grained dolerite retrieved at 4090.6 m depth (Zierenberg et al. 2017) provided detailed information on the rock composition (pyroxene 40%, plagioclase 55%, titanomagnetite 5%, grain size around 3 mm). The particles of the numerical dolerite model follow the same distribution. The heterogeneity of particles size is represented through the implementation of a particle-size distribution centered on 3 mm.

At the time of the study, macroscopic mechanical and thermal data were not available from the IDDP-2 cored samples. Thus, we chose a limited analogue rock with available macroscopic properties. Since petrographic characteristics affect mechanical properties, analogues are selected depending on the proximity in terms of rock petrographic characteristics as mineral composition and grain size. A North African gabbro, characterized by Keshavarz (2009), is retained as reference analogue. It contains almost 40% pyroxene and 60% plagioclase, with traces of other elements (among others magnetite). Laboratory tests performed on pressurized (stepwise up to 650 MPa) and heated (stepwise up to 600 °C) samples provide mechanical properties of the analogue rock covering

Table 3 Mechanical properties of the selected analogue rocks and of the numerical rock model

	Young's modulus (GPa)	Poisson's ratio	UCS (MPa)	UTS (MPa)	Cohesion (MPa)	Friction angle (°)
Analogue	85–90	0.18	225	12	68	43
Numerical model	87	0.17	214	15	61	35

UCS: uniaxial compressive strength, UTS: ultimate tensile strength

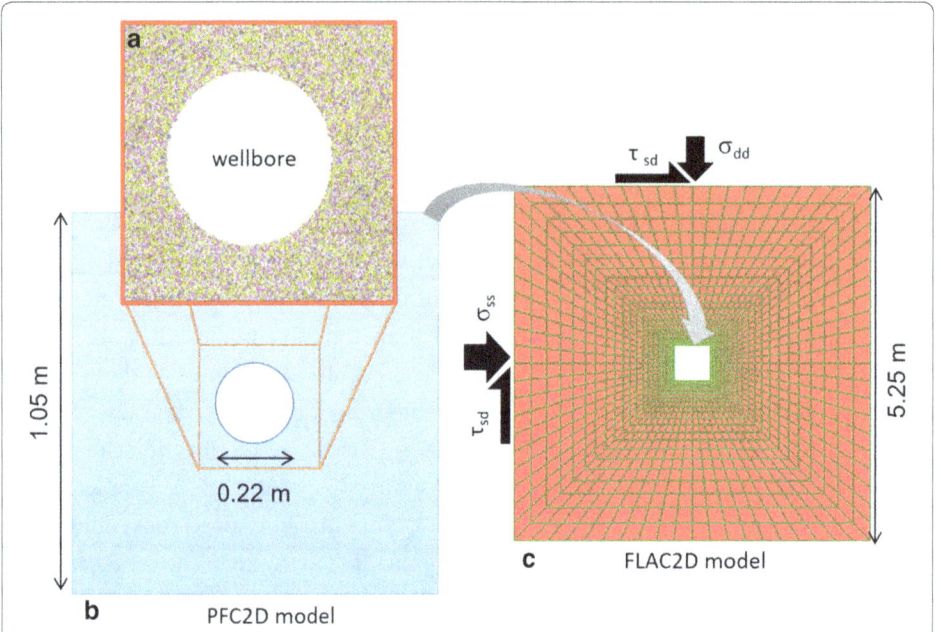

Fig. 8 FLAC/PFC2D-coupled model: **a** detail of the wellbore with particles of PFC2D; **b** PFC2D model (the entire extent of 1.05 by 1.05 m² —blue color—is discretized with particles, 141,500 in total); **c** FLAC model—mesh and stress state

conditions similar to those expected in the bottom of IDDP-2. As micro-gabbro/doler-ite have very low porosity [below 0.5% in the analogue rock (Keshavarz 2009), matrix porosity between 3.6 and 0.1% (no microporosity included) for the cored dolerite (Clau-dia Kruber, Equinor internal report in progress)], we assume that the pores can be seen as singularities in the rock matrix. The numerical rock model does not integrate the rock porosity. Therefore, in our numerical approach, no poroelastic effects are considered. The heat transfer process is thus limited to conduction between grains.

The range of values of the mechanical, thermal and thermo-mechanical micro-prop-erties (at the particle scale) has been first delimited according to a literature review on the properties of minerals (Simmons 1965; Carmichael 1989; Guéguen and Palciauskas 1992; Clauser and Huenges 1995). The properties must be physically consistent with the mineral phase characteristics and have to enable the reproduction of the macroscopic mechanical and thermo-mechanical behavior of the rock. The definitive calibration of the numerical particles and bonds particles is performed by fitting results of mechani-cal and thermal numerical tests (Uniaxial Compressive Strength—UCS, Ultimate Tensile

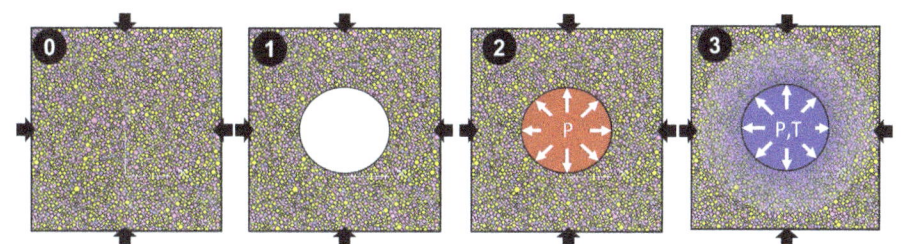

Fig. 9 Representations of the different steps of a simulation. (0) Application of the stress state on the numerical dolerite model. (1) Particles on the wellbore surface are removed to model drilling. (2) The hydraulic pressure is imposed in the wellbore. (3) The thermal loading is imposed. Black arrows symbolize the stress state. *P* represents the hydraulic pressure. *T* represents the thermal loading (cooling). The red color symbolizes the hot fluid, blue color the cold fluid

Strength—UTS, Triaxial and Thermal conductivity tests) with the macroscopic properties of the analogue rock (see final values of micro-properties in Table 2, and resulted macro-properties computed from numerical tests in Table 3).

Near-wellbore setup

The calculation setup concerns a 2D plane perpendicular to the wellbore axis at 4560 m MD depth. The definition of the near-wellbore setup needs to take into account an adequate extended area around the wellbore—at least three times the diameter of the wellbore (here 21.6 cm–8.5 in.) to limit the impact of boundary conditions on the numerical results. A significant number of particles are needed to build up such a large setup while keeping the size distribution close to the mineral size level (here close to an average of 3 mm). To push away the boundary conditions, the DEM near-wellbore model is embedded within a continuum-mechanics-based frame describing the region far away from the wellbore (FLAC Itasca Consulting Group Inc. 2002). The PFC2D simulation setup size is 1.05 m × 1.05 m, integrating more than 141,500 particles, and is embedded within a 5.25-m × 5.25-m FLAC mesh (Fig. 8). The coupling method between the continuum model and the discontinuous model is realized by an edge-to-edge approach for which the relevant overlapped elements are, respectively, segments of mesh in FLAC and a series of particles in PFC2D (Xiao and Belytschko 2004). A detailed description of the near-wellbore setup and of the PFC2D/FLAC coupled calculations is available in Shiu et al. (2011).

Simulation stepwise

The numerical simulation is performed stepwise with the aim to reproduce, as far as possible, the state of the rock in the vicinity of the wellbore before the thermal stimulation. Randomization is used for the construction of the numerical model. The radius of each particle is randomly drawn following the normal distribution $\mathcal{N}(1.37, 0.62)$ for pyroxene and plagioclase, and following $\mathcal{N}(1.25, 0.5)$ for titanomagnetite (based on cores observations). A periodic sample duplication process is used to build the numerical model faster.

Once the numerical setup is constructed and confined under a small confining pressure (which is always less than its corresponding in situ stress), it is loaded with its in situ stress which is presented in Fig. 9, step 0. We use the full-strain method (Itasca Consulting Group Inc. 2008c) for which a displacement increment is applied to each particle. Cycles are performed between two increments of displacement to reach a new mechanical equilibrium. Note that no contact breakage is allowed during the stress installation cycling. Thus, a pure elastic deformation is performed in this step. This method is very efficient when a large number of particles are included in the numerical model.

After the initial stress field is established, the borehole drilling is simulated by removing the particles located on the wellbore surface (Fig. 9 step 1). To avoid a sudden increase of the unbalanced forces of particles placed on the surface of the wellbore, leading to numerical instabilities, a force-reduction procedure is used at this step to release progressively the unbalanced forces of particles situated along the wellbore surface (Shiu et al. 2011). Note that this step is a very rough and simplified approximation of the drilling impact on the formation stability. On the one hand, the impact of the drilling bit at the excavation step is not considered. On the other hand, the pressure considered in the wellbore is assumed zero, due to limitation in the calculation procedure, which can lead to damage overestimation as pressure actually exists in the well during real drilling at ECDs (equivalent circulation densities) even above the static fluid column.

During the fluid injection step of the calculation schedule, the wellbore is subjected to a hydraulic pressure and to a thermal loading. The fluid injection is assumed to act only on particles forming the wellbore surface. A specific procedure (Itasca Consulting Group Inc. 2008c; Shiu et al. 2011) is used to detect a set of closed linked particles (connected by parallel bonds) around the wellbore. These particles are recorded in a specific list and will be referred to as the wellbore list in the following description. To simplify the numerical modeling setup, and to limit the computational time, the hydraulic pressure and the thermal loading are applied in two steps. The hydraulic pressure is applied first (Fig. 9 step 2) and the thermal loading (Fig. 9 step 3) takes place later (assuming that no significant thermal propagation occurs before the hydraulic pressure is fully installed on the wellbore surface). The underlying assumption is that the characteristic time for pressure effects is far shorter than the characteristic time for thermal effects. The list of the wellbore particles is updated automatically when cracks appear between particles in the wellbore list. Hence, once the cracks start propagating from the wellbore, the injection pressure and the fluid temperature can penetrate into the crack as well.

Results

Simulation of the drilling of the well

Among the four simulated stress states (cases A, B, C, D, defined in "Numerical approach" section), the drilling of the wellbore is the most critical for case C characterized by the highest 2D deviatoric stress in the near-wellbore area. Numerous cracks are observed in the rock with greater density in areas closer to the wellbore (Fig. 10). At the wellbore boundary, the cracks are connected, forming a slight caving (up to a depth of 3 cm with radial extension up to 10 cm) breakout in the direction of the 2D minimum stress. For the second highest deviatoric stress (case A), the number

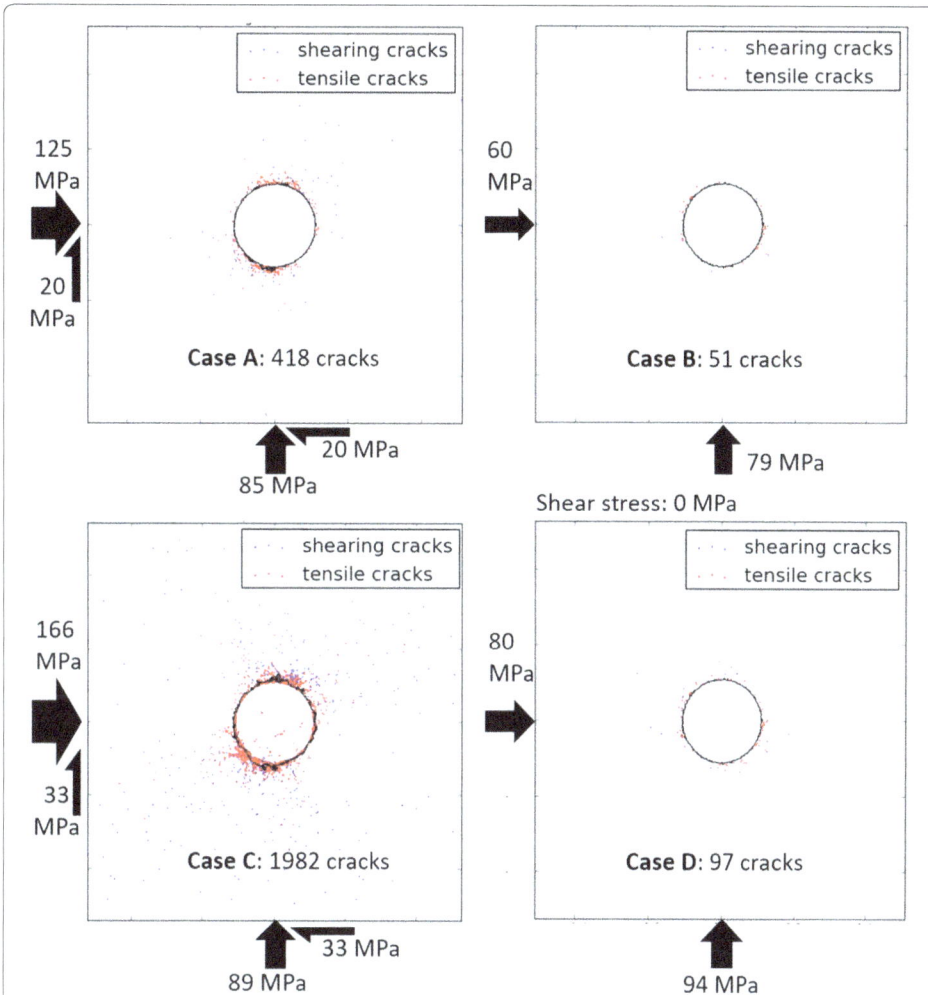

Fig. 10 Results of simulations, after drilling, for the four stress states (see "Numerical approach" section for definition of cases A, B, C and D). Each red point corresponds to the apparition of a tensile crack, each blue point corresponds to the occurrence of a shearing crack. The total number of cracks is mentioned in each subplot. The field of view for each subplot is 1 m × 1 m

Table 4 Results of the analytical approach for breakouts

Name	Principal stress in the 2D plane perpendicular to the well		Breakout characteristics			
			$P_{well} = 0$ MPa		$P_{well} = 34$ MPa	
	σ_{A*} (MPa)	σ_{B*} (MPa)	$\sigma_{90°, R=r}$ (MPa)	$r_{\theta=90°,}$ $\sigma_{\theta=UCS}$ (cm)	$\sigma_{90°, R=r}$ (MPa)	$r_{\theta=90°,}$ $\sigma_{\theta=UCS}$ (cm)
Case A	76	134	326	14.8	292	13.6
Case B	60	79	175	$<R$	141	$<R$
Case C	76	178	459	23.0	425	21.0
Case D	80	94	200	$<R$	166	$<R$

The computation of the borehole potential damage is linked to no to low wellbore fluid pressure (P_{well}). $\sigma_{90°, R=r}$ is the circumferential stress at the wellbore boundary in the direction of the minimal 2D stress, $r_{\theta=90°,}$ $\sigma_{\theta=UCS}$ is the distance from the wellbore center where the circumferential stress is equal to the UCS of the dolerite (214 MPa), in the direction of the minimal 2D stress, R is the radius of the borehole. Principal stresses in the cross section perpendicular to the well are obtained by diagonalization of the stress state (σ_{ss}, σ_{dd}, τ_{sd}) mentioned in Table 1 and are thus slightly different from horizontal principal stresses

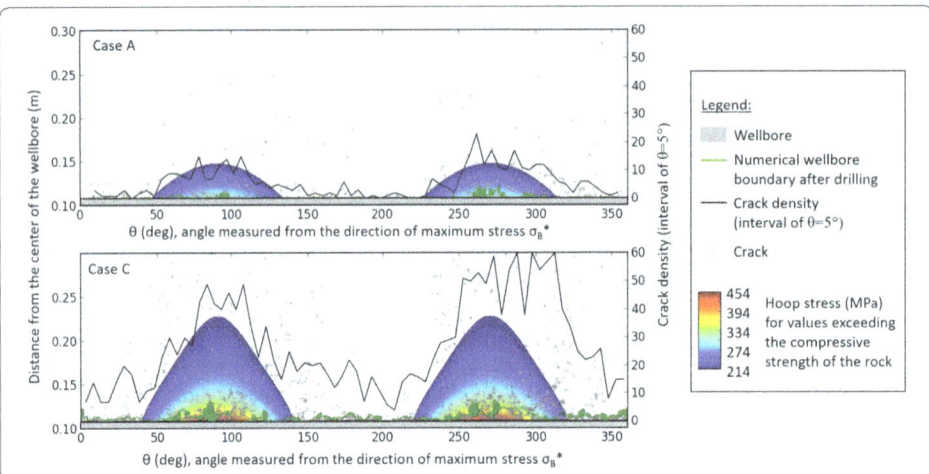

Fig. 11 Distance of the new wellbore boundary and of the cracks resulting from numerical simulations of drilling, and crack density within an interval of 5° for the stress states A and C in plot vs. the angle measured from the maximum stress direction θ. The colormap refers to area where the hoop stress exceeds the compressive strength according to the analytical solution

of cracks is four times less than for case C. It leads locally to a rock caving up to 1 cm. Other stress states (B and D) lead to a limited number of cracks.

These numerical results are compared with results obtained with the analytical Kirsch equations (see Appendix), presented in Table 4. The analytical solution predicts that breakout will occur for cases A and C for both zero fluid pressure and in case it is equal to 34 MPa.

There is good agreement between the analytical and numerical results. Most cracks in numerical results are in the area stress from analytical solution exceeds the rupture criterion. The ratios of cracks in this area compared to all cracks are, respectively, 88% and 89% for cases A and C for zero fluid pressure in the wellbore. Some differences between analytical and numerical results can nevertheless be noted (Fig. 11):

- In numerical simulation, cracks coalesce until creating a caved area; this area represents only a limited part of the area where the rupture criterion is exceeded in the analytical model.
- Cracks in the numerical model also occur outside of the area where the rupture criterion is exceeded according to the analytical solution.

Several possible explanations can be suggested to discuss these differences:

- As contact properties depend on the adjacent particle properties, the PB do not have all the same strength (the criterion of the analytical solution is the mean strength of the rock). Cracks may occur outside of the analytical breakout area when the strength of the PB is locally exceeded by the stress, even if it is lower than the criterion. Conversely, stronger bonds may resist in the numerical model, even if the analytical area predicts rupture.

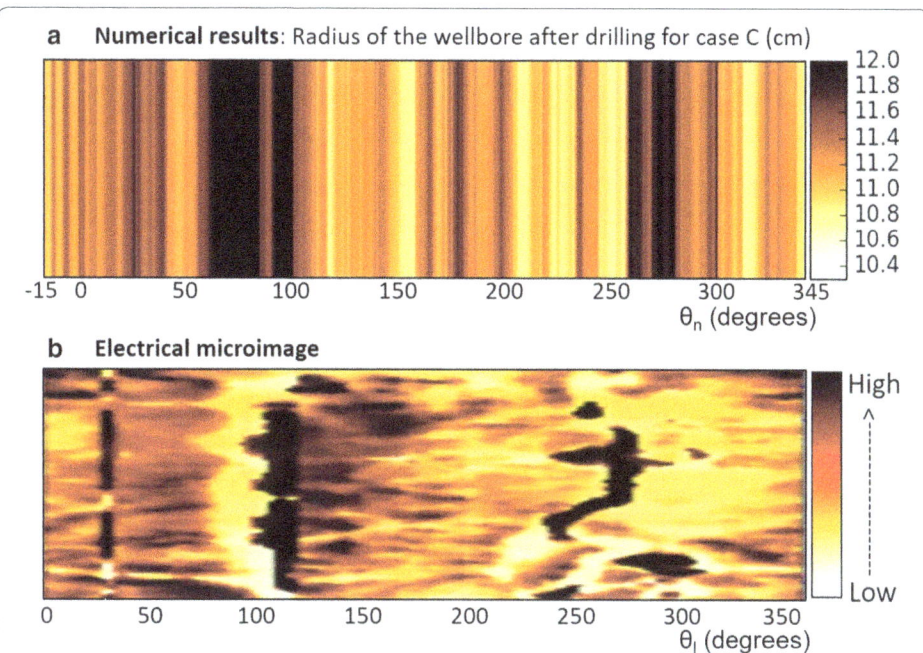

Fig. 12 Confrontation of the numerical results and of the logging images. **a** Numerical results were transformed to logging-like images. The color intensity corresponds to the distance between the boundary of the wellbore and the center of the wellbore, after drilling, for case C, with zero fluid pressure in the wellbore. θ_n is the angle measured from the direction of maximum stress. **b** Electrical microimage (extracted from the image presented in Fig. 2). The azimuthal origin for numerical results was adjusted manually for qualitative comparison (θ_l and θ_n scales have not the same origin)

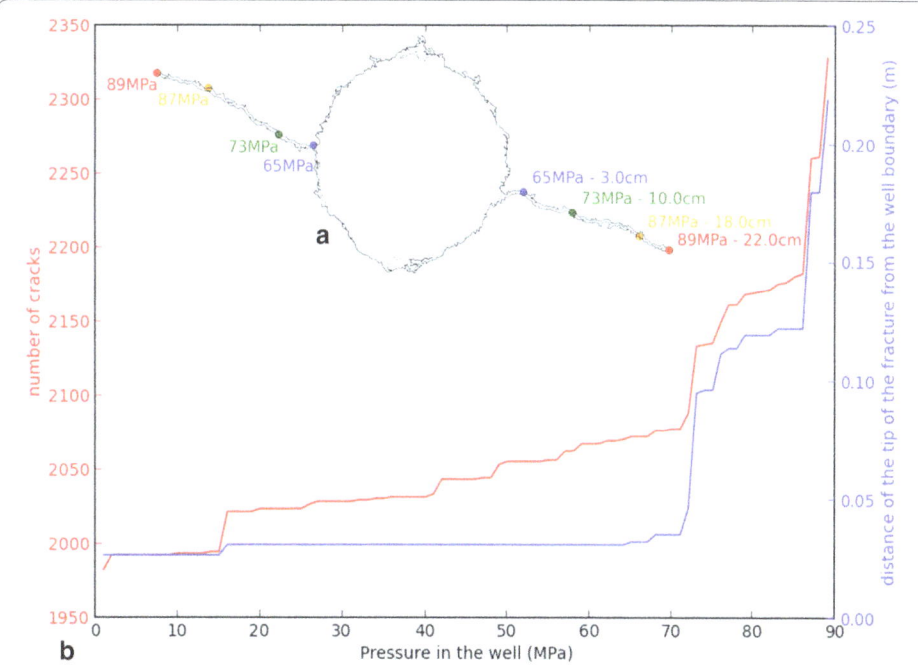

Fig. 13 Fracture propagation depending on the hydraulic pressure in the wellbore for case C. **a** Evolution of the wellbore shape—tip of the induced fracture is represented by a dot for different levels of pressure in the wellbore; **b** plot of the number of cracks (in red) and of the maximum distance of the tip of the fracture from the wellbore (in blue) depending on pressure in the well

Table 5 Results of the analytical approach for the computation of the breakdown pressure (P_{frac})

Name	Tectonics	Principal stresses		$P_{frac, UTS=15\,MPa}$
		σ_{A^*} (MPa)	σ_{B^*} (MPa)	
Case A	Intermediary normal/strike-slip fault ($\sigma_v = \sigma_H = \sigma_1$)	76	134	110
Case B	Normal fault—"low" horizontal isotropic stress ($\sigma_v = \sigma_1$)	60	79	116
Case C	Strike-slip fault ($\sigma_H = \sigma_1$ and $\sigma_v = \sigma_2$)	76	178	65
Case D	Normal fault—"high" horizontal isotropic stress ($\sigma_v = \sigma_1$)	80	94	161

Following Eq. 14 in Appendix with $\Delta T = 0$ and UTS $= 15$ MPa. Principal stresses in the cross section perpendicular to the wellbore axis are obtained by diagonalization of the stress state (σ_{ss}, σ_{dd}, τ_{sd}) mentioned in Table 1 and are thus slightly different from horizontal principal stresses

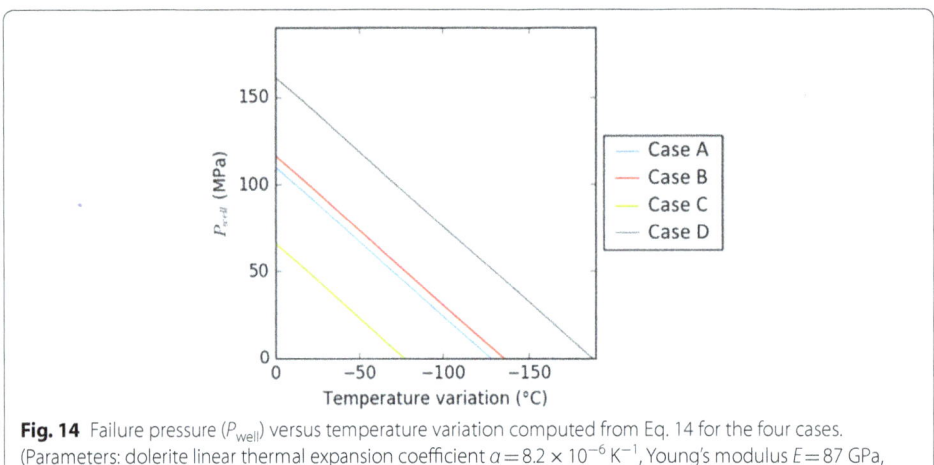

Fig. 14 Failure pressure (P_{well}) versus temperature variation computed from Eq. 14 for the four cases. (Parameters: dolerite linear thermal expansion coefficient $a = 8.2 \times 10^{-6}\,K^{-1}$, Young's modulus $E = 87$ GPa, Poisson ration $v = 0.17$ and UTS $= 15$ MPa.)

- Because of the heterogeneity of the properties of the particles and of the PB, stress local modification can occur. Thus, locally, a higher stress can lead to PB breaking, or a lower stress to PB integrity.
- The caving in the breakout area will affect the stress further—this case cannot be taken into account in the analytical solution.

The qualitative analysis of the in situ logging images of the RN-15/IDDP-2 reveals numerous features that may be interpreted as breakout. The results of the numerical simulation of the effect of the drilling in the strike-slip regime (case C) could potentially fit these observations (Fig. 12). High resistivity on logging images (black color) might correspond to cavings filled with fluid, thus matching with increased well radius in numerical results. To go beyond, it would be interesting to have better knowledge of the stress state and to have quantitative estimation of breakout caving (with calipers), thus enabling the comparison of caving dimensions (depth and lateral extension).

Impact of increased well pressure

In this section, we discuss the mechanical impact of increased well pressure, without thermal loading effects, with tensile failure as expected result.

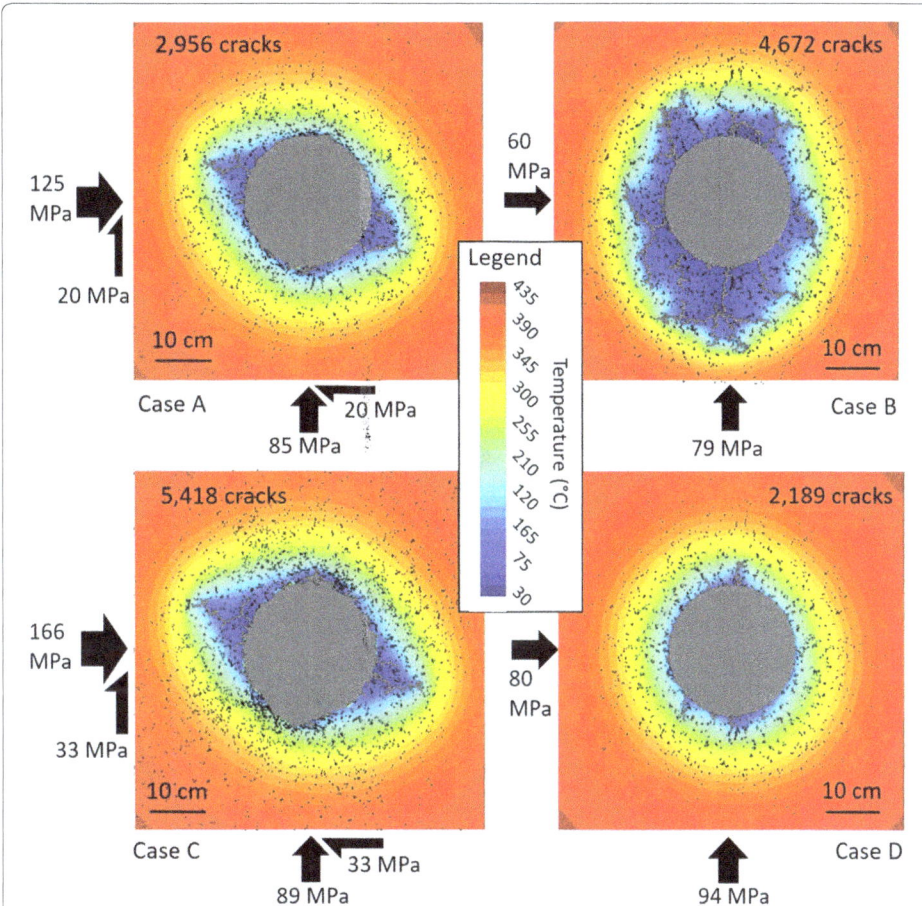

Fig. 15 Calculation results after 4 h of thermal loading for a heat transfer coefficient value of 1000 W m^{-2} K^{-1} (low thermal flux). The front color corresponds to the temperature (see legend). Each point corresponds to a crack (either tensile or shear crack). Grey uniform color corresponds to the area connected to the wellbore (penetration of hydraulic pressure and of thermal loading)

The numerical work focuses on case C (strike-slip regime). In the simulation, the pressure is applied by 1-MPa increment up to 90 MPa on the wellbore boundary. Figure 13 shows the result. The tensile fracture initiates for a well pressure close to 65 MPa, with 3 cm length into the rock matrix. Further stepwise increase of the pressure leads to progressive fracture depth. A pressure in the wellbore higher than 89 MPa is necessary for a wide propagation of the fracture.

For the sake of comparison, results obtained with the Kirsch equations (see Appendix) are presented in Table 5. The tensile failure appears for a pressure in the wellbore of 65 MPa, thus in good agreement with the numerical results. For the other stress states, a pressure in the well above 100 MPa is necessary to induce tensile failure.

In the RN-15/IDDP-2 wellbore, the pressure remained limited (below breakdown pressures computed in this section). Neither the analytical solution nor the numerical simulation results can explain the tensile fractures observed by the low well pressures. However, numerous features that may be interpreted as tensile fractures are observed in the image logs (see "Logging images" section and Fig. 2). Therefore, we study the effects due to the thermal cooling in the next section.

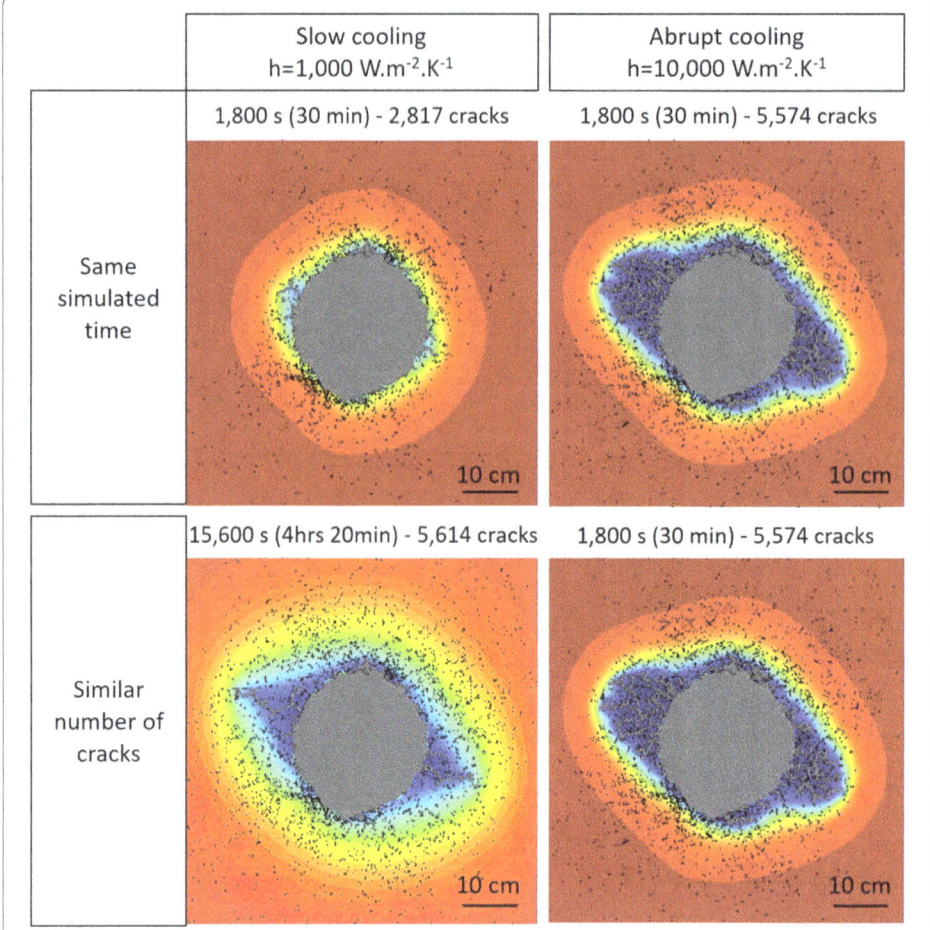

Fig. 16 Influence of the heat transfer coefficient on the temperature field and on crack apparition (legend: colormap is the same as in Fig. 15), for a stress state corresponding to case C (defined in "Numerical approach" section), with no hydraulic loading. Results on left correspond to a low heat transfer coefficient (1000 W·m^{-2} K^{-1}). Results on right correspond to a high heat transfer coefficient (10,000 W m^{-2} K^{-1}). On the top line, the comparison is made for the same duration of cool fluid injection (half-hour). On the bottom line, the comparison is made for a similar number of cracks (and different cooling durations)

Cooling effects on wellbore stability

The additional temperature term contributing to the hoop stresses in the analytical solution (Eq. 14 in Appendix) leads to tensile failure in all cases for a cooling larger than 190 °C (Fig. 14) without any need of fluid pressure in the well ($P_{well} = 0$ MPa). In RN-15/IDDP-2 well, the host rock temperature is estimated to be between 426 and 549 °C (see "Geological knowledge" section), which means that any downhole drilling fluid temperature below 236 °C is likely to induce tensile fractures. This is in good agreement with the numerous tensile fractures observed on image logs (see "Logging images" section and Fig. 2).

Numerical simulations are proposed here for an in-depth view of the failure under thermal loading. Our motivation is twofold: first, the DEM approach allows to take into account thermal effects at the microscale (notably the differential expansion of the grains of the rock) and thus the approach is indeed more detailed for simulating the thermal

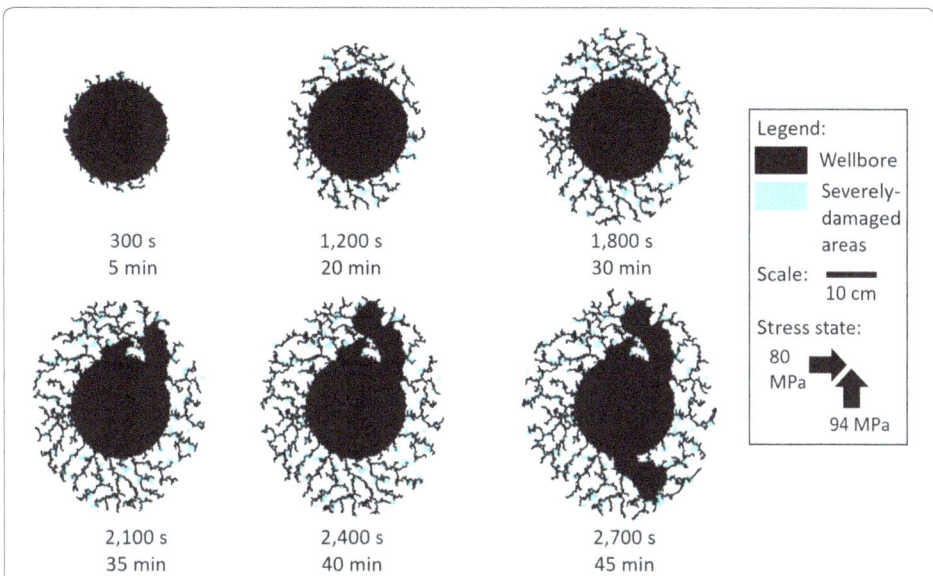

Fig. 17 Example of a separation of pieces of rock in the near-wellbore area, for the stress state case D, with high heat transfer coefficient value—$h = 10,000 \, \text{W m}^{-2} \, \text{K}^{-1}$, pressure in the wellbore $= 50$ MPa. In this figure, black color corresponds to the wellbore and to separated pieces of rock matrix, cyan color corresponds to unseparated but severely damaged areas (defined as area with more than 6 cracks cm^{-2})

effects. Second, the shape and size of damages can be retrieved and analyzed in comparison with observations. We investigate the impact of the thermal flux (through the heat transfer coefficient values) and of the pressure in the well. For the sake of comparison, and after a brief presentation of the stress state impact, a sensitivity analysis is presented.

Thermo-mechanical tensile failures depending on the stress state

Figure 15 shows the development of the cracks and of the induced fractures by the cracks' coalescence for the different stress states, in the absence of hydraulic pressure in the wellbore. Tensile fractures developed in all the four cases, as predicted by the analytical results. However, depending on the stress state, the shape, the propagation direction and the intensity of the damage differ. With the largest 2D deviatoric stresses (cases A and C), the fracture propagates in the direction of the 2D maximum stress σ_B^* (perpendicular to the direction of breakout caused by drilling, if any). In the most isotropic cases (cases B and D), fractures develop around the wellbore without preferential direction, following the path of least resistance defined by the local mineral distribution.

Impact of the thermal flux at the wellbore boundary

Figure 16 allows the comparison of the development of cracks in the rock depending on the thermal flux through extreme values of heat transfer coefficient for the stress state case C (strike-slip tectonic regime). With a low flux (low heat transfer coefficient value), the kinetics of the fracture development is drastically reduced. The second observation is less intuitive: besides kinetics, the cooling rate influences the shape of fractures. A fast cooling of the rock (high heat transfer coefficient, high thermal

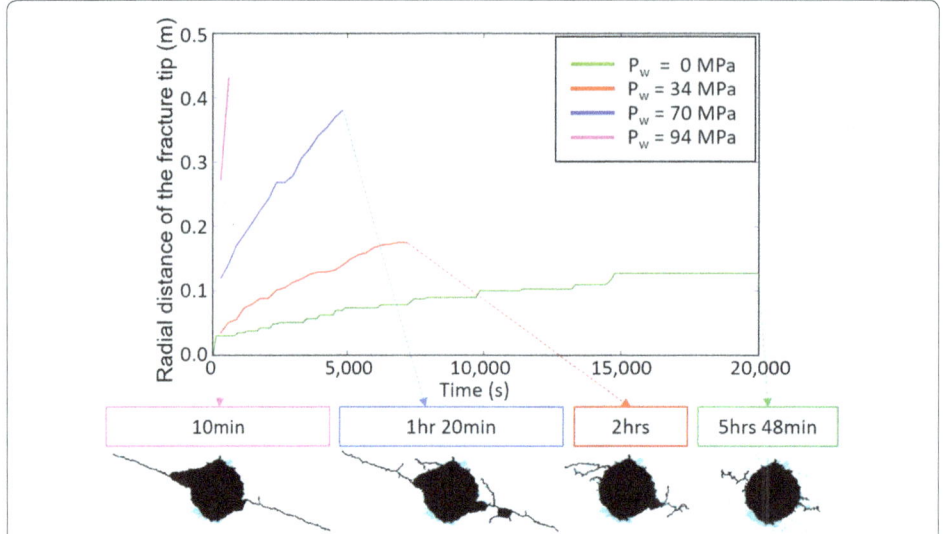

Fig. 18 Influence of the well pressure on fracture propagation (after both hydraulic pressure and thermal loading), for case stress state C, with $h = 1000\,W\,m^{-2}\,K^{-1}$. The four snapshots correspond to hydraulic pressure P_w of 94, 70, 34, 0 MPa. In this figure, black color corresponds to the wellbore and to separated pieces of rock matrix, cyan color corresponds to unseparated but severely damaged areas (more than 6 cracks cm^{-2}). Note that the fracture propagation is limited by the DEM model size, as a consequence, the radial distance of the tip cannot exceed 41.5 cm

gradient from the beginning of the cooling) creates several tortuous fractures, with dichotomy, while a slow cooling of the rock (low heat transfer coefficient value, progressive increase of the thermal gradient in the rock) allows the focusing of the crack within a single fracture.

Moreover, it should be noticed that the coalescence of cracks observed in the case of high thermal flux can lead to separation of larger pieces of rock as shown in Fig. 17. This may be interpreted as caving on the wellbore images. In this case, cavings are not the result of overly low ECD but are due to the thermal loading. These temperature-induced cavings tend to be oriented in the maximum principal stress direction, contrary to breakouts caused by overly low ECD. The layout of crack development and coalescence in the case of fast cooling may provide explanations on some logging features: one can observe that the logging image is not symmetric but has a tendency to develop larger on one azimuth than the systematic one (note that the logging image is on a shallower depth than result of numerical modelling).

Impact of the pressure in the wellbore

The hydraulic pressure has a significant influence on the damage in the near-wellbore area as illustrated in Fig. 18. With high well pressure, towards the range of the least compressive hoop stress, the induced fractures localize towards a distinct plane with preferential orientation. For a pressure lower than the theoretical breakdown pressure in the wellbore (65 MPa, see "Impact of increased well pressure""section), the induced fracture is more restrained, split and discontinuous. Features observed on the logging images correspond rather to discontinuous fractures than to very localized fractures. Since pressure in the wellbore was limited, this lends credibility to the model.

Discussion

The presented study may shed light on the effects of the wellbore drilling and of the thermal stimulation in a deep and very hot fine-grained rock. Drilling and pressurization impacts on the wellbore stability have been studied first. The formation of breakouts and induced tensile fracture have been successfully described by the calculation results, even though we consider total deconfinement of the wellbore during drilling and no dynamic processes as tools impact in a first approach. Slight differences can be explained by the level of greater detail included in the numerical approach compared to the analytical solution: grain heterogeneity in the rock matrix, caving processes allowed and not pre-defined. From both analytical solution and numerical results, a fluid pressure no less than 65 MPa is needed for inducing tensile fracture without considering thermal effects. During the drilling phase and after, the pressure in the bottom of the well is under the breakdown pressure. However, numerous induced fractures have been observed in the logging images; a thermal component appears to be necessary to explain the observations.

In the RN-15/IDDP-2 well, there is a drastic difference in temperature between the fluid in the well and formation, probably higher than 150 °C during the drilling phase and even higher during the thermal stimulation (up to 400 °C). From both analytical and DEM calculations, which take into account thermo-mechanical loadings, this constant thermal stimulation induces tensile fracture. Note that considering the high temperature difference, fluid pressure in the well is not necessary for fracture inducing.

Complementary to the analytical solution, DEM allows a more detailed study of the thermo-mechanical processes; beyond the "macro" thermo-mechanical processes, the impact of the differential behavior of the minerals composing the rock can be considered thanks to a modeling at the grain scale. In addition, this approach allows notably the quantification of the damage around the wellbore, the visualization of the pathway of the induced fractures.

Beyond the above-presented results fitting with the observations, and as for any model, it is important to keep in mind the limitations when analyzing modeling results. Among the model limitations, we can quote the two dimensionality, the matrix considered as impermeable, and the impossibility to generate intragranular cracks.

Other limitations of simulations come from the complexity of the model and from the difficulty to have well-characterized parameters to feed into the model. The long computational time (on average 4–5 weeks to simulate a few hours of thermal loading) makes these limitations more pronounced since the number of possible investigations is limited. The variety of rock behavior under thermal loading, depending on the different studied parameters as the stress state, the thermal flux or the pressure in the wellbore, illustrates the necessity of data acquisition to reduce uncertainties. Indeed, the efficiency of the thermal stimulation as well as the stability of the borehole during the drilling evaluation need a sound knowledge of the in situ conditions (thermal and mechanical properties of the rock, direction and magnitude of stress state among others), and the control of the thermal stimulation (depending notably on the flow rate, the temperature at surface, the pressure and the composition of the injected fluid). In addition to these influent uncertainties, the influence of the rock model should also be further investigated.

A second layer of uncertainties is introduced when comparing the modeling results with logging images, since these latter are also subject to uncertainties. Besides, note that the logging is performed more than 48 h after drilling while exposing the well to both thermocycling and pressure cycling; as a consequence, comparisons are mainly qualitative, but provide nonetheless a preliminary evaluation on the ability of the numerical approach to replicate successfully the observations.

Further investigations and numerical developments are needed to confirm the assumption and for a better understanding of the linked processes. Indeed, some limitations of the used version of the numerical code can lead to a misevaluation of the pathway and the propagation speed of the fractures: the energy of propagation of the fractures is not taken into account (Kanninen and Popelar 1985); the stability/instability of fracture growth in and out the zone of increased stress should be further investigated depending on the stimulation mechanism (either thermal or pressure effects). We have observed that the results are not accurately capturing the propagation of fractures into the far field once the close wellbore region is fractured when well pressures are larger than the minimum horizontal stress. The reason for that is found to be in the definition of the 2D plane strain cross section and the definition of the boundary conditions. The principal stress direction of the far field stresses is not aligned with the plane in which the 2D calculations are performed and also the axial stresses are not taken into the lower dimensional setup. For that reason, the system does not recognize that the out-of-plane, far-field stress is not a principal stress direction, but rather rotating with distance from the wellbore. This simplification overestimates the overall resistance against fracturing in the far field which in reality would be simply the minimum horizontal stress and rock resistance. Therefore, a fully 3D setup including all components of the stress tensor is proposed in future studies, which will shed light on stable vs unstable fracture growth beyond the close-wellbore region in the case of inclined wells exposed to temperature and pressure loading. This enhanced setup would then also enable a discussion of the paths of fractures propagating from the well into the far field eventually creating so-called "hackles".

Further developments are also in progress for interpreting the induced fractures in terms of injectivity and later on also for studying productivity gains. The goal of such future simulations will be to enable the numerical reproduction of transient productivity loss as the previously created fracture closes due to thermal expansion of the matrix.

Conclusion

A DEM using micro–macro approach is proposed to simulate the thermo-mechanical processes in the surrounding of the wellbore of RN-15/IDDP-2. The results of this approach—chosen in consistence with the observations of the field—are compared with the classical analytical solutions and with the logging images. Considering the numerical limitations, modeling approximations and assumptions is necessary for a relevant interpretation of the results. The most impacting are the following:

- The long computational time resulting in limited number of possible investigations in the parametric study;
- The two dimensionality of the model leading to a poor capture of the propagation of fractures into the far field;

Some of these limitations can be improved in future works, in particular, by considering a 3D setup.

Nonetheless, numerical results are consistent with the results of the analytical solutions. According to the numerical results, as well as to the analytical solution, and fitting with the observations in RN-15/IDDP-2, breakouts result from the drilling process—arguing for a quite high local deviatoric stress—and tensile fractures appear because of the high thermal loading. Overpressure in the wellbore speeds up the process.

Moreover, the numerical simulation allows a deeper investigation into the effect of the drilling and into the thermal stimulation. In particular, the impact of the differential behavior of the minerals composing the rock can be considered thanks to a modeling at the grain scale. In addition, this approach notably allows the quantification of the damage around the wellbore and highlights the caved areas and the pathway of the induced fractures in the near field.

As emphasized, a fresh aspect of this study is the consideration of the thermal flux at the wellbore boundary. We have shown that a high thermal flux between the fluid in the wellbore and the rock leads to tortuous pathways for induced fractures; In this case, pieces of rock can be separated from the rock mass. This could be one explanation for the observed induced fractures and cavings in the logging images, oriented perpendicular to the direction of breakouts due to low ECD.

Abbreviations

DEM: discrete element method; ECD: equivalent circulation density; EGS: enhanced/engineered geothermal system; IDDP: Iceland Deep Drilling Project; MD: measured depth; PB: parallel bond; PFC2D: particle flow code—2 dimensions; UCS: uniaxial compressive strength; UTS: ultimate tensile strength; VD: vertical depth.

List of symbols

α_L: linear thermal expansion coefficient ($1\ °C^{-1}$); ΔP: difference between the fluid pressure in the borehole and that in the formation ($P_{well}-P_p$) (MPa); Δt_{th}: thermal time step (s); ΔT: temperature difference between the mud and the rock (°C); θ: Azimuth measured from the direction of σ_B^* (°); ν: Poisson's ratio; σ_1: major principal stress (MPa); σ_2: middle principal stress (MPa); σ_3: minor principal stress (MPa); $\boldsymbol{\sigma_H}$: major horizontal stress (MPa); $\boldsymbol{\sigma_h}$: minor horizontal stress (MPa); σ_v: vertical stress (MPa); σ_A^*: 2D minimal principal stress component (MPa); σ_B^*: 2D maximum principal stress component (MPa); σ_{dd}: stress component parallel to the dip direction of the plan perpendicular to the well (MPa); σ_{ss}: stress component parallel to the strike direction of the plan perpendicular to the well (MPa); $\boldsymbol{\sigma_r}$: radial stress around the borehole (MPa); $\boldsymbol{\sigma_\theta}$: circumferential stress around the borehole (MPa); τ_{sd}: tangential shear stress component in the plane perpendicular to the well (MPa); τ_{r_θ}: tangential shear stress around the borehole (MPa); C_v: specific heat coefficient (J kg^{-1} °C^{-1}); E: Young's modulus (GPa); L_p: pipe length (m); m: mass of the heat reservoir (kg); P_{frac}: fracture pressure (MPa); P_{well}: well pressure (MPa); P_p: pore pressure of the formation (MPa); Q_p: power in the thermal pipe (W); r: distance from the center of the hole (m); R: radius of the borehole (m); R_{th}: thermal resistance (°C W^{-1} m^{-1}); T_i: temperature of the numerical particle i (°C); T_j: temperature of the numerical particle j (°C).

Authors' contributions
BRGM's authors performed modeling work and prepared the core of the manuscript. Equinor's authors provided data from wellbore logging and enriched the discussion part. HsOrka provided data from RN-15/IDDP-2. All authors read and approved the final manuscript.

Author details
[1] BRGM, 3 av. C. Guillemin, BP36009, 45060 Orléans Cedex 2, France. [2] Equinor ASA, Research and Technology, Rotvoll, Norway. [3] HsOrka, Svartsengi, 240 Grindavík, Iceland.

Acknowledgements
We would like to thank Kati Tänavsuu-Milkeviciene (Equinor) and Claudia Kruber (Equinor) who helped with the geology, mineralogy and porosity analyses; Théophile Guillon (BRGM) and Arnold Blaisonneau (BRGM) for fruitful discussion on modeling issues. The authors are grateful to the editor and to the two anonymous reviewers for their helpful comments and advice.

Competing interests
The authors declare that they have no competing interests.

Availability of data and materials
Not applicable (commercial code).

Funding
This study was part of the DEEPEGS project, which received funding from the European Union HORIZON 2020 research and innovation program under Grant agreement no. 690771.

Appendix: Simplified analytical approach for computation of stress development in the near-wellbore area

We propose to compute the analytical solution for the stress development in the near-wellbore area to evaluate the risk of breakout and breakdown. The proposed analytical solution is a simplified one: it requires the assumption of a wellbore parallel to principal stress that is not the case of the deep part IDDP-2 for the four stress states considered. For a cylindrical hole in a thick, homogeneous, isotropic, elastic plate subjected to effective minimum and maximum stresses (absolute values of minimum and maximum stresses are noted σ_A^* and σ_B^* hereafter), disregarding any thermal stresses, the following equations apply (Kirsch 1898 in Zoback et al. 1985):

$$\sigma_r = \frac{1}{2}\left(\sigma_B^* + \sigma_A^*\right)\left(1 - \frac{R^2}{r^2}\right) + \frac{1}{2}\left(\sigma_B^* - \sigma_A^*\right)\left(1 - 4\frac{R^2}{r^2} + 3\frac{R^4}{r^4}\right)\cos 2\theta + \Delta P\frac{R^2}{r^2},$$

(4)

$$\sigma_\theta = \frac{1}{2}\left(\sigma_B^* + \sigma_A^*\right)\left(1 + \frac{R^2}{r^2}\right) - \frac{1}{2}\left(\sigma_B^* - \sigma_A^*\right)\left(1 + 3\frac{R^4}{r^4}\right)\cos 2\theta - \Delta P\frac{R^2}{r^2}, \quad (5)$$

$$\tau_{r\theta} = -\frac{1}{2}\left(\sigma_B^* + \sigma_A^*\right)\left(1 + 2\frac{R^2}{r^2} - 3\frac{R^4}{r^4}\right)\sin 2\theta, \quad (6)$$

where σ_r is the radial stress, σ_θ is the circumferential stress, τ_{r_θ} is the tangential shear stress, R is the radius of the hole, r is the distance from the center of the hole, θ is the azimuth measured from the direction of $\sigma_B{}^*$ and ΔP is the difference between the fluid pressure in the borehole and the pore pressure (positive indicates overpressure in the borehole).

At the well boundary, when $r=R$, the set of equations becomes

$$\sigma_r = \Delta P, \tag{7}$$

$$\sigma_\theta = \sigma_B^* + \sigma_A^* - 2\left(\sigma_B^* - \sigma_A^*\right)\cos 2\theta - \Delta P, \tag{8}$$

$$\tau_{r\theta} = 0. \tag{9}$$

The most critical stresses at the well boundary, called hoop stresses, occur for σ_θ when θ equals 0° (minimal stress value, i.e., maximum tensile stress) and when θ equals 90° (maximal stress value, i.e., maximum compressional stress):

$$\sigma_{0°} = 3\sigma_A^* - \sigma_B^* - \Delta P, \tag{10}$$

$$\sigma_{90°} = 3\sigma_B^* - \sigma_A^* - \Delta P. \tag{11}$$

According to the analytical model, and for a Mohr–Coulomb strength criterion, damages occur if $\sigma_{0°}$ reaches the UTS or if $\sigma_{90°}$ exceeds the UCS. In the first case, drilling-induced tensile fractures develop in the direction of the maximum stress; the fluid pressure in the well has reached the so-called breakdown pressure. On the contrary, for $\theta = 90°$, damages appear in the form of breakout in the direction of the minimum stress; high pressure acts as a stabilizer.

Thermal effects can be integrated into the analytical model, by adding the thermal stress coefficient in the equations (Stephens and Voight 1982): $\alpha_L E \Delta T / (1 - \nu)$, where α_L is the linear coefficient of thermal expansion, E the Young's modulus, ΔT the temperature difference between the fluid in the wellbore and the rock (ΔT is negative for cooling), and ν the Poisson's ratio. The equations for σ_θ at the well boundary become

$$\sigma_{0°} = 3\sigma_A^* - \sigma_B^* - \Delta P + \frac{\alpha E \Delta T}{(1 - \nu)}, \tag{12}$$

$$\sigma_{90°} = 3\sigma_B^* - \sigma_A^* - \Delta P + \frac{\alpha E \Delta T}{(1 - \nu)}. \tag{13}$$

From these equations, it can be seen that the thermal effects favor the occurrence of induced fractures when cooling the wellbore ($\sigma_{0°}$ decreases since ΔT is negative for cooling, thus the failure may occur sooner). In the present case, in the absence of pore pressure, $\Delta P = P_{\text{well}}$ and thus the fracture pressure P_{frac} considering the thermal effect is

$$P_{\text{frac}} = 3\sigma_A^* - \sigma_B^* + \text{UTS} + \frac{\alpha E \Delta T}{(1 - \nu)}. \tag{14}$$

References

ALT advanced logic technology, acoustic borehole imagers and specifications ABI43. https://www.alt.lu/pdf/abi_2013. pdf. Accessed 26 Feb 2018.

Baujard C, Genter A, Dalmais E, et al. Hydrothermal characterization of wells GRT-1 and GRT-2 in Rittershoffen, France: implications on the understanding of natural flow systems in the rhine graben. Geothermics. 2017;65:255–68.

Batir J, Davatzes N, Asmundsson R (2012) Preliminary model of fracture and stress state in the Hellisheidi geothermal field, Hengill volcanic system, Iceland. In: Thirty-Seventh Workshop on Geothermal Reservoir Engineering. Palo Alto: Stanford University; 2012.

Carmichael RS. Practical handbook of physical properties of rocks and minerals. Boca Raton: CRC Press; 1989.

Clauser C, Huenges E. Thermal conductivity of rocks and minerals. Rock Phys Phase Relat. 1995;1(3):105–26.

Covell C. Hydraulic well stimulation in low-temperature geothermal areas for direct use. Thesis of 60 ETCS credits. Master of Science in Energy Engineering, Iceland School of Energy; 2016.

Davatzes N, Hickman S. Comparation of acoustic and electrical image logs from the Coso geothermal filed. In: Thirty-Seventh Workshop on Geothermal Reservoir Engineering. Palo Alto: Stanford University; 2005.

Deltombe JL, Schepers R. Combined processing of BHTV travel time and amplitude images. In: Proc. Int. Symp. Borehole geophysics for minerals, geotechnical and groundwater applications. Colorado: Golden; 2001. vol. 7, p. 29–42.

EGEC. 2016 EGEC geothermal market report, key findings, 6 edn; 2017.

Fjar E, Holt RM, Raaen AM, et al. Petroleum related rock mechanics, vol. 53, 2nd edn. New York: Elsevier Science; 2008.

Flores M, Davies D, Couples G, Palsson B. Stimulation of geothermal wells, can we afford it? In: Proceedings of the World Geothermal Congress, Antalya, Turkey, 2005.

Foulger GR, Du Z, Julian BR. Icelandic-type crust. Geophys J Int. 2003;155:567–90.

Friðleifsson GO, Bogason SG, Stoklosa AW, et al. Deployment of deep enhanced geothermal systems for sustainable energy business. In: Proceedings, European Geothermal Congress, Strasbourg; 2016.

Friðleifsson GO, Elders WA. The Iceland Deep Drilling Project geothermal well at Reykjanes successfully reaches its super-critical target. Geotherm Resour Counc Bull. 2017a;2017:30–3.

Friðleifsson GO, Elders WA. Successful drilling for supercritical geothermal resources at Reykjanes in SW Iceland. Geotherm Resour Counc Bull. 2017b;41:1095–107.

Friðleifsson GO, Elders WA, Zierenberg RA, et al. The Iceland Deep Drilling Project 4.5 km deep well, IDDP-2, in the seawater-recharged Reykjanes geothermal field in SW Iceland has successfully reached its supercritical target. Sci Drill. 2017;23:1–12.

Grant MA, Clearwater J, Quinao J, et al. Thermal stimulation of geothermal wells: a review of field data. In: Proceedings of the thirty-eighth workshop on geothermal reservoir engineering, Stanford University, Stanford; 2013.

Gudmundsson A. Dynamics of volcanic systems in Iceland: example of tectonism and volcanism at juxtaposed hot spot and Mid-Ocean Ridge systems. Annu Rev Earth Planet Sci. 2000;28:107–40.

Guéguen Y, Palciauskas V. Introduction à la physique des roches. Hermann edit; 1992.

Héðinsdóttir H. Mechanisms of injectivity enhancement in the thermal stimulation of geothermal wells. Master Thesis, Department of Earth Sciences ETH Zürich Geological Institute Engineering Geology, 2014.

Heidbach O, Tingay M, Barth A, et al. The World Stress Map database release, 2008.

Heidbach O, Rajabi M, Reiter K, Ziegler M. The WSM Team World Stress Map database release. 2016 GFZ data services, 2016.

Hokstad K, Tanavasuu-Milkeviciene K. Temperature prediction by multigeophysical inversion: application to the IDDP-2 well at Reykjanes, Iceland. Geotherm Resour Counc. 2017;41:1141–52.

Holl HG, Barton C. Habanero field—structure and state of stress. In: Proceedings World Geothermal Congress 2015, Melbourne; 2015.

Itasca Consulting Group Inc. PFC2D—Particle flow code in 2 dimensions, Version 4.0, Theory and background manual. Minneapolis: Itasca; 2008a.

Itasca Consulting Group Inc. PFC2D/3D (Particle flow code in 2/3 dimensions), Version 4.0, MN: ICG. Minneapolis; 2008b.

Itasca Consulting Group Inc. Fast lagrangian analysis of continua, 4th ed., 4.0 Minneapolis: Itasca; 2002.

Itasca Consulting Group Inc. PFC2D—Particle Flow Code in 2 Dimensions, Ver. 4.0, FISH Reference Manual. Minneapolis: Itasca; 2008c.

Kang Y, Yu M, Miska SZ, Takach N. Wellbore stability: a critical review and introduction to DEM. Soc Pet Eng. 2009. https://doi.org/10.2118/124669-MS.

Kanninen MF, Popelar CH. Advanced fracture mechanics. Oxford: Oxford University Press; 1985.

Karatela E, Taheri A, Xu C, Stevenson G. Study on effect of in situ stress ratio and discontinuities orientation on borehole stability in heavily fractured rocks using Discrete Element Method. J Pet Sci Eng. 2016;139:94–103.

Karson JA. Crustal accretion of thick, mafic crust in iceland: implications for volcanic rifted margins. Can J Earth Sci. 2016;53:1–11.

Keiding M, Lund B. Earthquakes, stress, and strain along an obliquely divergent plate boundary: Reykjanes Peninsula, southwest Iceland. J Geophys Res. 2009;114:1–16.

Keshavarz M. Contribution à l'étude expérimentale de l'endommagement mécanique et thermique de roches cristallines. Thèse de doctorat en Terre solide, sous la direction de Frédéric Pellet et de Benjamin Loret, à l'Université Joseph Fourier (Grenoble); 2009.

Khodayar M, Níelsson S, Hickson C, et al. The 2016 conceptual model of Reykjanes geothermal system, SW Iceland. Deliverable WD 4.1, DEEPEGS project, 2017.

Kirsch C. Die Theorie der Elastizität und die Bedürfnisse der Festigkeitslehre. Zeitschrift des Vereines deutscher Ingenieure. 1898;42:797–807.

Kranz RL. Microcracks in rocks: a review. Tectonophysics. 1983;100:449–80.

Kristjánsdóttir S. Microseismicity in the Krysuvik geothermal field, SW Iceland, from May to October 2009. Unpublished MS-thesis, University of Iceland, Reykjavik; 2013. p. 50.

Lu SM. A global review of enhanced geothermal system (EGS). Renew Sustain Energy Rev. 2018;81:2902–21.

MacKenzie, Adams. Atlas d'initiation à la pétrographie, éd. Dunod, Collection Sciences sup; 1999.

Menger S. New aspects of the Borehole televiewer decentralization correction. The log analyst 35 July/August: 14–20; 1994.

Pálmason G. Crustal structure of Iceland from explosion seismology, Science Institute, University of Iceland, Reykjavik; 1970.

Peter-Borie M, Blaisonneau A, Gentier S, et al. Study of thermo-mechanical damage around deep geothermal wells : from the micro-processes to macroscopic effects in the near well. In: Proceedings World Geothermal Congress 2015, Melbourne; 2015.

Peter-Borie M, Blaisonneau A, Gentier S, et al. A particulate rock model to simulate thermo-mechanical cracks induced in the near well by supercritical CO_2 injection. In: IAMG 2011 : the Annual Conference of the International Association for Mathematical Geosciences. Salzburg, Austria; 2011. p. 141–8.

Potyondy DO, Cundall PA. A bonded-particle model for rock. Int J Rock Mech Min Sci. 2004;41:1329–64.

Rajabi M, Tingay M, King R, Heidbach O. Present-day stress orientation in the Clarence-Moreton Basin of New South Wales, Australia: a new high density dataset reveals local stress rotations. Basin Res. 2016;29:622–40.

Santarelli F, Dahen D, Baroudi H, Sliman K. Mechanisms of borehole instability in heavily fractured rock media. Int J Rock Mech Min Sci Geomech Abstr. 1992;29:457–67.

Shiu W, Dedecker F, Rachez X, Peter-Borie M. Discrete modeling of near-well thermo-mechanical behavior during CO_2 injection. In: 2nd International FLAC/DEM symposium proceedings, 14-16 February, 2011, Melbourne; 2011.

Simmons G. Single crystal elastic constants and calculated aggregate properties, Progress. J Grad Res Center. 1965;34:273.

Stefanson A, Gislason TH, Sigurdsson O, Fridleifsson GO. The drilling of RN-15/IDDP-2 research well at Reykjanes in SW Iceland. Geother Resour Coun. 2017;41:512–22.

Stephens G, Voight B. Hydraulic fracturing theory for conditions of thermal stress. Int J Rock Mech Min Sci. 1982;19:279–84.

Tester JW, Anderson BJ, Batchelor AS, et al (2006) The future of geothermal energy. Impact of enhanced geothermal system (EGS) on the United States in the 21st Century, MIT-Massachusetts Institute of Technology, Cambridge; 2006.

Tingay M, Reinecker J, Müller B. Borehole breakout and drilling-induced fracture analysis from image logs, World Stress Map Project Guidelines: Image Logs; 2008.

Tulinius H. Estimation of Formation Temperature below 4000 m Depth in Well IDDP-2 Using Horner Plots. Prepared for DEEPEGS, Short report ÍSOR-17069; 2017.

Ulusay R, Türeli K, Ider MH. Prediction of engineering properties of a selected litharenite sandstone from its petrographic characteristics using correlation and multivariate statistical techniques. Eng Geol. 1994;38:135–57.

Wanne TS, Young RP. Bonded-particle modeling of thermally fractured granite. Int J Rock Mech Min Sci. 2008;45:789–99.

Weatherford international logging while drilling SinWave. https://www.weatherford.com/en/documents/brochure/products-and-services/drilling/lwd-sensors/. Accessed 26 Feb 2018.

Xiao SP, Belytschko T. A bridging domain method for coupling continua with molecular dynamics. Comput. Methods Appl. Mech. Eng. 2004;193:1645–69. https://doi.org/10.1016/j.cma.2003.12.053.

Yamamoto K, Shioya Y, Uryu N. Discrete element approach for the wellbore instability of laminated and fissured rocks. In: Proceedings of the SPE/ISRM Rock Mechanics Conference, 2002.

Yan C, Deng J, Yu B, et al. Borehole stability in high-temperature formations. Rock Mech Rock Eng. 2014;47:2199–209.

Ziegler M, Rajabi M, Hersir G, et al. Stress Map Iceland 2016. GFZ Data Services; 2016.

Zierenberg RA, Fowler A, Friðleifsson GÓ, et al. Preliminary description and alteration in IDDP-2 drill core samples recorved from Reykjanes Geothermal System, Iceland.; 2017.

Zoback MD, Moos D, Mastin L, Anderson RN. Well bore breakouts and in situ stress. J Geophys Res. 1985;90:5523–30.

5

Numerical analysis of the role of radiogenic basement on temperature distribution in the St. Lawrence Lowlands, Québec

Hejuan Liu[1,2], Bernard Giroux[1*] [ID], Lyal B. Harris[1], Steve M. Quenette[3,4] and John Mansour[4]

*Correspondence:
bernard.giroux@ete.inrs.ca
[1] Centre – Eau Terre
Environnement, Institut
national de la recherche
scientifique, 490 rue de la
Couronne, Québec, QC G1K
9A9, Canada
Full list of author information
is available at the end of the
article

Abstract

Regions with low or medium surface heat flow in stable cratonic areas, such as in eastern Canada, have received little attention for geothermal energy. In the presence of a high heat-producing basement overlain by a sedimentary cover, however, such areas might be prospective. Their potential will depend on various parameters such as heat production within the basement, thermal conductivity of sedimentary formations, and structural context. In this study, we aim at quantifying the importance of these parameters on temperature distribution at depth for a model representative of the St. Lawrence Lowlands, Québec, where locally anomalously higher heat flow has been observed and a 3D model of sedimentary cover is available. Scenarios involving physical properties from neighbouring Grenvillian domains are considered: Portneuf–Mauricie domain with radiogenic heat production of 0.94–5.83 $\mu W\, m^{-3}$, 0.02–4.13 $\mu W\, m^{-3}$ for the Morin Terrane, and 0.34–1.96 $\mu W\, m^{-3}$ for the Parc des Laurentides domain. The impact of radiogenic heating on temperature distribution at depth was simulated using the Underworld2 numerical modeling code. Results show that at 5 km depth, the range of temperature difference is 22 °C for all modeled scenarios. In addition, the benefit of the thermal blanket effect of the sedimentary cover can be significant, but depends strongly on the contrast in thermal conductivity between the basement and the cover, as well as on the structural context, and less on heat production in the basement. Finally, depth of the 120 °C isotherm varies by up to 1 km for the scenarios considered; carefully assessing the boundary conditions therefore, appears critical in an exploration context.

Keywords: Deep geothermal potential, St. Lawrence Lowlands, Radiogenic Grenvillian basement, Temperature distribution, Underworld2 numerical modeling

Introduction

Active volcanic or magma-related systems possessing a high heat flow at plate margins have received attention for geothermal energy because of their high temperature at shallow depths (Capuno et al. 2010). Regions with low or medium surface heat flow in stable cratonic areas, such as in eastern Canada, have received much less attention. In some such regions, there, however, may still be potential to develop high-temperature "Hot Dry Rock" geothermal resources (summarized by Hillis et al. 2004), especially when there is a high heat-producing basement, e.g. granites rich in radiogenic elements (Costain et al. 1980). The Cooper Basin in Australia is an example of such "Hot Dry Rock" geothermal exploration where high heat producing (3.8–8.7 $\mu W\ m^{-3}$) basement granites, and thick sedimentary rocks with low thermal conductivity act as a "thermal blanket", give rise to temperature anomalies in the range of 240 °C at 3.5 km depth (Beardsmore 2005; Hillis et al. 2004; Horspool et al. 2012). There has, however, been little previous study of whether there is potential for "Hot Dry Rock" geothermal resources in eastern Canada.

In Canada, the utilization of geothermal energy has increased significantly during the last decade, although it is mainly limited to low enthalpy systems (Raymond et al. 2015; Thompson 2010). Compared to other countries, such as the USA, Indonesia, Philippines, Mexico, and New Zealand, the development and application of geothermal power in Canada lags because of the lack of federal and provincial support (Thompson et al. 2015). There is, however, great potential for developing geothermal power in western Canada, especially in British Columbia and SW Yukon (Grasby et al. 2012; Thompson et al. 2015) which are along the Pacific Ring of Fire. In contrast, the Canadian Shield of eastern Canada and the Appalachian orogen on its SE margin have an overall low heat flow of less than 60 mW m^{-2} (Drury et al. 1987; Fou 1969; Grasby et al. 2012; Jaupart et al. 1998; Majorowicz and Minea 2012; Mareschal et al. 2000; Misener et al. 1951; Pinet et al. 1991; Saull et al. 1962), making the utilization of geothermal power from deep geothermal resources less attractive.

Grasby et al. (2012) concluded that there is little potential for high-temperature geothermal resources in the province of Québec because of the generally low heat generation (0.7–0.8 $\mu W\ m^{-3}$) and high thermal conductivity (> 3 W $m^{-1}\ K^{-1}$) of the Precambrian basement rocks, resulting in low geothermal gradients. However, recent evaluations in parts of the St. Lawrence Lowlands have suggested differently, especially south of Montréal and Trois–Rivières, in the Gaspésie Peninsula, and on Anticosti Island (Majorowicz and Grasby 2010; Majorowicz and Minea 2012, 2013; Minea and Majorowicz 2011, 2012; Perry et al. 2010; Raymond et al. 2012). The local elevated temperatures at depth suggested in these studies are likely due to the presence of radiogenic crystalline gneissic basement. For example, in the N Appalachians ca. 230 km NE of Québec City, Gicquel et al. (2015) present a correlation between higher thermal maturation in Appalachian sedimentary rocks (calculated from vitrinite reflectance data,) where they overlie radiogenic migmatitic Grenvillian paragneisses interpreted from aeromagnetic data. Some basement granitoids or alkaline and carbonatite intrusions in the sedimentary cover (e.g., in Oka, Ford et al. 2001) may result in local, high radiogenic heat flow in St. Lawrence Lowlands, and N Appalachian orogen. Preliminary interpretations of aeromagnetic and gravity data suggest the possibility for lateral offset of Grenvillian basement rocks along ductile shear zones that parallel the St. Lawrence River (Gicquel et al. 2015); therefore, direct extrapolation of basement domains in the Grenville Province beneath the St. Lawrence Lowlands is not yet possible. Our research has focused on the general problem of whether radiogenic rocks in basement to the St. Lawrence Lowlands in the region south of Trois–Rivières (Fig. 1) may play a significant role in affecting the temperature distribution at depth and whether thermal modeling may help constrain the nature of basement in this area.

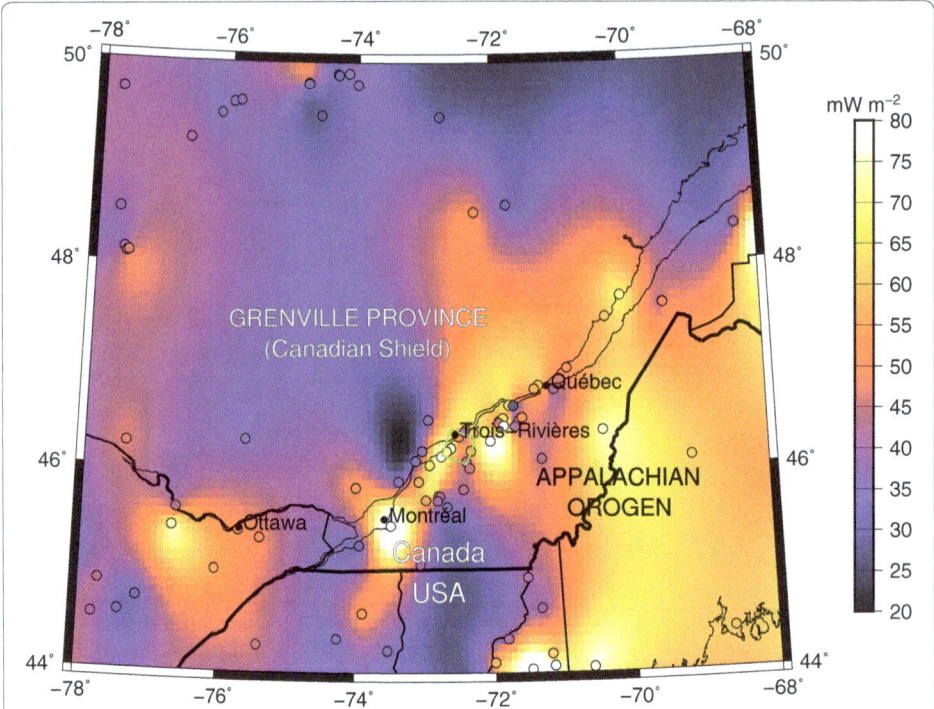

Fig. 1 Surface heat flow in southwestern Québec. Open circles show the data points of heat flow values based on Perry et al. (2010) and references therein; Bédard et al. (2016); SMU Geothermal Laboratory (2018); Jean-Claude Mareschal, pers. comm. Green box is the study area

This area was selected, as it was the subject of prior research evaluating the potential for the CO_2 storage in the St. Lawrence Lowlands in which petrophysical parameters of the sedimentary rocks were measured and present-day stress field calculated (Bédard et al. 2013b; Konstantinovskaya et al. 2012; Malo et al. 2014), and a 3D GOCAD model (Bédard et al. 2013a) and a structural interpretation of 2D seismic line M2002 (Castonguay et al. 2010) that crosses the area are available. Although a broad overview of depth to basement and main faults has been documented by Thériault et al. (2005), the detailed lithological composition of the Grenville Province basement underlying the Cambro-Ordovician platform sequence in this region is not well constrained.

A sensitivity study was undertaken using the Underworld2 geothermal modeling software (Mansour et al. 2018; Moresi et al. 2007; Quenette et al. 2015) on small scale 3D model, using the area S of Trois–Rivières as representing a geological setting typical of the St. Lawrence Lowlands to investigate the impacts of: (1) a homogeneous basement with properties representative of the different basement domains (each with different concentrations of uranium, thorium, and potassium and variable thermal conductivity) that may occur beneath the Saint Lawrence Lowlands; (2) intrusions within basement; (3) the thermal blanket effect of sedimentary cover; and (4) boundary conditions, especially the temperature at the bottom boundary of the model.

Geological setting

St. Lawrence Lowlands

The St. Lawrence Lowlands in southern Québec is composed of Cambrian to Ordovician sedimentary rocks that overly a Mesoproterozoic Grenvillian metamorphic and intrusive basement (geology of basement domains is discussed in the next section). The St. Lawrence Lowlands basin developed during the intracratonic rifting of the Rodinia supercontinent related to the opening of the Iapetus ocean during the Neoproterozoic to Early Cambrian (Hersi et al. 2003; Tremblay et al. 2003). During the Cambrian to early Ordovician, it evolved into an oceanic basin (Kumarapeli 1985) characterized by a syn-rift and passive continental margin succession (Lavoie et al. 2014). During the Middle-to-Late Ordovician onset of Taconian thrusting in the initial stage of Appalachian orogenesis, it evolved into a foreland basin characterized by the deposition of carbonates, shales, and clastic sequences (Bernstein 1991; Lavoie et al. 2009, 2014; Tremblay et al. 2003). Deposition in the St. Lawrence Lowlands ended following the emplacement of large ophiolitic nappes onto Laurentia and the accretion of volcanic arcs and continental blocks induced by the closure of the Iapetus Ocean and convergence between Laurentia and Gondwana Taconian Orogeny (De Souza et al. 2014; Lavoie and Chi 2010; Rimando and Benn 2005; Rocher and Tremblay 2001).

From bottom to top, the sedimentary sequence of the St. Lawrence Lowlands is: Potsdam (PO), Beekmantown (BK), Trenton–Black River–Chazy (Tr–BR–Ch), Utica and Sainte-Rosalie–Lorraine–Queenston (SR–LO–QT) groups which are mainly composed of sandstones, dolomites, limestone, shale, and siltstone, respectively (Clark 1972; Globensky 1987; Lavoie et al. 2009); see Figs. 2 and 3.

Densities and radiogenic element concentrations of rocks representative of the overlying sedimentary formations are listed in Table 1. Utica shales have the highest concentrations of radiogenic elements, with uranium, thorium and potassium in the range of 2.59–3.41 ppm, 2.9–9.8 ppm, and 0.29–3.42%, respectively. Dolomites from the Beekmantown group also have high concentration of radiogenic elements. There is only one measurement for uranium and thorium in the uppermost cap rock (SR–LO–QT groups), which is relatively high but likely not representative. Thermal conductivities of the sedimentary rocks show that the sandstones of the Potsdam group have the highest values in the range 4.77–6.9 $W\,m^{-1}\,K^{-1}$, followed by the dolomites of the Beekmantown group in the range of 2.7–4.24 $W\,m^{-1}\,K^{-1}$ while for other groups with an average value less than 3.0 $W\,m^{-1}\,K^{-1}$.

Grenvillian basement

The Grenville Province of western Québec is composed of high-grade metamorphic terranes accreted to the SE Laurentian margin (Davidson 1984; Dufréchou et al. 2014; Rivers 1997, 2015). Grenvillian rocks outcropping N of our study area comprise the generally N–S trending Portneuf–Mauricie and Parc des Laurentides domains on the E margin of the Mékinac–Taureau dome, along with an eastern extremity of the Morin terrane on the dome's southern and eastern margins (Nadeau and Brouillette 1994, 1995; Nadeau et al. 2008; Sappin et al. 2009), as portrayed in Fig. 4.

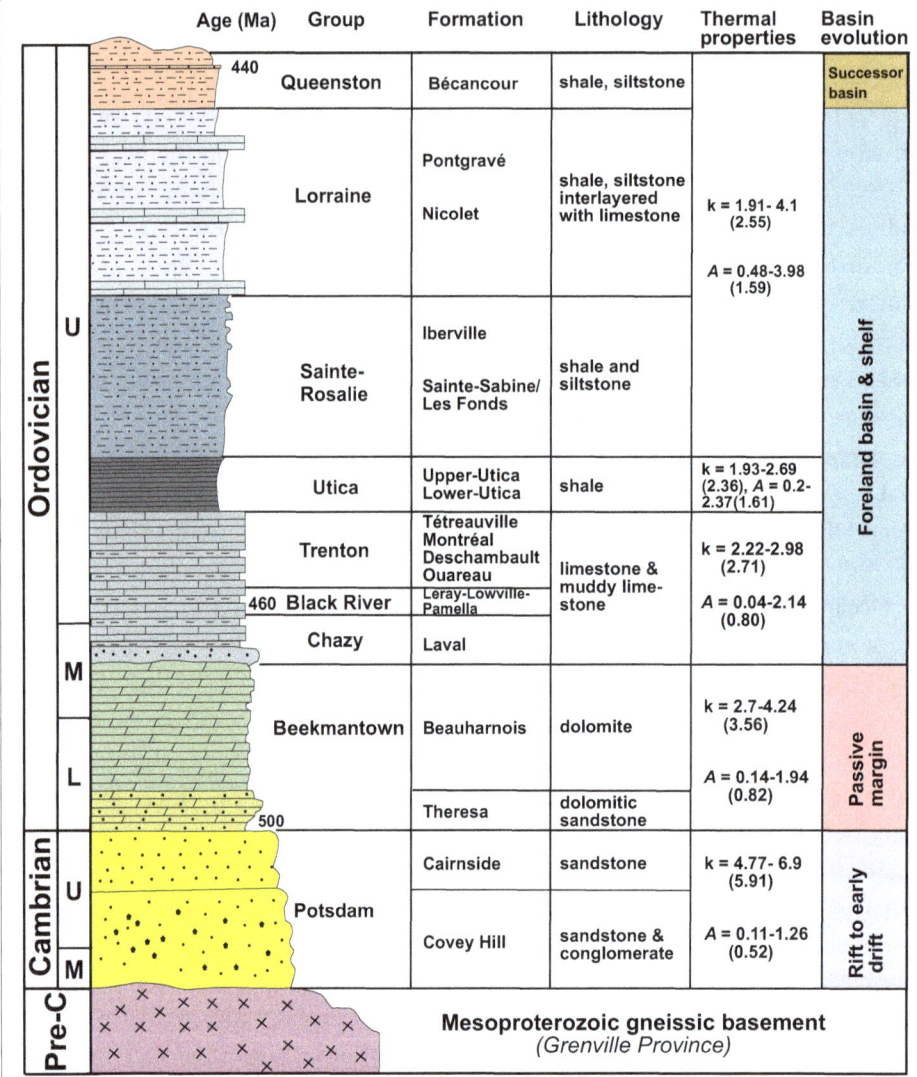

Fig. 2 Stratigraphy of St. Lawrence Lowlands. Stratigraphy of St. Lawrence Lowlands, with lithology, thermal and reservoir properties (thermal conductivity k in W m^{-1} K^{-1} and radiogenic heat production A in μW m^{-3}, values of the minimum, maximum and mean are in parentheses; Perm = permeability). Basin evolution is modified after (Comeau et al. 2012), combining data obtained from Bédard et al. (2016), Globensky (1987), Hersi et al. (2003), Hofmann (1972), Nasr (2016), Pinti et al. (2011), Rocher and Tremblay (2001)

The concentration of uranium, thorium, and potassium of ten basement core samples of orthogneiss, granite ± pegmatite, and quartzite from wells located in or around the region of interest is in the range of 0.67–4.73 ppm, 0.6–22 ppm 1.34–5.8%, respectively (Table 2).

Portneuf–Mauricie (PM) domain

The Portneuf–Mauricie domain (PM) region comprises (1) metasedimentary and meta-volcanic rocks of the Montauban group, including amphibolite facies quartzofeldspathic gneiss, quartzite, marble, and calc-silicate rocks and andesitic to rhyolitic volcanics and (2) the La Bostonnais Complex plutons, including the de Lapeyrère and Gagnon

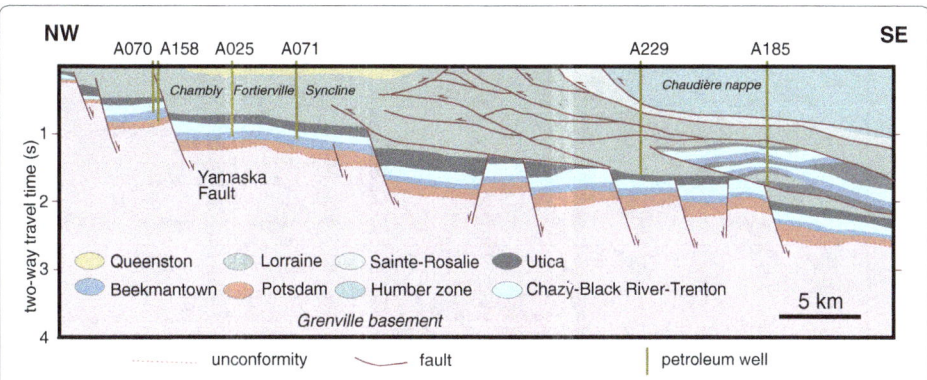

Fig. 3 Interpretation of seismic profile M2002. Interpretation of seismic profile M2002 crossing the modeled region (simplified from Castonguay et al. (2010). Bottom hole temperatures after Harrison correction in °C for wells A070, A229 and A185 are shown in the profile

Table 1 Density, concentration of radiogenic elements, and thermal conductivity of sedimentary rocks of the St. Lawrence Lowlands, with the minimum, maximum values and average in parentheses

Groups	ρ_r (kg m^{-3})	C_U (ppm)	C_{Th} (ppm)	C_K (%)	k (W m^{-1} K^{-1})
SR–LO–QT siltstone	2540–2720 (2602)	3.1	10.2	0.05–0.8 (0.24)	1.91–4.1 (2.55)
Utica shale	2700–2710 (2705)	2.59–3.41 (2.9)	2.9–9.8 (7.4)	0.29–3.42 (2.33)	1.93–2.46 (2.36)
TR-BR-Ch limestone	2630–2700 (2680)	0.2–6.6 (1.6)	2.4–4.7 (3.55)	0.04–2.66 (0.67)	2.22–2.98 (2.71)
Beekmantown dolomite	2640–2810 (2717)	0.07–3.08 (2.0)	3–14 (7.5)	0.19–6.26 (2.08)	2.7–4.24 (3.56)
Potsdam sandstone	2540–2640 (2602)	0.4–0.45 (0.43)	1.38–1.4 (1.39)	0.05–2.3 (0.32)	4.77–6.9 (5.91)

Densities of sedimentary rocks are based on measurements of 32 samples by Nasr (2016); averages are in parenthesis. Radiogenic element data are from Owen and Greenough (2008), Pinti et al. (2011), Rivard et al. (2002), Vautour et al. (2015) as well as new ICP–MS and ICP–OES measurements. Thermal conductivity data of sedimentary rocks are from Nasr (2016) and Perozzi et al. (2016)

gabbronorites, granite, quartz monzonite, quartz syenite, and pegmatite which contain high concentrations of K_2O and Th (Corrigan 1995; Gautier 1993; Nadeau et al. 1992; Sappin 2012). Montauban group rocks are interpreted as part of a 1.45 Ga arc–back–arc complex into which the La Bostonnais complex plutons were intruded between 1.40 and 1.37 Ga after deformation and metamorphism of the Montauban group (Nadeau and van Breemen 1994; Sappin 2012; Sappin et al. 2009, 2014). Physical properties of Portneuf–Mauricie domain rocks are listed in Table 3. There are limited thermal conductivity data in the literature; therefore, five additional measurements were performed using a KD2 Pro probe and lab thermal properties analyzer.

Morin Terrain (MT) and Mékinac Taureau domain

The Morin Terrane in our study area (the Shawinigan domain of Corrigan and van Breemen (1997)) is characterized by Grenville Supergroup paragneisses and quartzite along with granitic and granodioritic gneiss and metavolcanics. The main area of exposure of the Morin Terrane W of our study area underwent regional ca. 1280 Ma deformation and metamorphism prior to diapiric emplacement of 1165–1135 Ma anorthosite–mangerite–charnockite–granite (AMCG) suite intrusions (Corrigan and van Breemen 1997; Corriveau and van Breemen 2000; Emslie and Hunt 1990;

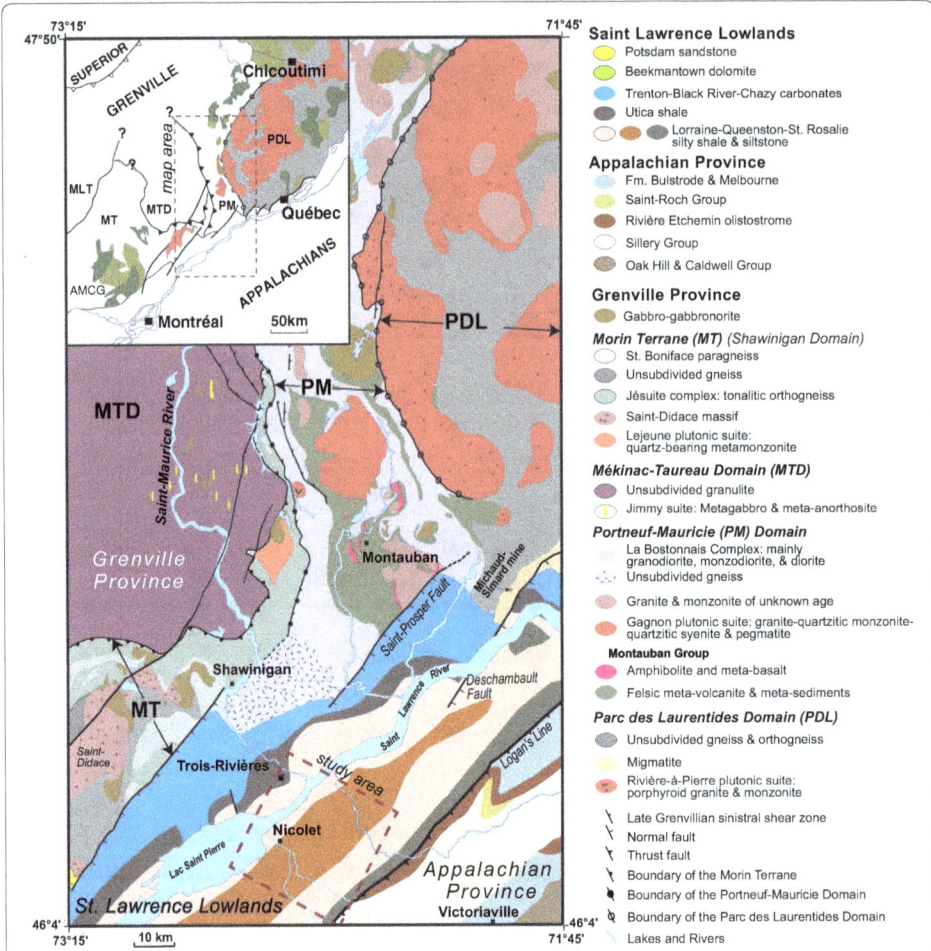

Fig. 4 Simplified regional geology. Geology of the Grenville Province is taken from Nadeau and Brouillette (1994, 1995); St. Lawrence Lowlands and Appalachian Province geology is from SIGÉOM (2018). Dashed maroon square indicates the study area. Inset map is modified after (Charlier et al. 2010)

Table 2 Concentration of radiogenic elements for basement rocks obtained from wells located south of Trois–Rivières from ICP–MS and ICP–OES analyses undertaken in our study

Well no.	Long.	Lat.	Lithology	C_U (ppm)	C_{Th} (ppm)	C_K (%)
A027	46.3186	−72.4857	Granitic gneiss	4.73	11.3	3.93
A126	46.1539	−72.6774	Pegmatite	2.19	22	5.80
A128	46.3356	−72.6103	Granite	1.58	6.7	3.92
A129	46.2895	−72.7064	Quartzite	0.67	1.4	1.34
A130	46.3068	−72.6997	Granitic gneiss	0.47	0.6	2.23
A167	46.6216	−71.7040	Granite	1.01	3.2	2.63
A175	46.8181	−71.3353	Granite	1.96	9.0	4.02
A186	46.1850	−72.6439	Pegmatite	1.59	5.9	3.91
A197	46.1814	−72.6714	Granite	0.87	5.2	6.59
A222	46.1819	−72.3201	Granitic gneiss	0.92	3.4	2.05

Table 3 Density, radiogenic element concentration and thermal conductivity data for rocks of the Portneuf–Mauricie domain

Basement rock	ρ_r (kg m^{-3})	C_U (ppm)	C_{Th} (ppm)	C_K (%)	k (W m^{-1} K^{-1})
Amphibolite (12)	2860–3090 (3010)	0.5–3 (2.2)	3–6 (3.8)	0.19–0.92 (0.41)	2.27–2.32 (2.3)
Basalt (3)	–	–	3	0.49–3.14 (1.47)	2.42–4.91 (3.32)
Diorite (10)	2730–2970 (2840)	0.5–2.2 (1.33)	0.55–5.2 (4.08)	0.11–4.75 (1.64)	3.35–3.92 (3.64)
Gabbro-gab-bronorite (21)	2600–3530 (2995)	0.05–0.6 (0.35)	0.2–3.6 (1.45)	0.1–2.32 (0.47)	1.84–4.72 (3.02)
Granite (22)	2620–2760 (2670)	0.5–4.9 (1.41)	1.2–56 (6.18)	0.5–5.59 (3.6)	1.7–4.0 (2.76)
Granitic gneiss (37)	2530–3160 (2650)	0–22 (3.33)	0.8–55.9 (9.35)	0.2–4.9 (2.12)	1.53–5.88 (2.24)
Iherzolite (4)	–	0.2	0.4–0.7 (0.55)	0.14–0.19 (0.17)	–
Migmatite (4)	–	0.5–4.5 (1.85)	0.2–14.4 (5.18)	0.2–1.3 (1.0)	–
Pyroxenite (2)	–	0–0.2 (0.1)	0–0.2 (0.1)	0.08–0.09 (0.085)	–
Paragneiss (46)	2580–3350 (2790)	0–3.9 (2.4)	0–14 (7.09)	0–3.73 (1.83)	1.53–5.88 (2.24)
Pegmatite (2)	2470–2810 (2630)	640.4–1202 (921.2)	1141–1151 (1146)	–	2.28–2.98 (2.46)
Quartzite (5)	2580–3350 (2790)	3	3–11 (6.6)	0.48–1.73 (1.42)	6.08–6.38 (6.23)
Quartz monzo-nite (9)	2600–2760 (2700)	0.5–1.8 (1.0)	3.0–3.2 (3.03)	1.48–4.52 (3.64)	3.35–3.92 (3.64)
Websterite (3)	–	0.05–0.2 (0.12)	0.05–0.7 (0.42)	0.11–0.54 (0.28)	–

Data sources: densities are from Emslie and Ermanovics (1975), Feininger and Goodacre (1995), Kearey (1978), Paradis (2004). Radiogenic element concentration data are from 163 samples from the SIGÉOM (2018) database. Thermal conductivity data are from Mareschal (2018), Jessop et al. (2005). Average values are in parenthesis. Number of samples for radiogenic element concentration is given in parenthesis after rock type

Martignole and Schrijver 1977). The domal structure of the Mékinac Taureau domain (Fig. 4), interpreted as being exhumed during crustal extension by Soucy La Roche et al. (2015), has subsequently been attributed from geophysical data to the presence of an underlying, contemporaneous AMCG body by Dufréchou (2017), which our unpublished geophysical data support. Sinistrally oblique extension along the Tawachiche shear zone (TSZ) on the E margin of the Morin Terrane, marking the mapped contact with the Portneuf–Mauricie domain, occurred between 1065 and 1035 Ma (Soucy La Roche et al. 2015). The Shawinigan norite, the Lejeune complex and the St. Didace complex intrude Morin Terrane granulites on the S margin of the Mékinac–Taureau dome between 1080 and 1056 Ma (Nadeau et al. 2008; Nadeau and van Breemen 2001; Soucy La Roche 2014; Soucy La Roche et al. 2015).

Paradis (2004) presents density data for granite (2630–2710 kg m^{-3}, average = 2668.3), quartzite (2640–2730 kg m^{-3}, average = 2685), charnockite (2700–2750 kg m^{-3}, average = 2733), anorthosite (2670–2830 kg m^{-3}, average = 2726) and paragneiss (2930–3100 kg m^{-3}, average = 2990); the overall bulk density of the Morin Terrane is in the range 2580–3350 kg m^{-3}, with an average of 2760 kg m^{-3}. Table 4 contains concentration of radiogenic elements data collected from the SIGÉOM (2018) and Mareschal (2018) databases, Mareschal and Jaupart (2013), Peck (2012), as well as field and laboratory measurements performed in this study. Radiogenic

Table 4 **Concentration of radiogenic elements for outcropping rocks of the Morin Terrane (including samples from outside our study area)**

Rock type	C_U (ppm)	C_{Th} (ppm)	C_K (%)
Anorthosite (7)	–	–	0.07–1.0
Amphibolite (23)	0.2–17.7	1.7–35.1	0.9–4.5
Diorite (4)	1.7–4.5	3.6–38	1.8–5.0
Granite (7)	0.9–13.9	2.2–31	0.5–6.29
Granitic gneiss (36)	0.2–3.95	0.25–35.5	0.9–5.9
Granodiorite gneiss (4)	0.54–1.35	0.6–4.7	2.71–3.35
Mangerite (1)	0.5	0.2	0.36
Marble (20)	2.1–3.4	–	0.11–7.91
Mixed gneiss (91)	0–9.2	0.1–35.9	0–5.8
Pyroxenite (4)	0.5–3.4	–	0.01
Quartz-monzonite (1)	0.5	–	5.96
Quartzite (10)	1.7–2.4	–	0.05–0.22
Pegmatite (5)	0.3–59.3	11.2–421.5	2.8–7.1

Table 5 **Concentration of radiogenic elements of rocks from Rivière-à-Pierre suite in the Parc des Laurentides domain Data sources: (SIGÉOM 2018; Higgins and van Breemen 1996)**

Rock type	C_U (ppm)	C_{Th} (ppm)	C_K (%)
Amphibolite (1)	0.8	3	0.78
Charnockite (5)	–	–	2.94–4.25
Gabbro (2)	0.5–0.7	3	2.25–4.57
Granite (16)	0.5–1.9	2.2–14	1.37–5.48
Granitic gneiss (7)	0.5–1.8	3–4.1	0.99–3.87
Mangerite (5)	0.5–1.0	2.1–2.2	2.17–3.44
Quartz-monzonite (7)	0.5–1.2	3–10	4.23–5.14
Quartzite (1)	1.7	3	4.26
Bulk rock	0.5–1.9 (1.18)	2.1–14 (4.05)	0.78–5.48 (3.55)

element concentrations are U = 0.01–4.5 (average = 1.67) ppm, in Th = 0.2–31 (average = 10.7) ppm and K = 0.01–7.91% (average: 2.43%). Thermal conductivities range between 2.36 and 4.58 $W m^{-1} K^{-1}$, with an average value of 2.89 $W m^{-1} K^{-1}$ (Jessop et al. 2005). Additional measurements for nine samples obtained from the Shawinigan domain vary between 2.017 and 3.924 $W m^{-1} K^{-1}$, with an average value of 2.648 $W m^{-1} K^{-1}$.

Parc des Laurentides domain

The Parc des Laurentides domain (PDLD) is mainly composed of the quartz–monzonite, porphyritic granite, mangeritic–charnockitic–granitic and granodioritic orthogneiss (Hébert and Nadeau 1995; Paradis 2004; Sappin 2012). Its western boundary separated from the Portneuf–Mauricie domain was obliterated by large granitic intrusions of the 1058 ± 1 Ma Rivière-à–Pierre suite (Hébert and Nadeau 1995).

The bulk density of the PDLD lies between 2575 and 2775 kg m^{-3}, with a mean density of 2680 kg m^{-3} (Paradis 2004). The radioelement data for rocks from the Rivière-à–Pierre suite are shown in Table 5. Uranium, thorium, and potassium concentrations are

0.5–1.9 ppm, 2.1–14 ppm, and 0.78–5.48%, respectively. The thermal conductivity of 3 samples obtained from the PDLD is in the range of 3.02–5.15 W m^{-1} K^{-1}, with an average value of 3.8 W m^{-1} K^{-1} (Jessop et al. 2005).

Method

Model definition

In the absence of fluid flow, the heat equation with a radiogenic source term is (Jaupart and Mareschal 2011)

$$\rho_r C_p \frac{\partial T}{\partial t} = \nabla \cdot (k \nabla T) + A, \tag{1}$$

where T is temperature (°C), t is time (s), k is thermal conductivity (W m^{-1} K^{-1}), ρ_r is density (kg m^{-3}), C_p is specific heat (J kg^{-1} K^{-1}) and A is heat production (W m^{-3}) calculated from the concentration of radiogenic elements. Symbol $\nabla \cdot$ is the divergence operator and ∇T the temperature gradient. In the following, we assume a steady-state regime. The time-dependence term, therefore, vanishes, which yields

$$\nabla \cdot (k \nabla T) = -A. \tag{2}$$

Heat production is considered using the following empirical function (Rybach 1976, 1988):

$$A = \rho_r (9.52 C_u + 2.56 C_{Th} + 3.48 C_K) \times 10^{-5}, \tag{3}$$

where C_u is the uranium content (ppm), C_{Th} is the thorium content (ppm), and C_K is the potassium content (%). Thermal conductivity is also assumed to be temperature dependent. The correction of Sekiguchi (1984) was used to take into account this dependence. According to Lee and Deming (1998), this correction is applicable to arbitrary rocks over the temperature range 0–300 °C. It is given by

$$k = \left(\frac{T_0 T_m}{T_m - T_0} \right) (k_0 - k_m) \left(\frac{1}{T} - \frac{1}{T_m} \right) + k_m, \tag{4}$$

where T is the *in situ* temperature (K), k_0 is the thermal conductivity at laboratory temperature, T_0 is the laboratory temperature (K), k_m and T_m are calibration coefficients, respectively, equal to 1.8418 W m^{-1} K^{-1} and 1473 K.

The geothermal module of the 3D geodynamic modeling package Underworld2 (Mansour et al. 2018; Moresi et al. 2007) was used to solve Eq. (2) numerically for temperature distribution. Underworld2 includes a finite-element solver for Eulerian meshes that embed a set of Lagrangian integration points or particles. These particles allow storing material properties and can be moved to follow the deformations of geological materials, through the use of a particle-in-cell method, to map the finite-element integrals in the system. When solving the steady-state heat equation with the Underworld2 module, a standard Gaussian quadrature module is used in place of the particle-in-cell method because there is no deformation.

The area of interest is represented by a simplified 30 km × 30 km × 10 km geological model representative of the sedimentary configuration of the St. Lawrence Lowlands south of Trois–Rivières (Fig. 1). This model was set up based on available seismic and well-log data and comprises the Grenvillian basement and the overlying

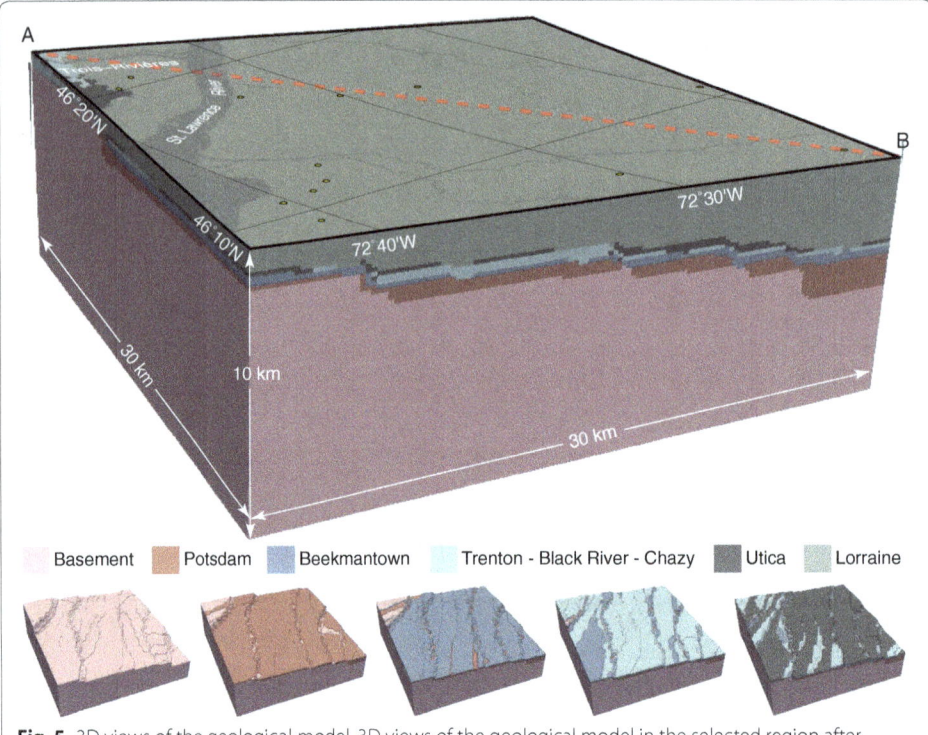

Fig. 5 3D views of the geological model. 3D views of the geological model in the selected region after discretization. The red dashed line represents the location of profile **A–B** that crosses the model from shallower (**A**) to deeper basement (**B**)

Cambrian-Ordovician sedimentary sequence (see Fig. 5). To perform the sensitivity analysis, various structures are inserted in this model, as described below.

To perform the calculations, the model was discretized in 128×128 cells in the horizontal directions, and in 64 cells in the vertical direction. The bottom boundary condition was established based on geothermal gradient data in the area as well as bottom hole temperature and surface heat flow data available inside the area of the model. For the St. Lawrence Lowlands, Bédard et al. (2016) found that the geothermal gradient is in the range of 14–35 °C km^{-1}, with an average of 24.3 ± 4.9 °C km^{-1} (Fig. 6). Within the model limits, 9 bottom hole temperature data are available and 15 heat flow measurements are found (Bédard et al. 2016). For each modeled scenarios described below, a series of modeling runs were done with heat flow imposed at the bottom boundary, with values ranging from 50 to 66 mW m^{-2}. The value that yielded a temperature distribution giving the best fit with the 24.3 °C gradient, the bottom hole temperature data and the surface heat flow data was selected. The fit is defined as the sum of the Relative Root Mean Square Error (rRMSE) of all three variables, with a lower weight (1/2) given to the rRMSE of the gradient because it was obtained with data outside the modeling domain. In all cases, the temperature at the top boundary is specified to be 10 °C and no-flow Neumann conditions are applied on the sides. Physical properties of the lithological units have been assigned based on the data presented in the geology section. Due to the size of the cells in the numerical mesh, the lithology of each formation must be simplified. Therefore, the main lithology of the Potsdam Group is sandstone, dolomite for the Beekmantown Group, limestone for the Trenton–Black River–Chazy Groups

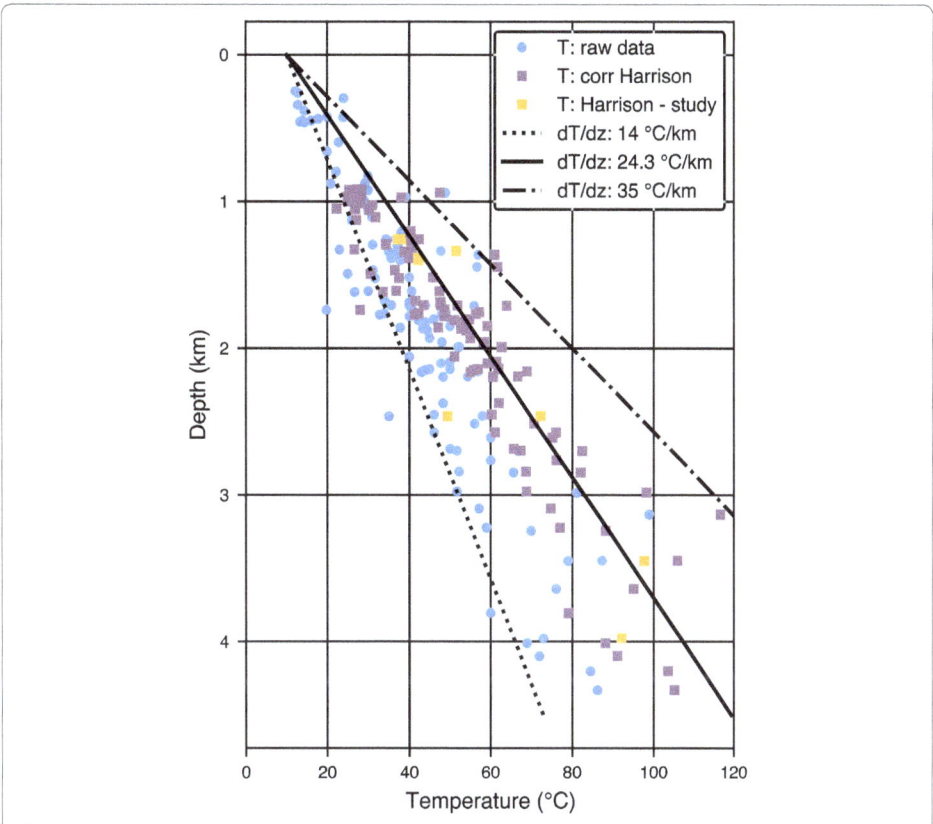

Fig. 6 Temperature–depth relationship. Temperature–depth relationship based on 124 bottom hole temperature data (Bédard et al. 2016) from petroleum wells located in the St. Lawrence Lowlands. The yellow squares are the temperature data after Harrison correction in the modeled region of this paper

(or TR–BR–Ch), shale for the Utica Group, siltstone for the top three groups of Sainte–Rosalie, Lorraine and Queenston (i.e. SR–LO–QT), respectively. For Precambrian basement, the influence of various rock types will be studied, as described next.

Modeled scenarios

In the study area, the basement is buried by a sedimentary cover varying in thickness between 0.5 and 6.5 km (see Fig. 3). As previously mentioned, a detailed interpretation of the basement structure is not possible due to scarcity of available data. Therefore, we begin with a simple homogeneous basement with thermal properties obtained from rocks outcropping in the Portneuf–Mauricie domain, which will be referred to as the base case in the simulation study. In the five scenarios presented in this paper, only differences in basement characteristics are considered. For four of the five scenarios, a homogeneous basement is considered, with input data summarized in Table 6. Each scenario is described in more detail below.

Model 1: PM domain basement

From data listed in Table 3, using Rybach's empirical function (Eq. 3), heat production in the basement is equal to $0.94\,\mu\mathrm{W}\ m^{-3}$. For this model, a heat flow value of $59\,\mathrm{mW}\ m^{-2}$

Table 6 Physical properties of Grenvillian rocks in the PM domain (Model 1), Shawinigan domain (Model 2), PDLD (Model 3) and from well cores (Model 4); average values in parentheses

Basement	ρ_r (kg m^{-3})	C_U (ppm)	C_{Th} (ppm)	C_K (%)	A (μW m^{-3})	k (W m^{-1} K^{-1})
Model 1	2620–3050	0.05–6.0	0.5–56	0.05–5.59	0.94	2.39–3.87
	(2790)	(1.61)	(4.65)	(1.64)		(3.0)
Model 2	2580–3350	0–17.7	0–68.1	0.01–7.91	1.42	2.02–3.92
	(2760)	(1.91)	(10.38)	(2.92)		(2.65)
Model 3	2575–2775	0.5–1.9	2.1–14	0.78–5.48	0.91	3.02–5.15
	(2680)	(1.18)	(4.05)	(3.55)		(3.8)
Model 4	2580–2620	0.67–4.73	0.6–22	1.34–5.8	1.18	2.26–2.71
	(2598)	(1.60)	(6.87)	(3.64)		(2.54)

at the bottom boundary gave the best fit to the data, with the sum of rRMSE equal to 36%.

Model 2: Morin Terrane basement

Based on data in Table 4, the calculated heat production for the Morin Terrane is 1.42 μW m^{-3}. In this case, a bottom boundary condition of 52 mW m^{-2} was obtained (sum of rRMSE equal to 35%).

Model 3: Parc des Laurentides domain basement

An average heat production value of 0.91 μW m^{-3} is obtained using the empirical function of Rybach (1988). In this model, the mean concentration of uranium, thorium, and potassium is, respectively, 1.18 ppm, 4.05 ppm and 3.55%. Heat flow at the bottom boundary is in this case 62 mW m^{-2} (sum of rRMSE equal to 39%).

Model 4: basement based on well data

The parameters of the fourth model are computed from measurements taken on 10 samples obtained from cores collected from wells in the modeled region (Table 2) for which the average bulk density is 2598 kg m^{-3} (Nasr 2016). With the values given in Table 2 for this model, the heat production is 1.18 μW m^{-3}. For this model, 53 mW m^{-2} at the bottom boundary gave the best fit to the data (sum of rRMSE equal to 37%).

Models 5 and 6: intrusions in the basement

Two models are used to evaluate the influence of a relatively "hot" intrusion on temperature distribution. The shape of the intrusion is arbitrarily set to a vertical cylinder that crosses the basement in the center of the model. The top of the intrusion is set at 2.4 km, at the interface between the basement and the Potsdam sandstones. In the first model, the radius of the intrusion is 1 km and its volume is approximately 2.5×10^{10} m^3, whereas in the second case the radius is 2 km, for a volume of 9.5×10^{10} m^3. Physical properties of the model unit parameters are given in Table 7. The radiogenic element concentrations for the host rock are assumed to be the same as in Model 1. For the intrusion, properties of granites of the Gagnon pluton (Fig. 4) are used. The bottom boundary condition is the same as for Model 1.

Table 7 Parameters used for the heterogeneous basement model

Unit	C_U (ppm)	C_{Th} (ppm)	C_K (%)	k (W m^{-1} K^{-1})	A (μW m^{-3})
Host rock	1.61	4.65	1.64	3.0	0.94
Intrusion	4.9	56	5.59	2.24	5.53

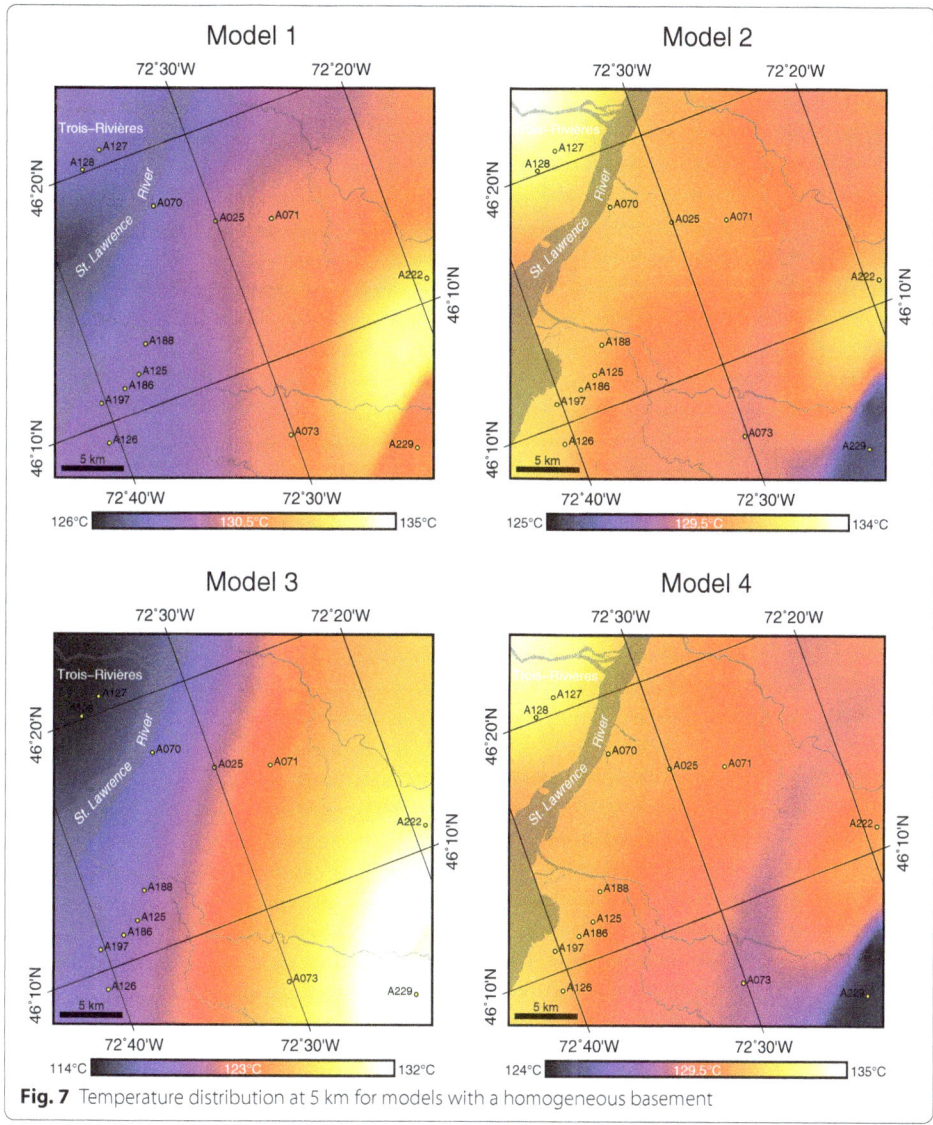

Fig. 7 Temperature distribution at 5 km for models with a homogeneous basement

Results and discussion

Homogeneous basement

Figure 7 shows the temperature at 5 km for the different homogeneous basement models. The first observable trend is the relative temperature in the Trois–Rivières vicinity (where basement rocks are predominant), which is colder for Models 1 and 3, and warmer for Models 2 and 4, this for lower heat production in Models 1 and 3 but for lower heat flow at the bottom for Models 2 and 4. Thus, not surprisingly, the effect of

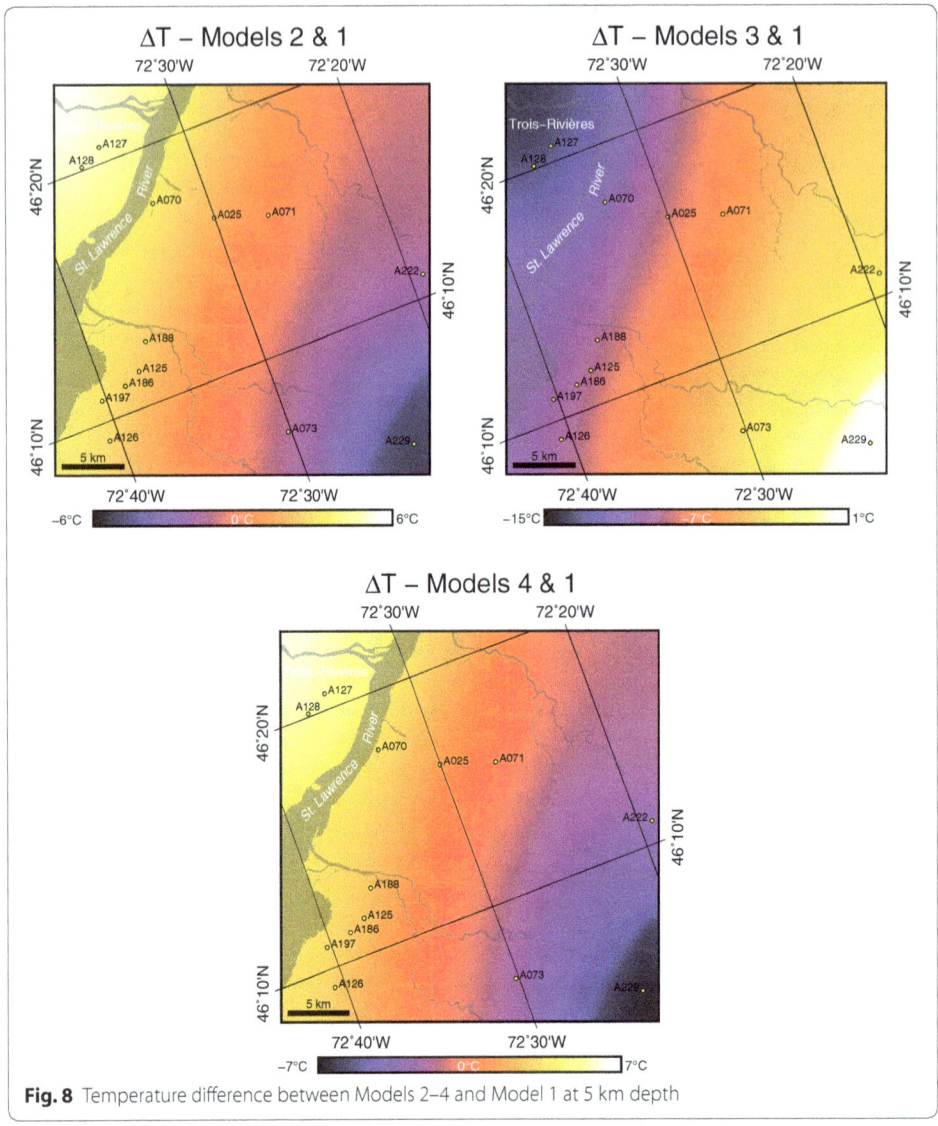

Fig. 8 Temperature difference between Models 2–4 and Model 1 at 5 km depth

heat production is counterbalanced by heat flow at the bottom boundary: models with high heat production have a better fit to the available data with lower heat flow at the bottom. On the other hand, Models 1 and 4 have the highest temperature, followed by Model 2 and Model 3. This might appear surprising, because basement rocks in Model 2 have the highest value of heat production and Model 3 the highest value of heat flow at the bottom boundary. However, Model 2 also has the second lowest thermal conductivity and Model 3 has the highest thermal conductivity (respectively, 3 and 3.8 W m^{-1} K^{-1} for Models 1 and 3 vs 2.7 and 2.5 W m^{-1} K^{-1} for Models 2 and 4). These observations thus show that the thermal conductivity of the basement rocks has a significant impact on temperature distribution, regardless of heat production. It is, therefore, equally important to accurately determine both parameters to correctly assess the geothermal potential of a given area.

The difference between Models 2–4 and Model 1 at 5 km depth is shown in Fig. 8. The maximum difference with Model 1 is 15 °C (Model 3), with a range of 22 °C (−15 to

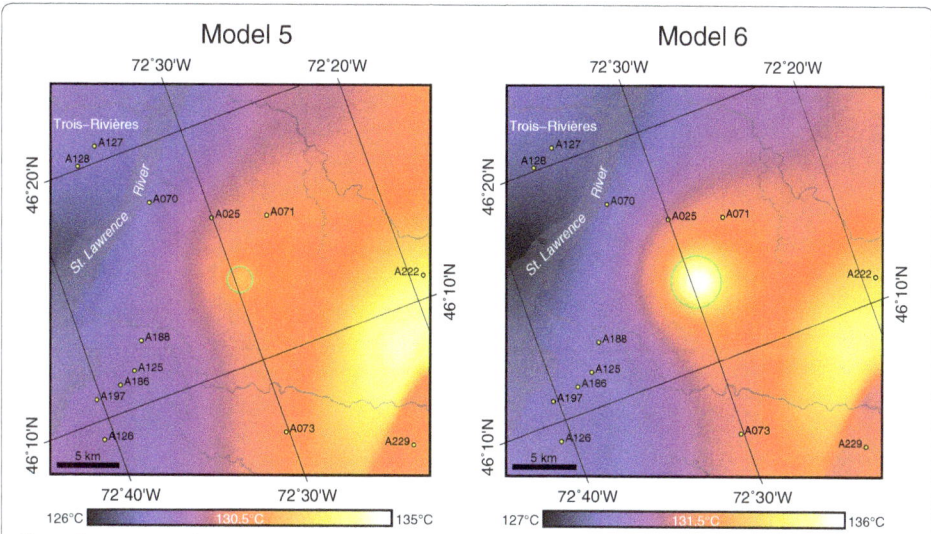

Fig. 9 Temperature distribution at 5 km for models with intrusions. Temperature distribution at 5 km for models with a 1 km radius vertical intrusion (left) and 2 km radius vertical intrusion (right) in the basement. Green circles are the surface projection of the vertical intrusion

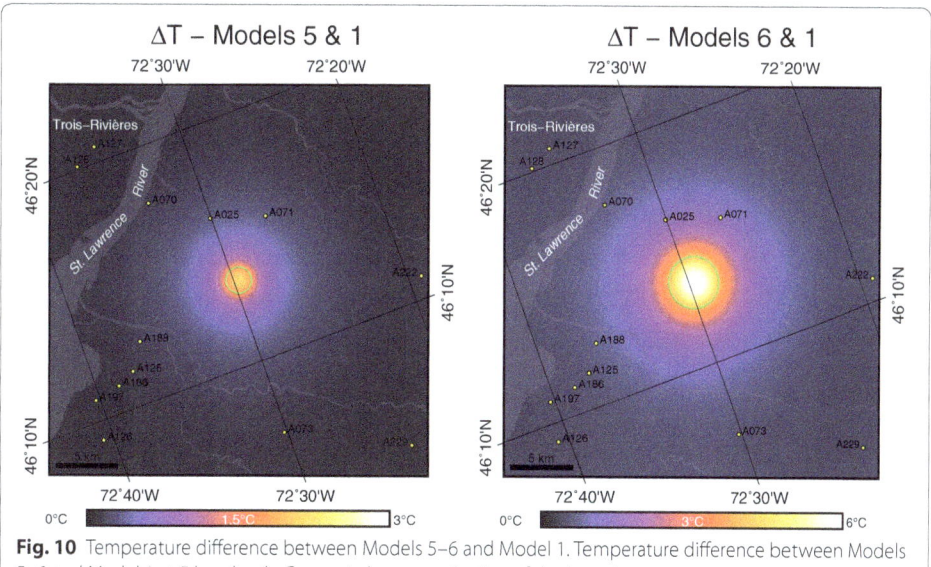

Fig. 10 Temperature difference between Models 5–6 and Model 1. Temperature difference between Models 5–6 and Model 1 at 5 km depth. Green circles are projection of the intrusion

7 °C) if all scenarios are considered. Such a difference is important in the context of electricity production, because it can imply drilling deeper than expected may be required to meet specific temperature requirements.

Effect of an intrusion in the basement

As shown in Figs. 9 and 10, the influence of a relatively hot intrusion is rather limited, both spatially and in terms of temperature increase. For the smaller intrusion, the temperature increase at 5 km depth is only 3 °C. Moreover, although the larger intrusion is four times bigger than the smaller, the temperature increase is only 6 °C. In Addition,

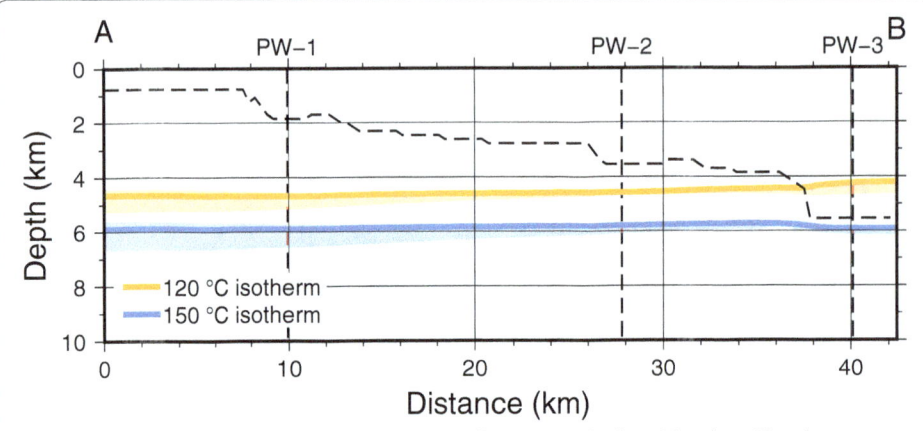

Fig. 11 Depth of 120 °C and 150 °C isotherms along profile A–B. Depth of 120 °C and 150 °C isotherms along profile A–B for the modeled scenarios. Lines are isotherm for Model 1 and light shadow areas indicate the min–max range given by Models 2, 3, and 4. The black dashed line is the depth to basement profile and PW-1, PW-2, and PW-3 are virtual wells used below

the spatial extent of the anomaly is mostly limited to the location of the intrusion. It thus appears that "hot" intrusions are of limited interest as geothermal exploration targets.

Effect of bottom boundary condition

At this point in the study, we examine the choice of the bottom boundary condition. In addition to the basement parameters, the boundary condition obviously plays a significant role in affecting the temperature distribution at depth, but to what quantitative extent? Figure 11 shows the 120 °C and 150 °C isotherms as function of depth for Model 1 (thick lines) and the range of depths for all models (shaded areas). There is a significant difference in isotherm depth between the difference models, with up to a km in depth difference in the northern part. In a given geothermal project, a poor estimation of this boundary condition would likely have dramatic impact on drilling costs.

Figure 12 shows a comparison between the simulated temperature for Model 1 under different bottom boundary conditions, with the measured temperature obtained from the drill stem tests in wells A126, A188, A197 and A222. This figure also indicates that the choice of the boundary condition value has a strong influence on the modeled temperatures at depth. This emphasizes the importance of having high quality data to calibrate the model.

"Thermal blanket" effect of the sedimentary rocks

Because sedimentary rocks usually have a relatively low thermal conductivity, they behave as a "thermal blanket", preserving the heat in the rocks at depth (Beardsmore 2005). We examine this effect taking Models 1, 2 and 3 as examples. Figure 13 shows temperature profiles at three virtual wells located along profile AB, as illustrated in Figs. 5 and 11. Figure 13 shows that the temperature gradient in the sedimentary rocks is stronger than in the basement for Models 1 and 3, and especially for Model 3 in which the thermal conductivity of the basement is relatively higher. The opposite can be observed for Model 2, which has the lowest thermal conductivity but the highest heat

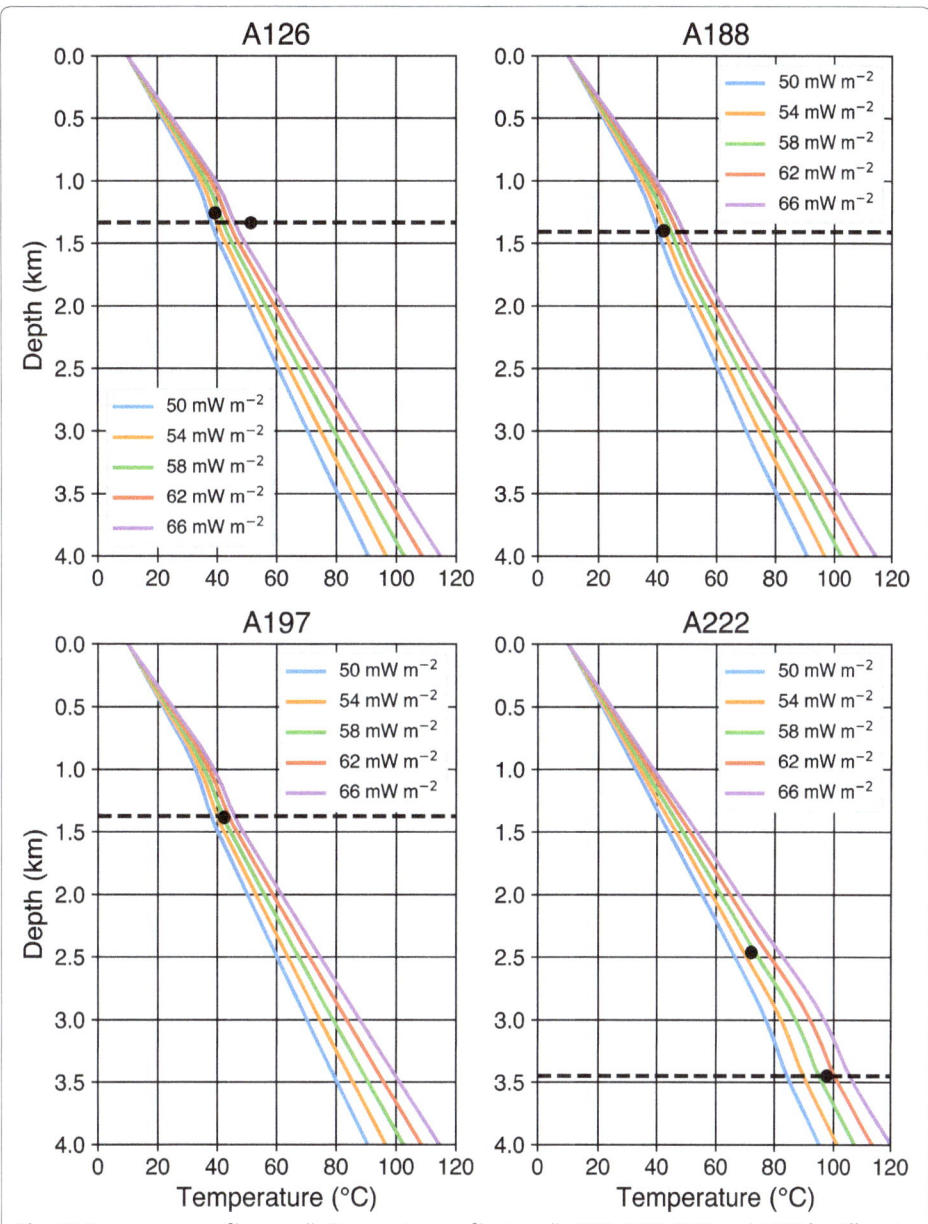

Fig. 12 Temperature profiles in wells. Temperature profiles in wells A126, A188, A197, and A222 for different heat flow conditions at the bottom boundary of Model 1 (dashed line indicates basement depth; red solid circles represent temperature after Harrison's correction)

production. Thus, the thermal blanket effect is strongly controlled by the contrast in thermal conductivity. The inflection point varies in depth depending on thickness of the sedimentary cover, with an advantage at well PW-3. However, although the difference in sedimentary cover thickness is almost 4000 m between PW-1 and PW-3, the difference in inflection point depth is about 1500 m, i.e., the so-called blanket effect is not as strong as might be expected at first. This can be attributed to the fact that the topography of the basement is irregular, with a large "step" not far from well PW-3 (see Fig. 11, and that lateral effects influence the thermal response. Through this case study, it appears that the benefit of the insulating effect of the sedimentary cover can be significant, but that it

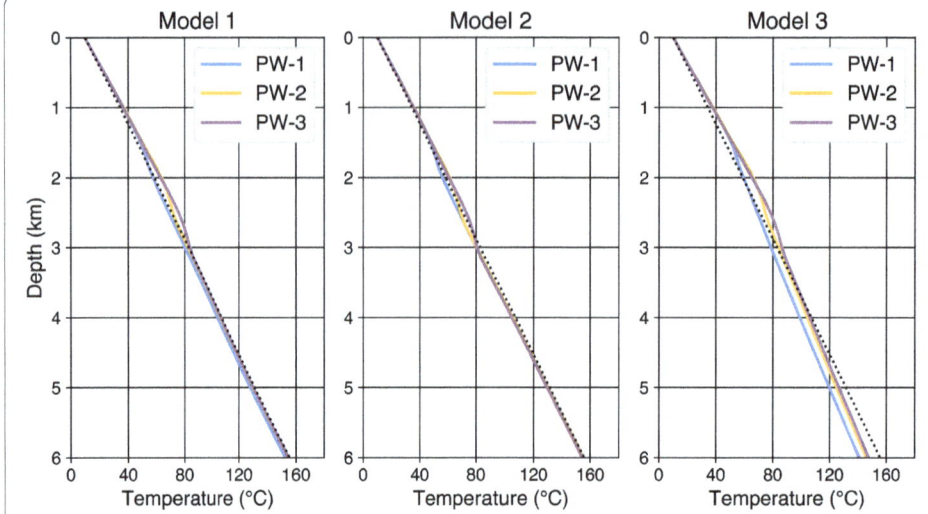

Fig. 13 Temperature changes along three vertical profiles. Temperature changes along three vertical profiles in Fig. 11 showing the thermal blanket effect of the sedimentary rocks (the sedimentary cover is thinnest at PW-1 and thickest at PW-3). Dashed lines show a 24.3 °C/km gradient

strongly depends on the contrast in thermal conductivity between the basement and the cover, as well as on the structural context.

Surface heat flow

Surface heat flow values computed from borehole data in the study area are given by Bédard et al. (2016). These data are compared with vertical heat flow computed from the results of the modeling with Underworld2. Figure 14 shows maps of surface heat flow for the six modeled scenarios, together with 18 data points from Bédard et al. (2016). The general agreement is best for the model with a basement representative of the Parc des Laurentides domain (Model 3), followed by models corresponding to the Portneuf–Mauricie domain (Models 1, 5, and 6). More specifically, the rRMSE between data points from Bédard et al. (2016) and modeled surface heat flow is 13% for Model 1, 15% for Model 2, 11% for Model 3, 17% for Model 4, 13% for Model 5 and 13% for Model 6. These results should, however, be considered carefully, as the average of the borehole data (76 mW m^{-2}) is much higher than the interpolated value that can be seen in the map of Fig. 1 ($\approx$ 55 mW m^{-2}), and Model 3 also has the worst rRMSE fit with the bottom hole temperature and regional gradient.

Comparisons between Model 6 and Model 1 reveal that, although the top of the large intrusion is 2.4 km below the surface, it produces sufficient heat to alter the temperature gradient up to the surface, and thus affects the heat flow at the surface. It is interesting to note, however, that the heat flow is lower for Model 6, which is counterintuitive. It is found that the presence of the intrusion slightly decreases the temperature gradient in the sedimentary cover, and thus decreases the heat flow at the surface, as shown in Fig. 15. Interpretation of temperature at depth from surface heat flow data is, therefore, a delicate matter.

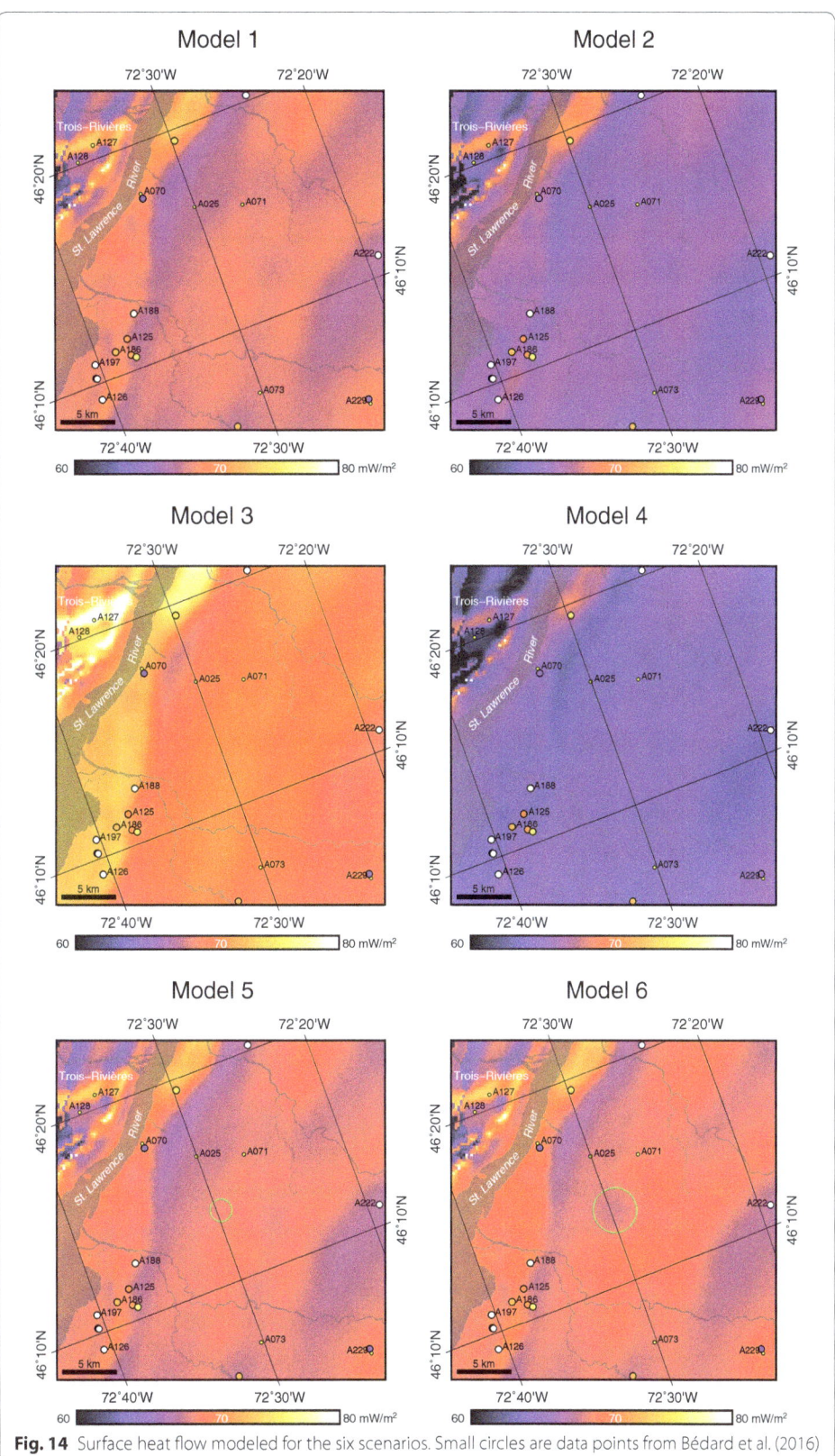

Fig. 14 Surface heat flow modeled for the six scenarios. Small circles are data points from Bédard et al. (2016)

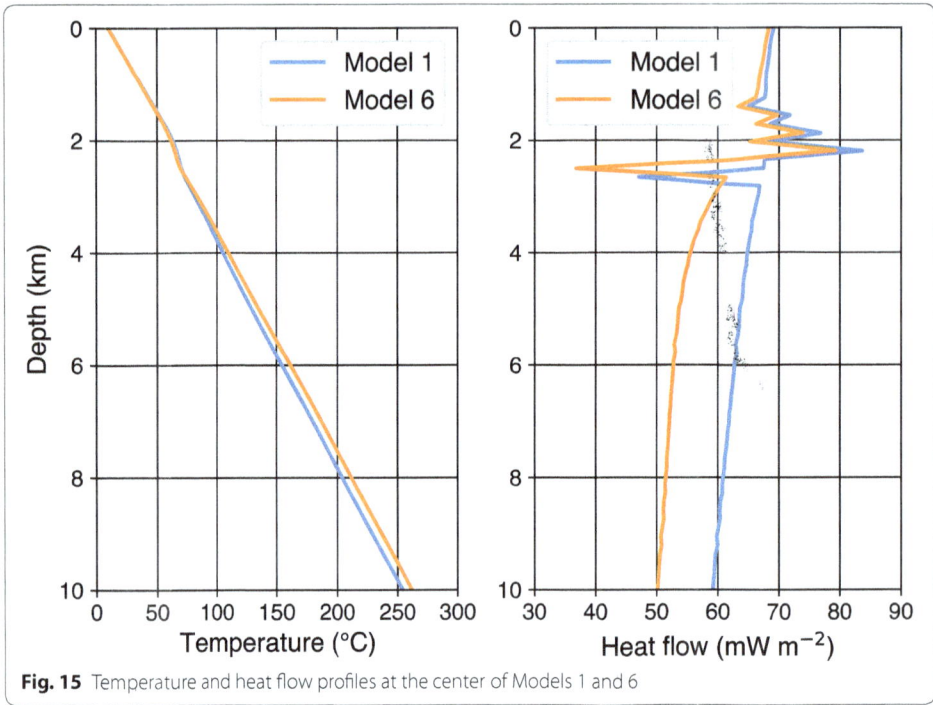

Fig. 15 Temperature and heat flow profiles at the center of Models 1 and 6

Conclusions

The main conclusions that can be drawn from this study are:

1. Statistical analysis of radiogenic element concentrations suggests that the radiogenic heat production south of Trois–Rivières is in the range of 0.34–3.24 $\mu W\ m^{-3}$, 0.94–5.83 $\mu W\ m^{-3}$ for the Portneuf–Mauricie domain, 0.02–4.13 $\mu W\ m^{-3}$ for the Morin Terrane, especially its Shawinigan domain, 0.34–1.96 $\mu W\ m^{-3}$ for the Parc des Laurentides domain.

2. The distribution of temperature is affected by many factors, e.g., thermal conductivity, density, and concentration of radioelements in sedimentary rocks and basement, thickness of sedimentary cover, and structural context. For the modeled scenarios, temperature is highest in Models 1 and 4 (respectively, PM and well data basement), because it has the highest thermal conductivity, followed by Model 2 (Morin Terrane) and by Model 3 (PDLD basement). At a depth of 5 km, the maximum difference with Model 1 is 15 °C (with Model 3), and the range of difference is 22 °C (−15 to 7 °C) if all scenarios are considered.

3. Thick Cambro-Ordovician sedimentary rocks in this region may limit heat loss to the surface by virtue of the so-called "thermal blanket" effect. The benefit of this insulating effect can be significant, but it strongly depends on the contrast in thermal conductivity between the basement and the cover, as well as on the structural context, and less on heat production in the basement.

4. The boundary condition value imposed at the bottom boundary of the model plays a significant role in controlling the distribution of temperature. Depending on basement rock parameters, values in the range [52–62] mW m^{-2} yield comparable fit to the regional temperature gradient, bottom hole temperature data and surface heat

flow data. Moreover, heat production values in the models are counterbalanced by heat flow at the bottom boundary.

5 The effect of hot intrusions in the basement is relatively modest and localized in terms of temperature increase. The presence of intrusions underneath the sedimentary cover may also decrease the heat flow at the surface, and caution should be taken when using surface heat flow as an exploration variable.

The sensitivity of model outcomes to small changes in input parameters shows that better constrained input parameters are required to establish with certainty the nature of basement to our study area. Computed and measured heat flow models for Portneuf–Mauricie domain basement beneath our study area are, however, compatible with the results of preliminary geophysical interpretations that ca. 35 km dextral transcurrent displacement may have occurred on late Grenvillian ductile shear zones beneath the northern St. Lawrence platform (Gicquel, unpublished data; Gicquel et al. 2015). Model results can, nevertheless, be applied to adjacent areas where the nature of basement domains is better defined.

A limitation of the present study lies in the assumption that, except for the models with intrusions, the basement is homogeneous. The input parameters, e.g., concentration of radiogenic elements, thermal conductivity, and density, used in these simulation models are derived from measurements made on different types of rocks, both from the St. Lawrence Lowlands and the geological domains surrounding the city of Trois–Rivières, and exhibit a broad range of values. Besides this, some parameters are poorly constrained due to the limited number of available data. Therefore, the selected scenarios do not encapsulate the variability that is likely encountered in reality. For this reason, work is underway to generate models with stochastic distributions of physical properties that are representative of the spatial variability observed in the outcropping rocks for the different domains considered in the study. This will in turn allows establishing ranges of possible temperatures at depth for the different scenarios.

Abbreviations
AMCG: anorthosite–mangerite–charnockite–granite; BK: Beekmantown; MT: Morin Terrane; PDLD: Parc des Laurentides domain; PM: Portneuf–Mauricie; PO: Potsdam; SR–LO–QT: Sainte–Rosalie–Lorraine–Queenston; Tr–BR–Ch: Trenton–Black River–Chazy; TSZ: Tawachiche shear zone.

Authors' contributions
HL gathered data, prepared the model, ran Underworld2 simulations and wrote the first draft of the manuscript. BG reviewed the model and the numerical simulations, ran additionnal Underworld2 simulations, and revised the manuscript. LBH reviewed the manuscript and completed the section on Geological setting. SMQ and JM contributed to the elaboration of the numerical model. All authors read and approved the final manuscript.

Author details
[1] Centre – Eau Terre Environnement, Institut national de la recherche scientifique, 490 rue de la Couronne, Québec, QC G1K 9A9, Canada. [2] State Key Laboratory of Geomechanics and Geotechnical Engineering, Institute of Rock and Soil Mechanics, Chinese Academy of Sciences, Wuhan 430071, China. [3] Monash eResearch Centre, Monash University, Clayton, VIC 3800, Australia. [4] School of Mathematical Sciences, Monash University, Clayton, VIC 3800, Australia.

Acknowledgements
We would like to thank the reviewers for their constructive comments. We would also like to show our appreciation to Jean-Claude Mareschal at the Université du Québec à Montréal, Karine Bédard, Aurélie Gicquel, Félix-Antoine Comeau, Jasmin Raymond, Emmanuelle Millet, Maher Nasr, and Jean-Philippe Drolet from INRS-ETE. Vasile Minea from Hydro-Québec gave us much help with the collection of basic geothermal data, and Jennifer Levett from Mira Geoscience Ltd. of Montréal gave us invaluable help in technical support of GOCAD software, to Louis Moresi, Fabio Capitanio, Julian Giordani, Jerico Revote, Owen Kaluza and Ben Mather for their guidance in using the Underworld2 software. We also give our thanks to Adrien Bouffard at the Ministère de l'Énergie et des Ressources naturelles (MERN) who helped in collecting the basement rock samples.

Competing interests
The authors declare that they have no competing interests.

Funding
Research was funded by the Natural Sciences and Engineering Research Council of Canada and IREQ/Hydro Québec (Grant RDCPJ 451432-13).

References

Beardsmore GR. Thermal modeling of the Hot Dry Rock geothermal resource beneath GEL99 in the Cooper Basin, South Australia. In: World Geothermal Congress 2005. 2005. http://www.geothermal-energy.org/pdf/IGAstandard/WGC/2005/1634.pdf. Accessed 22 Dec 2018.

Bédard K, Comeau F-A, Malo M. Modélisation géologique 3D du basin des Basses-Terres du Saint-Laurent. Research Report INRSCO2-2013-V1.5, Institut national de la recherche scientifique. 2013. http://espace.inrs.ca/1643/1/R001439.pdf. Accessed 22 Dec 2018.

Bédard K, Comeau F-A, Millet E, Raymond J, Malo M, Gloaguen E. Évaluation des ressources géothermiques du bassin des Basses-Terres du Saint-Laurent. Research Report INRS-1659, Institut national de la recherche scientifique. 2016. http://espace.inrs.ca/4845/1/R001659.pdf. Accessed 22 Dec 2018.

Bédard K, Malo M, Comeau F-A. CO_2 geological storage in the province of Québec, Canada. Capacity evaluation of the St. Lawrence Lowlands basin. Energy Procedia. 2013;37:5093–100.

Bernstein L. The Lower Ordovician Beekmantown Group, Québec and Ontario. Ph.D. thesis, Université de Montréal, Montréal. Canada. 1991.

Capuno VT, Maria RBS, Minguez EB. Tiwi geothermal field, Philippines: 30 years of commercial operation. In: Proceedings World Geothermal Congress 2010, Bali, Indonesia. 2010.

Castonguay S, Dietrich J, Lavoie D, Laliberte JY. Structure and petroleum plays of the St. Lawrence Platform and Appalachians in southern Quebec: insights from interpretation of MRNQ seismic reflection data. Bull Can Petrol Geol. 2010;58(3):219–34.

Charlier B, Namur O, Malpas S, de Marneffe C, Duchesne JC, Vander Auwera J, Bolle O. Origin of the giant Allard Lake ilmenite ore deposit (Canada) by fractional crystallization, multiple magma pulses and mixing. Lithos. 2010;117(1–4):119–34.

Clark TH. Région de Montréal area. Geological Report RG-152, Ministère des Richesses Naturelles, Québec. 1972.

Comeau F-A, Bédard K, Malo M. Les régions de Nicolet et de Villeroy: état des connaissances pour le stockage géologique du CO_2. Research Report R1332, INRS-ETE. 2012. http://espace.inrs.ca/532/1/R001332.pdf. Accessed 22 Dec 2018.

Corrigan D. Mesoproterozoic evolution of the south-central Grenville orogen: structural, metamorphic, and geochronologic constraints from the Mauricie transect. Ph.D. thesis, Carleton University; 1995. https://curve.carleton.ca/df84deaf-2f8e-4ba2-9c7e-58c98d5faff6. Accessed 22 Dec 2018.

Corrigan D, van Breemen O. U-Pb age constraints for the lithotectonic evolution of the Grenville province along the Mauricie transect, Quebec. Can J Earth Sci. 1997;34(3):299–316.

Corriveau L, van Breemen O. Docking of the Central Metasedimentary Belt to Laurentia in geon 12: evidence from the 1.17-1.16 Ga Chevreuil intrusive suite and host gneisses, Quebec. Can J Earth Sci. 2000;37(2):253–69.

Costain JK, Glover L, Sinha AK. Low-temperature geothermal resources in the eastern United-States. Eos Trans Am Geophys Union. 1980;61(1):1–3.

Davidson A. Identification of ductile shear zones in the south-western Grenville Province of the Canadian Shield. In: Kroner A, Greiling R, editors. Precambrian Tectonics Illustrated. Stuttgart: Schweizerbart'sche Verlagsbuchhandlung; 1984. p. 263–79.

De Souza S, Tremblay A, Ruffet G. Taconian orogenesis, sedimentation and magmatism in the southern Quebec-northern Vermont Appalachians: Stratigraphic and detrital mineral record of Iapetan suturing. Am J Sci. 2014;314(7):1065–103.

Drury MJ, Jessop AM, Lewis TJ. The thermal nature of the Canadian Appalachian crust. Tectonophysics. 1987;133:1–14.

Dufréchou G. Aeromagnetic signature of an exhumed double dome system in the sw Grenville Province (Canada). Terra Nova. 2017;29(6):363–71.

Dufréchou G, Harris LB, Corriveau L. Tectonic reactivation of transverse basement structures in the Grenville orogen of SW Quebec, Canada: insights from gravity and aeromagnetic data. Precambrian Res. 2014;241:61–84.

Emslie RF, Ermanovics IF. Major rock units of the Morin Complex, southwestern Quebec. Paper 74–48, Geological Survey of Canada. 1975. https://doi.org/10.4095/102555

Emslie RF, Hunt PA. Ages and petrogenetic significance of igneous mangerite-charnockite suites associated with massif anorthosites, Grenville Province. J Geol. 1990;98(2):213–31.

Feininger T, Goodacre AK. The eight classical Monteregian hills at depth and the mechanism of their intrusion. Can J Earth Sci. 1995;32(9):1350–64.

Ford KL, Savard M, Dessau J-C, Pellerin E, Charbonneau BW, Shives BK. The role of gamma-ray spectrometry in radon risk evaluation: a case history from Oka, Quebec. Geosci Can. 2001;28(2):59–64.

Fou JTK. Thermal conductivity and heat flow at St. Jerôme, Quebec. M.Sc. thesis, McGill University. 1969.

Gautier E. Géochimie et pétrologie du complexe de La Bostonnais et du gabbro du Lac Lapeyrère. M.Sc. thesis, Université Laval; 1993.

Gicquel A, Harris LB, Groulier PA. Interpretation of new aeromagnetic data and implications for geothermal power from a suite of radiogenic units beneath the Appalachian sedimentary cover. In: Québec Mines, Québec, Canada; 2015. http://gq.mines.gouv.qc.ca/documents/examine/DV201506/DV201506.pdf. Accessed 22 Dec 2018.

Globensky Y. Géologie des Basses-Terres du Saint-Laurent. Report MM85-02, Ministère de l'Énergie et des Ressources du Québec. 1987.

Grasby SE, Allen DM, Bell S, Chen Z, Ferguson G, Jessop A, Kelman M, Ko M, Majorowicz J, Moore M, Raymond J, Therrien R. Geothermal energy resource potential of Canada. Open File 6914, Geological Survey of Canada. 2012. https://doi.org/10.4095/291488.

Hébert C, Nadeau L. Géologie de la région de Talbot (Portneuf), carte 2213, échelle 1:50000. Étude 95-01, Ministère des Ressources naturelles, Québec. 1995.

Hersi OS, Lavoie D, Nowlan GS. Reappraisal of the Beekmantown group sedimentology and stratigraphy, montreal area, southwestern Quebec: implications for understanding the depositional evolution of the Lower-Middle Ordovician Laurentian passive margin of eastern Canada. Can J Earth Sci. 2003;40(2):149–76.

Higgins MD, van Breemen O. Three generations of anorthosite-mangerite-charnockite-granite (AMCG) magmatism, contact metamorphism and tectonism in the Saguenay-Lac-Saint-Jean region of the Grenville Province, Canada. Precambrian Res. 1996;79(3–4):327–46.

Hillis R, Hand M, Mildren S, Morton J, Reid P, Reynolds S. Hot Dry Rock geothermal exploration in Australia: Application of the in situ stress field to Hot Dry Rock geothermal energy in the Cooper Basin. In: PESA Eastern Australasian Basins Symposium II, Adelaide, South Australia; 2004. p. 415–423.

Hofmann HJ. Stratigraphy of the Montreal area. Excursion Guidebook B-O3, 24th International Geological Congress. 1972.

Horspool N, Lescinsky DT, Kirkby AL, Meixner AJ. The Cooper Basin Region 3D Map Version 2: Thermal modeling and Temperature Uncertainty. 2012. http://pid.geoscience.gov.au/dataset/ga/74291. Accessed 22 Dec 2018.

Jaupart C, Mareschal JC, Guillou-Frottier L, Davaille A. Heat flow and thickness of the lithosphere in the Canadian Shield. J Geophys Res Solid Earth. 1998;103(B7):15269–86.

Jaupart C, Mareschal J-C. Heat Generation and Transport in the Earth. Cambridge: Cambridge University Press; 2011. p. 464.

Jessop AM, Allen VS, Bentkowski W, Burgess M, Drury M, Judge AS, Lewis T, Majorowicz J, Mareschal JC, Taylor AE. The Canadian geothermal data compilation. Open File 4887, Geological Survey of Canada. 2005.

Kearey P. An interpretation of the gravity field of the Morin anorthosite complex, southwest Quebec. Geol Soc Am Bull. 1978;89(3):467–75.

Konstantinovskaya E, Malo M, Castillo DA. Present-day stress analysis of the St. Lawrence Lowlands sedimentary basin (Canada) and implications for caprock integrity during CO_2 injection operations. Tectonophysics. 2012;518:119–37.

Kumarapeli PS. Vestiges of Iapetan rifting in the craton west of the northern Appalachians. Geosci Can. 1985;12(2):54–9.

Lavoie D, Chi GX. Lower Paleozoic foreland basins in eastern Canada: tectono-thermal events recorded by faults, fluids and hydrothermal dolomites. Bull Can Petrol Geol. 2010;58(1):17–35.

Lavoie D, Pinet N, Castonguay S, Dietrich J, Giles P, Fowler M, Thériault R, Laliberté JY, Peter S, Hinds S, Hicks L, Klassen H. Hydrocarbon systems in the Paleozoic basins of eastern Canada. Open File 5980, Geological Survey of Canada. 2009.

Lavoie D, Rivard C, Lefebvre R, Sejourne S, Theriault R, Duchesne MJ, Ahad JME, Wang B, Benoit N, Lamontagne C. The Utica Shale and gas play in southern Quebec: geological and hydrogeological syntheses and methodological approaches to groundwater risk evaluation. Int J Coal Geol. 2014;126:77–91.

Lee YM, Deming D. Evaluation of thermal conductivity temperature corrections applied in terrestrial heat flow studies. J Geophys Res Solid Earth. 1998;103(B2):2447–54.

Majorowicz J, Grasby SE. Heat flow, depth-temperature variations and stored thermal energy for enhanced geothermal systems in Canada. J Geophys Eng. 2010;7(3):232–41.

Majorowicz J, Minea V. Geothermal energy potential in the St. Lawrence River area, Quebec. Geothermics. 2012;43:25–36.

Majorowicz J, Minea V. Geothermal anomalies in the Gaspésie Peninsula and Madeleine Islands, Québec. Trans GRC. 2013;37:1–12.

Malo M, Aznar J-C, Bédard K, Claprood M, Comeau F-A, Konstantinovskaya E, Giroux B, Moutenet J-P, Tran Ngoc TD. Rapport synthèse 2008-2013, Chaire de recherche sur la séquestration géologique du CO_2. Research Report 1492, INRS-ETE. 2014. http://espace.inrs.ca/2092/1/R001492.pdf. Accessed 22 Dec 2018.

Mansour J, Kaluza O, Giordani J, Beucher R, Farrington R, Kennedy G, Moresi L, Velic M, Beall A, Sandiford D. underworldcode/underworld2: v2.6.0b. 2018. https://doi.org/10.5281/zenodo.1475861.

Mareschal JC. Données géothermiques. 2018.

Mareschal JC, Jaupart C. Radiogenic heat production, thermal regime and evolution of continental crust. Tectonophysics. 2013;609:524–34.

Mareschal JC, Jaupart C, Gariepy C, Cheng LZ, Guillou-Frottier L, Bienfait G, Lapointe R. Heat flow and deep thermal structure near the southeastern edge of the Canadian Shield. Can J Earth Sci. 2000;37(2):399–414.

Martignole J, Schrijver K. Anorthosite-farsundite complexes in the southern part of the Grenville Province. Geosci Can. 1977;4(3):137–43.

Minea V, Majorowicz J. Assessment of enhanced geothermal systems potential in Québec, Canada. In: AAPG/SPE/SEG HEDBERG Research Conference Enhanced Geothermal Systems, Napa, California. 2011. http://www.searchandd iscovery.com/pdfz/abstracts/pdf/2011/hedberg-california/abstracts/ndx_minea.pdf.html Accessed 22 Dec 2018.

Minea V, Majorowicz J. Preliminary assessment of deep geothermal resources in Trois–Rivières area, Québec. Trans Geotherm Resour Council. 2012;36:709–15.

Misener AD, Thompson LGD, Uffen RJ. Terrestrial heat flow in Ontario and Quebec. Trans Am Geophys Union. 1951;32(5):729.

Moresi L, Quenette S, Lemiale V, Meriaux C, Appelbe B, Muhlhaus HB. Computational approaches to studying non-linear dynamics of the crust and mantle. Phys Earth Planet Inter. 2007;163(1–4):69–82.

Nadeau L, Brouillette P. Carte structurale de la région de La Tuque (SNRC 31P), Province de Grenville, Québec. Open File 2938, Geological Survey of Canada. 1994.

Nadeau L, Brouillette P. Carte structurale de la région de Shawinigan (SNRC 31I), Province de Grenville, Québec. Open File 3012, Geological Survey of Canada. 1995.

Nadeau L, van Breemen O. Do the 1.45–1.39 Ga Montauban group and the La Bostonnais complex constitute a Grenvillian accreted terrane. In: Geological Association of Canada, Programs with Abstracts, vol. 19, p. 81; 1994.

Nadeau L, van Breemen O. U-Pb Zircon age and regional setting of the Lapeyrère gabbronorite, Portneuf–Mauricie region, south central Grenville Province, Quebec. Current Research 2001-F6, Geological Survey of Canada. 2001. https://doi.org/10.4095/212673.

Nadeau L, Brouillette P, Hébert C. Arc magmatism, continental collision and exhumation: the mesoproterozoic evolution of the south-central Grenville Province, Portneuf–Mauricie region, Quebec, Field Trip B3. 2008. http://www.friendsoftheGrenville.org/fog_2010.pdf. Accessed 22 Dec 2018.

Nadeau L, van Breemen O, Hébert C. Géologie, âge et extension géographique du groupe de Montauban et du complexe de La Bostonnais. Report DV 92-03, Ministère de l'énergie et des Ressources, Québec. 1992.

Nasr M. Évaluation des propriétés thermiques de la séquence des Basses-Terres du Saint-Laurent : Mesures au laboratoire et approche diagraphique. M.Sc. thesis, Institut National de la Recherche Scientifique—Centre Eau Terre Environnement. 2016. http://espace.inrs.ca/4637/. Accessed 22 Dec 2018.

Owen JV, Greenough JD. Influence of Potsdam sandstone on the trace element signatures of some 19th-century American and Canadian glass: Redwood, Redford, Mallorytown, and Como-Hudson. Geoarchaeology. 2008;23(5):587–607.

Paradis N. Modélisation gravimétrique et magnétique des intrusions gabbronoritiques de Lapeyrère et d'édouard, région de Portneuf–Mauricie, Province de Grenville, Québec. M.Sc. thesis, Institut National de la Recherche Scientifique—Centre Eau Terre Environnement. 2004.

Peck WH. Reconnaissance geochronology and geochemistry of the Mont-Tremblant gneiss of the Morin terrane, Grenville Province, Quebec. Geosphere. 2012;8(6):1356–65.

Perozzi L, Raymond J, Asselin S, Gloaguen E, Malo M, Bégin C. Simulation géostatistique de la conductivité thermique: application à une region de la communauté métropolitaine de Montréal. Research Report R1663, Institut National de la Recherche Scientifique—Centre Eau Terre Environnement. 2016. http://espace.inrs.ca/3374/1/R001663.pdf. Accessed 22 Dec 2018.

Perry C, Rosieanu C, Mareschal JC, Jaupart C. Thermal regime of the lithosphere in the Canadian Shield. Can J Earth Sci. 2010;47(4):389–408.

Pinet C, Jaupart C, Mareschal JC, Gariepy C, Bienfait G, Lapointe R. Heat-flow and structure of the lithosphere in the eastern Canadian Shield. J Geophys Res Solid Earth. 1991;96(B12):19941–63.

Pinti DL, Beland-Otis C, Tremblay A, Castro MC, Hall CM, Marcil JS, Lavoie JY, Lapointe R. Fossil brines preserved in the St. Lawrence Lowlands, Quebec, Canada as revealed by their chemistry and noble gas isotopes. Geochim Cosmochim Acta. 2011;75(15):4228–43.

Quenette S, Xi YF, Mansour J, Moresi L, Abramson D. Underworld-GT applied to Guangdong, a tool to explore the geothermal potential of the crust. J Earth Sci. 2015;26(1):78–88.

Raymond J, Malo M, Comeau FA, Bédard K, Lefebvre R. Assessing the geothermal potential of the St. Lawrence Lowlands sedimentary basin in Quebec, Canada. In: IAH 2012 Congress, Niagara Falls, Canada; 2012. http://grrebs.ete.inrs.ca/wp-content/uploads/2014/10/2012_Raymond_Assessing-the-geothermal-potential-of-the-St-Lawrence-Lowlands-sedimentary-basin-in-Quebec.pdf. Accessed 22 Dec 2018.

Raymond J, Malo M, Tanguay D, Grasby S, Bakhteyar F. Direct utilization of geothermal energy from coast to coast: a review of current applications and research in Canada. In: World Geothermal Congress, Melbourne, Australia; 2015.

Rimando RE, Benn K. Evolution of faulting and paleo-stress field within the Ottawa Graben, Canada. J Geodyn. 2005;39(4):337–60.

Rivard P, Ollivier JP, Ballivy G. Characterization of the ASR rim—application to the Potsdam sandstone. Cem Concr Res. 2002;32(8):1259–67.

Rivers T. Lithotectonic elements of the Grenville Province: review and tectonic implications. Precambrian Res. 1997;86(3–4):117–54.

Rivers T. Tectonic setting and evolution of the Grenville orogen: an assessment of progress over the last 40 years. Geosci Can. 2015;42(1):77–124.

Rocher M, Tremblay A. L'effondrement de la plate-forme du Saint-Laurent : ouverture de Iapetus ou de l'Atlantique ? Apport de la reconstitution des paléocontraintes dans la région de Québec (Canada). Comptes Rendus de l'Académie des Sciences-Series IIA-Earth and Planetary Science. 2001;333(3):171–8.

Rybach L. Radioactive heat-production in rocks and its relation to other petrophysical parameters. Pure Appl Geophys. 1976;114(2):309–17.

Rybach L. In: Hanel R, Rybach L, Stegena L editors. Determination of heat production rate. Dordrecht: Kluwer Academic Publishers; 1988. p. 125–142.

Sappin A-A. Pétrologie et métallogénie d'indices de Ni-Cu-éléments du Groupe du Platine du domaine de Portneuf–Mauricie, Québec (Canada). Ph.D. thesis, Université Laval. 2012. https://doi.org/10.1139/E09-022.

Sappin A-A, Constantin M, Clark T. Les indices de Ni-Cu±EGP du domaine de Portneuf–Mauricie dans la Province de Grenville: un exemple de minéralisation magmatique mise en place dans un environnement d'arc magmatique. Open File 7610, Geological Survey of Canada. 2014. https://doi.org/10.4095/293908

Sappin A-A, Constantin M, Clark T, van Breemen O. Geochemistry, geochronology, and geodynamic setting of Ni-Cu +/- PGE mineral prospects hosted by mafic and ultramafic intrusions in the Portneuf–Mauricie domain, Grenville Province, Quebec. Can J of Earth Sci. 2009;46(5):331–53.

Saull VA, Clark TH, Doig RP, Butler RB. Terrestrial heat flow in the St. Lawrence Lowland of Québec. CIM Bull. 1962;65:63–6.

Sekiguchi K. A method for determining terrestrial heat-flow in oil basinal areas. Tectonophysics. 1984;103(1–4):67–79.

SIGÉOM: SIGÉOM Système d'information géominière du Québec. 2018. http://sigeom.mines.gouv.qc.ca. Accessed 22 Dec 2018.

SMU Geothermal Laboratory: SMU Node of the National Geothermal Data System. 2018. http://geothermal.smu.edu/gtda/. Accessed 22 Dec 2018.

Soucy La Roche R. Histoire tectono-métamorphique de la zone de cisaillement Taureau orientale et ses implications pour l'exhumation de la croute moyenne dans la Province de Grenville. M.Sc. thesis, Université du Québec à Montréal. 2014.

Soucy La Roche R, Gervais F, Tremblay A, Crowley JL, Ruffet G. Tectono-metamorphic history of the eastern Taureau shear zone, Mauricie area, Québec: implications for the exhumation of the mid-crust in the Grenville Province. Precambrian Res. 2015;257:22–46.

Thériault R, Laliberté JY, Brisebois D, Rheault M. Fingerprinting of the Ottawa-Bonnechère and Saguenay grabens under the St. Lawrence Lowlands and Québec Appalachians: prime targets for hydrocarbon exploration. In: Geological Association of Canada, vol. 65. Halifax, Nova Scotia; 2005.

Thompson A. Geothermal development in Canada: Country update. In: World Geothermal Congress, Bali, Indonesia. 2010. http://www.geothermal-energy.org/pdf/IGAstandard/WGC/2010/0116.pdf. Accessed 22 Dec 2018.

Thompson A, Bakhteyar F, Van Hal G. Geothermal industry development in Canada-country update. In: World Geothermal Congress, Melbourne, Australia; 2015.

Tremblay A, Long B, Masse M. Supracrustal faults of the St. Lawrence rift system, Quebec: kinematics and geometry as revealed by field mapping and marine seismic reflection data. Tectonophysics. 2003;369(3–4):231–52.

Vautour G, Pinti DL, Méjean P, Saby M, Meyzonnat G, Larocque M, Castro MC, Hall CM, Boucher C, Roulleau E, Barbecot F, Takahata N, Sano Y. 3H/3He, 14C and (U-Th)/He groundwater ages in the St. Lawrence Lowlands, Quebec, Eastern Canada. Chem Geol. 2015;413:94–106.

6

Doped bentonitic grouts for implementing performances of low-enthalpy geothermal systems

Marco Viccaro[1,2]* iD

*Correspondence:
m.viccaro@unict.it
[1] Dipartimento di Scienze
Biologiche Geologiche e
Ambientali – Sezione di
Scienze della Terra, Università
degli Studi di Catania, Corso
Italia 57, 95129 Catania, Italy
Full list of author information
is available at the end of the
article

Abstract

Mechanical (flexural and uniaxial compressive strengths) and physical (thermal conductivity) properties of two new bentonitic grouts doped with 5 and 10% of graphite powder are discussed in this work and evaluated for their potential use in low-enthalpy geothermal applications. The same tests have been also conducted on a pure starting material (bentonitic grout) already present on the market and used to seal geothermal probes into boreholes. Experimental data show that the addition of 5 and 10% of graphite powder positively alters the mechanical properties of the doped bentonitic grouts, i.e., both flexural and uniaxial compressive strengths increased with respect to those of the pure material. Thermal conductivity also improved up to 60% in the doped bentonitic grouts. A simple analysis of the cost/benefit ratio suggests, however, that the bentonitic grouts doped with 5% of graphite powder is more suitable and competitive for a launch on the market and utilization as a sealing material in boreholes aimed at low-enthalpy geothermal installations. Implementation of thermal properties of the grout material implicates a reduction of the total borehole length of 15–20%.

Keywords: Low-enthalpy geothermal energy, Flexural strength, Uniaxial compressive strength, Thermal conductivity, Borehole, Bentonitic grout, Graphite powder

Background

Energy is a fundamental condition for the social growth of the community, being an essential ingredient for almost all human activities and services. During the past decade, the energy policy has been working towards more environmental-friendly renewable power plants in order to achieve maximum advantages and/or fewer impacts from the exploitation of fossil-fuel resources (Kalaiselvam and Parameshwaran 2014). Efforts have been concentrated mainly in the field of applied solar and wind energy (generally promoted by many subventions), and geothermal energy played comparatively a less prominent role among renewable sources. Although anomalous geothermal fluxes useful for exploitation are present in specific geological settings, there is a worldwide incredible potential related to low-enthalpy geothermal energy that is, at present, scarcely considered in many countries (Banks 2012 and references therein). In this regard, primary reasons are the lack of scientific-technical background in the field of low-enthalpy geothermal resources, coupled with the absence of adequate basic information of small-to-medium scale subsoil characteristics that are essential to appropriately manage feasibility studies.

It is worth to note that the majority of low-enthalpy geothermal systems have negligible environmental impact, chiefly because energy capture systems are placed into the ground and they can be fully integrated into the subsurface if planned during the construction phase of buildings (cf. also Florides and Kalogirou 2007). Low-enthalpy geothermal systems can, therefore, become of strategic importance in the near future, especially if technological efforts will point towards an improvement of their cost/benefit ratio in comparison with other renewable energy systems (Kalaiselvam and Parameshwaran 2014 and references therein). Indeed, the ability to produce energy for heating in winter, cooling in summer and heating water makes low-enthalpy geothermal energy the ideal alternative to other traditional systems. The advantage is given by the fact that a single installation includes functions normally delegated to different devices (e.g., boilers, air conditioners, etc.). Also, heating and cooling systems from low-enthalpy geothermal energy seem to be the most acceptable solution in circumstances where solar thermal or photovoltaic are prevented (e.g., scarce exposure to solar radiation, intense urbanization, protected natural/archeological sites, historical city centers, etc.).

Low-enthalpy geothermal systems are based on a rather simple technology of heat exchange with the subsoil (e.g., Florides and Kalogirou 2007). Beside the chemical–physical characteristics of rocks in the subsurface (which anyhow remain a paramount factor), the total efficiency of the installation is determined by the following: (1) efficiency of the Ground Source Heat Pump, which allows the transfer of heat from the subsoil upward to the surface; (2) efficiency of thermal exchange between probes installed in boreholes and rocks in the subsurface; (3) chemical-physical characteristics of the bentonitic grout to seal probes into boreholes; (4) efficiency of the heat distribution system into the environment at the surface. During the past years, technological development has been pointing with increasing interest towards new materials and solutions able to enhance the performances of one or more of these components, with particular reference to the composition of the grout used as sealing material in the boreholes (e.g., Smith and Perry 1999; Alrtimi et al. 2009; Wang et al. 2011; Delaleux et al. 2012; Desmedt et al. 2012; Lee et al. 2012; Borinaga-Treviño et al. 2013; Kim et al. 2013; Erol and Francois 2014; Indacoechea-Vega et al. 2015; Blazquez et al. 2017). Some innovative solutions have also found applications in other related fields of interest of bentonitic grouts, such as thermal storage (e.g., Sari 2016) or sealing of radioactive wastes (Tang et al. 2008; Jobmann and Buntebarth 2009).

Sealing the boreholes with mixtures of bentonitic grout should guarantee either perfect hydraulic insulation (e.g., among aquifers intercepted during the perforation) or good heat exchange between the geothermal probes and the subsoil. The permeability, generally on the order of 10^{-10} cm/s, and the flow ability of the sealing material during the fresh state of the grout are also important for the hydraulic sealing of the aquifers and for good performances during pumping into the borehole. In this regard, Fleuchaus and Blum (2017) put into evidence that the connection of aquifers by leaky annular space grouting can be the cause of damages to the installation. The use of bentonitic cements, more than pure cements, also ensures elasticity of the sealing structure, avoiding damages due to contractions ascribable to anthropic and natural solicitations. This becomes a crucial factor especially for low-enthalpy geothermal installations placed in areas affected by recurrent freeze–thaw cycles that cause circumferential effective stresses (Anbergen et al. 2014; Erol and Francois 2015). Furthermore, the use of bentonitic cements is beneficial also for installation in intensely urbanized areas (e.g., vibrations due to heavy traffic, subway, etc.) or in geologically and tectonically areas with high vulnerability, which can be frequently subjected to elevated risks (seismic, hydrogeological, etc.) producing continuous mechanical stress on the technological system. Pure bentonite, however, has very poor thermal conductivity [$\ll 1$ W/m K; e.g., Tang et al. (2008); Jobmann and Buntebarth (2009); Kim et al. (2015); Sari (2016)]. This means that its massive utilization in the mixture produces significant reduction of the thermal conductivity and, therefore, of the thermal exchange into the borehole between probes and subsoil. In this regard, some authors have already investigated the possibility to improve the thermal exchange by enhancing the thermal conductivity of the sealing grout in the boreholes by adding components (e.g., quartz, sands, carbon-based components, etc.) at high thermal conductivity in the mixture (Delaleux et al. 2012; Lee et al. 2012; Erol and Francois 2014; Indacoechea-Vega et al. 2015). In these instances, however, the mechanical properties of the obtained grouts were not verified or the experimentally determined thermal conductivity is too high (>6 W/m K) and far from common values observable on Earth (average thermal conductivity of rocks in the subsurface is 3.2 W/m K; Eppelbaum et al. 2014 and references therein). However, experimentation and production of new materials should always ensure exceptional resistance and flexibility of the grout in the borehole, together with thermal conductivity capable to assure balanced thermal exchange with the subsoil. Indeed, thermal conductivity needs to be maintained at values compatible with those of the surrounding subsurface in order to reduce the occurrence of thermal short-circuits in the borehole with the consequent loss of performance of the whole low-enthalpy geothermal system. This is even more valid for areas characterized by subsoils with very low thermal conductivity (e.g., 0.3–2.3 W/m K for clays, silt, water saturated and dry sands; Eppelbaum et al. 2014 and references therein), which can become the main resistance of the whole exchange system. All these aspects lead to the final consideration that a compromise needs to be found to obtain commercial products characterized by good mechanical and physical properties.

This work presents results coming from the experimental production of two original bentonitic grouts bearing an additive with high thermal conductivity (pure graphite powder). With respect to previous literature, this study is aimed at finding an optimum grout mixture with enhanced properties from both mechanical and thermal standpoints. Improvement of mechanical properties (flexural and uniaxial compressive strengths) together with increase of thermal conductivity makes these two mixtures highly suitable for their use as effective sealing grout into boreholes for geothermal probe installations, having also the advantage of implementing the thermal exchange between the technological system and the subsoil.

Methods

A bentonitic grout already launched in the market has been used as starting material. The grout mixture is composed of fine-grained (<0.1 mm) cement with addition of 4% of bentonite. Declared density of the fresh starting grout for pumping ranges between

1.49 kg/dm^3 and 1.65 kg/dm^3, depending on the water/solid ratio used to prepare the mixture. Declared fluidity for 1000 ml of suspension is 45 s (Marsh Cone Test with nozzle of 10 mm following the UNI 11,152:2005 normative for draining). For the investigated purposes, technical specifications declared for this bentonitic mixture report flexural strength ~ 2 N/mm^2 and uniaxial compressive strength ~ 3 N/mm^2 after 28 days (following the DIN 18136:2003–2011 normative) and thermal conductivity $\geq$ 1 W/mK. The additive used for the experimental production of the new bentonitic grouts is pure graphite powder, with grain size analogous to that of the starting material (i.e., grain size $\leq$ 0.50 µm) and thermal conductivity > 50 W/mK (cf. Clauser and Huenges 1995). Mechanical and physical properties have been determined on three typologies of samples: (1) starting material; (2) starting material plus 5% of graphite powder; (3) starting material plus 10% of graphite powder. The starting material has been also included in mechanical and physical tests, just to minimize the possible analytical bias and to evaluate true differences among the three mixtures. All samples have been prepared at the accredited laboratories of SIDERCEM srl (Misterbianco, Sicily) by using a colloidal grout mixer. Mixtures have been prepared following the UNI EN 196-1:2016 normative in 160 × 40 × 40 prismatic molds with proportions reported in Table 1. Drying of samples has been performed at temperature of + 20 °C and humidity ranging between 95 and 99% into rooms able to monitor continuously the environmental parameters.

Flexural and uniaxial compressive strengths have been determined by a TECNOTEST KE 72 instrument equipped with 4 reading channels and a digital dynamometer able to perform tests for compression, flexure, indirect tensile, paving blocks, breaking in load/deformation control, elastic module up to 999 cycles of loading/unloading for evaluations of creeping or ductility of various building materials. Measurements of flexural strength have been performed on 1 sample of pure material after 14 days and 2 samples after 28 days. The same number of samples has been investigated for the starting material doped with 5 and 10% of graphite powder after 14 and 28 days. Measurements of uniaxial compressive strength have been performed on 2 samples of pure material after 14 days and 4 samples after 28 days. The same number of samples has been investigated for the starting material doped with 5 and 10% of graphite powder.

Experimental determinations of thermal conductivity have been conducted on 4 different samples of starting material and 10 different samples of the doped bentonitic grouts (5 samples for each mixture with 5 and 10% of graphite powder). Measurements have been performed through a heat flow meter Shotherm QTM Showa Denko at the laboratories of Technical Physics of the DIEEI of the University of Catania following the UNI EN 12664:2002. This instrument uses the non-stationary hot-wire method at

Table 1 Components used for the preparation of the bentonitic grout mixtures with various proportions of pure graphite powder as additive (G0% no graphite powder; G5% mixture with 5% of graphite; G10% mixture with 10% of graphite)

Component/weight (g)	Mixture G0%	Mixture G5%	Mixture G10%
Starting material	1000	950	900
Pure graphite powder	0	50	100
Water	600	600	600

Data refer to weights of components reported in grams

constant heat flow and operates in the temperature range from -10 to $+200$ °C with precision and reproducibility in the order of 2%. The procedure allows the heat transfer from the instrument to the sample with consequent temperature increase (fixed at maximum 20 °C) through generation of coaxial cylindrical isotherms. Heating time is 45 min, whereas measurement duration is 180 s. This method ensures no significant changes of the material properties during measurements, as demonstrated in other studies performed on similar materials (Fukai et al. 2000; Karaipekli et al. 2007; Lee et al. 2010).

Results

Mechanical properties

Flexural and compressive strengths of the starting bentonitic grout and of grouts with the graphite powder additive at 5 and 10% after 14 and 28 days have been reported in Tables 2 and 3. Flexural strength (R_f) has been determined from the relation:

$$R_f = \left(1.5F_f/b^3\right) \times l, \tag{1}$$

where F_f is the strength applied at the center of the prismatic sample, b^3 is the side dimension of the resistant section, l is the distance between the two supports.

Compressive strength has been determined as follows:

$$R_c = F_c/ab, \tag{2}$$

where F_c is the breaking load, a and b the sides of the resistant section.

Each typology of material displays an increase of both flexural and uniaxial compressive strengths with time, i.e. with measurements performed at 14 and 28 days (Tables 2, 3). Tests performed on the starting material confirm values of the flexural and uniaxial compressive strengths after 28 days declared in the technical specification of the product, being, respectively, 1.99 N/mm^2 (average on 2 samples) and 3.33 N/mm^2 (average on 4 samples). Both the bentonitic grouts with 5 and 10% of graphite powder additive

Table 2 Flexural strengths (R_f) at the fracture point obtained on the starting and doped (+ 5 and + 10% of graphite powder) bentonitic grouts

	F_f (N)	R_f (N/mm^2)
Pure starting material		
June 30, 2016 (14 days)	490	1.84
July 14, 2016 (28 days)	500	1.88
July 14, 2016 (28 days)	560	2.10
Starting material + 5%		
June 30, 2016 (14 days)	930	3.49
July 14, 2016 (28 days)	980	3.68
July 14, 2016 (28 days)	1080	4.05
Starting material + 10%		
June 30, 2016 (14 days)	820	3.08
July 14, 2016 (28 days)	1060	3.98
July 14, 2016 (28 days)	1000	3.75

F_f indicates the load applied at the center of the sample at the fracture point. Dates indicate times between preparation and realization of the mechanical tests (at 14 and 28 days)

Table 3 Uniaxial compressive strengths (R_c) at the fracture point obtained on the starting and doped (+ 5 and + 10% of graphite powder) bentonitic grouts

	F_c (N)	R_c (N/mm^2)
Starting material		
June 30, 2016 (14 days)	4370	2.73
June 30, 2016 (14 days)	4680	2.93
July 14, 2016 (28 days)	4850	3.03
July 14, 2016 (28 days)	4870	3.04
July 14, 2016 (28 days)	5860	3.66
July 14, 2016 (28 days)	5740	3.59
Starting material + graphite 5%		
June 30, 2016 (14 days)	9020	5.64
June 30, 2016 (14 days)	8830	5.52
July 14, 2016 (28 days)	9830	6.14
July 14, 2016 (28 days)	9620	6.01
July 14, 2016 (28 days)	10,150	6.34
July 14, 2016 (28 days)	10,230	6.39
Starting material + graphite 10%		
June 30, 2016 (14 days)	8150	5.09
June 30, 2016 (14 days)	8910	5.57
July 14, 2016 (28 days)	11,520	7.2
July 14, 2016 (28 days)	11,320	7.08
July 14, 2016 (28 days)	10,630	6.64
July 14, 2016 (28 days)	10,750	6.72

F_c indicates the load applied at the fracture point. Dates indicate times between preparation and realization of the mechanical tests (at 14 and 28 days)

display increase of the flexural and uniaxial compressive strengths with respect to the starting material, either after 14 days from the production or after 28 days (Tables 2, 3). On the whole, measurements put into evidence comparable flexural strengths for samples with the graphite powder additive at 5 and 10% after 28 days (average 3.87 N/mm^2 for both; Table 2), whereas slight differences have been observed for what concerns the obtained uniaxial compressive strengths and their evolution during the drying process (Table 3). Specifically, the bentonitic grout with 5% graphite powder additive shows, on average, uniaxial compressive strength at 5.58 N/mm^2 after 14 days and increase at 6.22 N/mm^2 after 28 days. The bentonitic grout with 10% graphite powder additive exhibits, on average, lower values of uniaxial compressive strength after 14 days (5.33 N/mm^2) with respect to the bentonitic grout with 5% of graphite powder additive, but values are the highest after 28 days (6.91 N/mm^2).

Thermal conductivity

Thermal conductivity of the starting bentonitic grout and mixtures with the graphite powder additive at 5 and 10% on samples after 28 days has been reported in Table 4. The experimental values of thermal conductivity (λ) have been calculated as a function of the temperature–time curve (T-τ) and the heat flow at the input (q) through the following equation:

$$\lambda = [q \times \ln(\tau_2/\tau_1)]/[4\pi \, (T_2 - T_1)] \tag{3}$$

Table 4 Thermal conductivity obtained on the starting and doped (+ 5 and + 10% of graphite powder) bentonitic grouts

	λ (W/m K)
Starting material	
1st measurement	1198
2nd measurement	1207
3rd measurement	1171
4th measurement	1186
Starting material + graphite 5%	
1st measurement	1861
2nd measurement	1819
3rd measurement	1940
4th measurement	1872
5th measurement	1896
Starting material + graphite 10%	
1st measurement	1917
2nd measurement	1963
3rd measurement	2013
4th measurement	1928
5th measurement	2045

Equation (3) can be implemented according to electrical current and voltage following the relation:

$$\lambda = K \times [I^2 \times \ln(\tau_2/\tau_1)]/[4\pi \ (V_2-V_1)] -H, \tag{4}$$

where K and H are constants for the instrument calibration, I the electrical current and $(V_2 - V_1)$ the electric potential.

Measurements performed on the starting material confirm thermal conductivity as declared in the technical specifications, being the obtained $\lambda = 1.191$ W/m K (Table 4; average of 4 repeated measurements). Bentonitic grouts doped with graphite powder at 5 and 10% present increase of the thermal conductivity up to ca. 60%. Specifically, the average thermal conductivity of the bentonitic grout at 5% is 1.878 W/m K and further increases at 1.973 W/m K for the mixture with 10% of graphite powder (Table 4; average of 5 repeated measurements for both).

Discussion

Optimum doping rate defined by mechanical vs. thermal properties

The experimental results on bentonitic grouts presented in this work have been only evaluated for mechanical and physical properties useful for geothermal applications, which means that experimental materials have not any claim in order to be used as structural concretes for construction and/or building materials. Results emphasize the importance of graphite powder as additive in bentonitic grouts. Indeed, graphite powder is basically inert, so it does not alter the drying/compaction process of the bentonitic grout. Furthermore, small addition (5–10%) of graphite powder improves the flexural and uniaxial compressive strengths of the bentonitic grout after 14–28 days (times commonly assumed for structural concretes), having also the advantage of increasing its final thermal conductivity up to 60%. Improvement of the investigated mechanical properties suggests that the experimental bentonitic grouts doped with graphite powder can be even more resistant to various possible damages after the installation (e.g., freeze–thaw cycles, natural vs. anthropic vibrations, etc.). At identical water/solid ratios, density of the fresh doped grout ($\sim$ 1.7 kg/dm^3 for the mixture at 5% and $\sim$ 1.8 kg/dm^3 for the mixture at 10%) is not significantly modified with respect to that of the fresh starting material ($\sim$ 1.5 kg/dm^3). Fluidity has been evaluated through the Marsh Cone Test (UNI EN 445:2007, nozzle of 10 mm) following indications of the UNI 11152:2005 normative during the draining process. Declared fluidity for 1000 ml of the fresh starting material for pumping is 45 s, whereas measured fluidities of the fresh doped grouts are 65 and 85 s for the mixture at 5 and 10% of graphite powder, respectively. These values are well in the range (45–100 s) of other optimized bentonitic grouts generally found on the market, which therefore ensure good performances during the pumping process of the grout into the borehole.

It is worth noting that the amount of graphite powder added to the bentonitic grout affects the final cost of the experimental product. As a consequence, choice of a mixture with minimal quantities of additive is certainly fundamental to guarantee a competitive commercial strategy, even maintaining improvement of the mechanical and physical properties of the final product. Looking at the flexural and uniaxial compressive strengths of the starting material and those of the experimental bentonitic grouts with 5 and 10% of graphite powder, a significant difference between the starting material and both the doped bentonitic grouts can be observed (Fig. 1). However, the experimental mixtures doped at 5 and 10% by graphite powder display final values that, although slightly different, appear rather clustered within the same order of magnitude (Fig. 1). Similar trends are also recognizable for thermal conductivity (Fig. 2). Reasons to explain comparable values for the grouts doped at 5 and 10% are not clear. Addition of graphite can bring to two antagonistic effects: (1) graphite particles have high thermal conductivity, which finally increases that of the bentonite-graphite mixture and (2) at the same time, graphite addition could increase the porosity of the whole mixture, having opposite effect on the overall thermal conductivity. These two effects are, therefore, in competition: for a doping rate of 5% the first probably prevails, whilst the second starts to become important for the mixture at doping rate of 10%. This suggests that the mechanical vs. thermal optimum is at the doping rate of 5% of graphite.

Definition of the cost/benefit ratio is also paramount for the ultimate launch in the market of a competitive experimental product. Simple commercial considerations lead to the idea that utilization of bentonitic grouts doped at 10% with graphite powder probably does not satisfy the cost/benefit ratio. Following evaluations have been, therefore, conducted considering only the bentonitic mixture with 5% of graphite powder as additive. Pure graphite powder has rather elevated commercial cost, estimable in 0.0035 €/gm throughout the European market at the retail level. However, industrial supply of graphite powder may reasonably undergo to favorable discount up to 30–40% depending on the requested amounts. The additional cost due to the graphite powder additive can be, therefore, estimable in ca. 1000–1200 € per ton of bentonitic grout. Principal benefits resulting from the improved thermal conductivity of the bentonitic mixture are consequent in sizing of the total thermal exchange surface between the geothermal probes and the subsoil, which is quantifiable taking into account the probe cross-section (diameter)

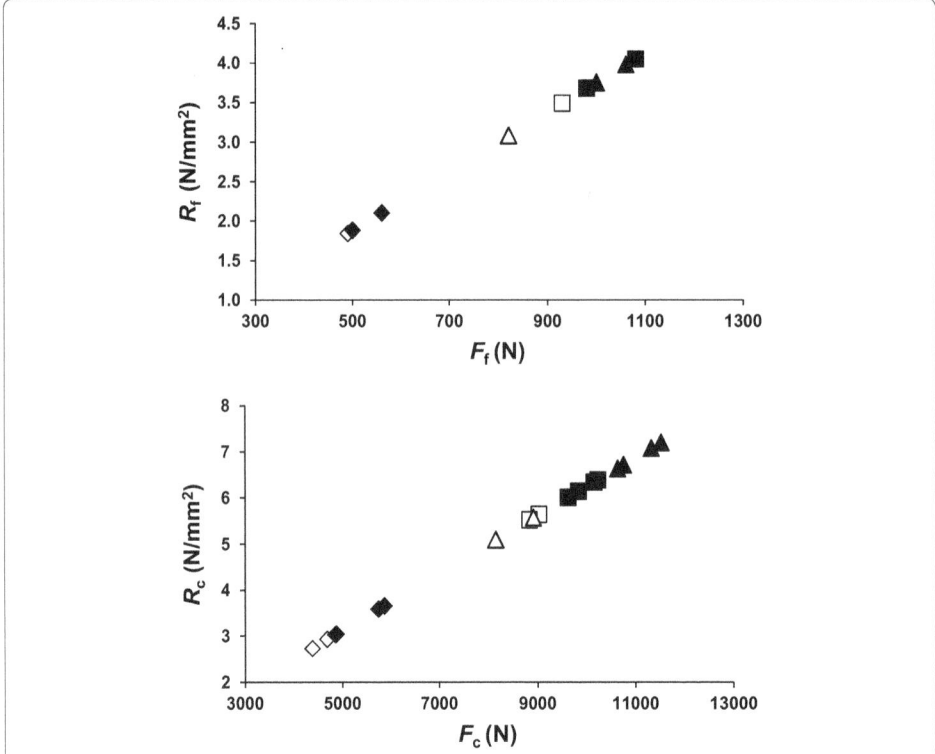

Fig. 1 Diagram reporting the flexural (R_f) and uniaxial compressive (R_c) strengths at the fracture point for the starting and doped (+ 5 and + 10% of graphite powder) bentonitic grouts as a function of the load applied to the sample (F_f and F_c). Symbols are as follows: diamonds for the starting grout; squares for the grout + 5% of graphite powder; triangles for the grout + 10% of graphite powder. Open symbols are for tests at 14 days, filled symbols for those at 28 days

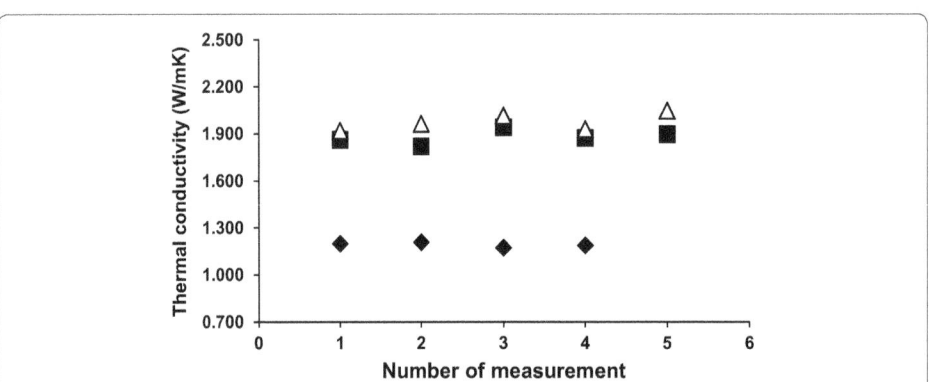

Fig. 2 Diagram of the repeated measurements of thermal conductivity obtained on the starting and doped (+ 5 and + 10% of graphite powder) bentonitic grouts. Symbols are as follows: diamonds for the starting grout; squares for the grout + 5% of graphite powder; triangles for the grout + 10% of graphite powder

and its total length (function of linear meters of perforation). Abatement of costs are, therefore, due to considerable reduction of the amount of bentonitic grout used to seal the borehole, in a way that: (1) fixing the borehole depth, the probe cross-section can be reduced (common probes diameters are 20, 25, 32 and 40 mm); (2) fixing the probe cross-section, the final depth of perforation can be shortened. In the former instance, savings principally come from lower costs of perforation, geothermal probes and their related cables/hoses, whereas in the latter chiefly from reduction of linear meters of the boreholes.

Effects on the total borehole length reduction

Calculations of the borehole length reduction can be performed through the general equations for sizing of thermal exchangers in the ground accepted by the ASHRAE (American Society of Heating, Refrigerating and Air-Conditioning Engineers), which were developed by Ingersoll et al. (1954) and later implemented by Kavanaugh and Rafferty (1997). Length of the borehole needed for heating (L_h) and cooling (L_c) can be, respectively, obtained from the following equations:

$$L_h = Q_a \cdot R_{ga} + Q_{g,hD} \cdot \left[R_b + \left(PLF_{m,hD} \cdot R_{gm} \right) + \left(R_{gd} \cdot F_{sc} \right) \right] / t_g - [(t_{wi} + t_{wo})/2)]_{hD} - t_p \tag{5}$$

$$L_c = Q_a \cdot R_{ga} + Q_{g,cD} \cdot \left[R_b + \left(PLF_{m,cD} \cdot R_{gm} \right) + \left(R_{gd} \cdot F_{sc} \right) \right] / t_g - [(t_{wi} + t_{wo})/2)]_{cD} - t_p, \tag{6}$$

where: Q_a is the average heat flux exchanged during 1 year (expressed in W); R_{ga} is the equivalent thermal resistance per length unit of the ground (annual pulse, expressed in m K/W); $Q_{g,hD}$ and $Q_{g,cD}$ are the project heat output at the ground side during the heating and cooling seasons, respectively (expressed in W); R_b is the equivalent thermal resistance per length unit of the heat exchanger, corresponding to the thermal exchange between the heat transfer fluid and the borehole surface (expressed in m K/W); $PLF_{m,hD}$ and $PLF_{m,cD}$ are the monthly part load factors during the heating and cooling seasons, respectively; R_{gm} is the equivalent thermal resistance per length unit of the ground (monthly pulse expressed in m K/W); R_{gd} is the equivalent thermal resistance per length unit of the ground (daily pulse expressed in m K/W); F_{sc} is the loss factor due to possible thermal short-circuits in the exchanger between input/output pipes; t_g is the ground temperature (in °C) not disturbed by the exchanger; t_{wi} and t_{wo} are the input and output temperatures (in °C) of the transfer fluid during the heating and cooling seasons; t_p is the penalty temperature (in °C), which evaluates the interference between exchangers.

Results presented in this work put into evidence the effect of changing the parameter R_b in the two Eqs. (5) and (6) considering that:

$$R_b = R_{pp} + R_{gr}, \tag{7}$$

where R_{pp} is the total thermal resistance of pipes containing the heat transfer fluid and R_{gr} is the thermal resistance of the sealing grout (i.e., pipe/ground interface), both expressed in m K/W. The thermal resistance of the sealing grout (R_{gr}) can be calculated through the relation provided by Remund (1999), which is valid for a single input/output circuit as follows:

$$R_{gr} = 1/S_b \cdot \lambda_{gr}, \tag{8}$$

where S_b is the short-circuit factor and λ_{gr} is the thermal conductivity of the sealing grout. The short-circuit factor (S_b) can be calculated as follows:

$$S_b = b_0 \cdot \left(d_b / D_{po} \right)^{\beta 1}, \tag{9}$$

where β_0 and β_1 are coefficients depending on the geometry of input/output pipes into the borehole (cf. Remund 1999 for the list of values), whereas d_b is the borehole diameter and D_{po} that of the pipe (i.e., the probe diameter). Sensitivity in changing λ_{gr} on the total sizing of the low-enthalpy geothermal installation (here evaluated in terms of borehole length) can be, therefore, calculated combing the relations (8) and (9). Indeed, the total thermal exchange between the heat transfer fluid and the borehole surface (i.e., the equivalent thermal resistance R_b) is strictly dependent on the thermal conductivity of the sealing material that operates at the probe/ground interface.

Fixing the technical configuration of the thermal exchange system (i.e., geometry of the input/output probe pipes, borehole and probe diameters), changes of λ_{gr} from 1.191 W/m K (average of the pure starting grout material) up to 1.878 W/m K (average of the grout material doped with 5% of graphite powder) are able to reduce the final R_{gr} from 0.12 down to 0.07 m K/W. Although small, these variations have important effects on the final calculation of the total borehole length either in heating or cooling conditions (L_h and L_c in Eqs. 5 and 6, respectively) due to changes of the R_b value. A simple calculation can be obtained by fixing all the environmental parameters and the related energetic budgets. As a mere example, decreasing of the equivalent thermal resistance R_b applied to a small edifice (surface ~ 150 m^2, for which ~ 200 linear meters of borehole are needed) can entail a total borehole length reduction on the order of 30–40 m (ca. 15–20%). Of course, the larger is the sizing of the low-enthalpy geothermal installation the higher is the economic benefit due to reduction of the total borehole length.

Conclusions

In this study, two bentonitic grouts doped with 5 and 10% of pure graphite powder have been considered for what concerns their flexural and uniaxial compressive strengths, together with thermal conductivity. Experimental results have put into evidence good perspectives for their possible use in low-enthalpy geothermal applications, especially as sealing material of probes into boreholes. Both the experimental bentonitic mixtures display improved mechanical and physical properties, although the cost/benefit ratio appears more favorable for starting materials doped with low percentages (5%) of graphite powder as additive. Importance of such experimental studies consists in finding new materials that, through implementation of the thermal exchange between the technological system and the subsoil, are able to make low-enthalpy geothermal installations even more competitive in the renewable energy market.

Authors' contributions
MV conceived the work, performed the mechanical and physical tests, elaborated the data and packed the whole paper. The author read and approved the final manuscript.

Author details
[1] Dipartimento di Scienze Biologiche Geologiche e Ambientali – Sezione di Scienze della Terra, Università degli Studi di Catania, Corso Italia 57, 95129 Catania, Italy. [2] Istituto Nazionale di Geofisica e Vulcanologia – Sezione di Catania, Osservatorio Etneo, Piazza Roma 2, 95125 Catania, Italy.

Acknowledgements
MV is pleased to thank Dr. Giuseppe Belfiore at EarTherm for his technical support and availability of materials. Dr. Giuseppe Cristaldi at Sidercem is greatly acknowledged for the preparation of bentonitic grout mixtures and his supervision during mechanical tests. MV is also grateful for the availability of the Technical Physics Lab at DIEEI of the University of Catania. Three anonymous reviewers are greatly acknowledged for their helpful suggestions, which have finally led to improvement of the paper.

Competing interests

Not applicable.

Funding

This work has been supported by research funds granted to Marco Viccaro by the University of Catania (FIR 2014 cod. 2F119B and PRA 2016-18 cod. 22722132120) and by EarTherm (Spin-Off Enterprise of the University of Catania).

References

Alrtimi AA, Rouainia M, Manning DAC. Thermal enhancement of PFA-based grout for geothermal heat exchangers. Appl Therm Eng. 2009;54(2):559–64.

Anbergen H, Frank J, Muller L, Sass I. Freeze-thaw cycles on borehole heat exchanger grouts: impact on the hydraulic properties. Geotech Test J. 2014;37:639–51.

Banks D. An introduction to thermogeology: ground source heating and cooling. Hoboken: Wiley-Blackwell; 2012. p. 526.

Blazquez CS, Martin AF, Nieto IM, Garcia PC, Sanchez Perez LS, Gonzalez-Aguilera D. Analysis and study of different grouting materials in vertical geothermal closed-loop systems. Renew Energy. 2017;114:1189–200.

Borinaga-Treviño R, Pascual-Muñoz P, Castro-Fresno D, Del Coz-Diaz JJ. Study of different grouting materials used in vertical geothermal closed-loop heat exchangers. Appl Therm Eng. 2013;50(1):159–67.

Clauser C, Huenges E. Thermal conductivity of rocks and minerals. In: Ahrens TJ, editor. Rock physics and phase relations—a handbook of physical constants, vol. 3. Washington, D.C: AGU Reference Shelf; 1995. p. 105–26.

Delaleux F, Py X, Olives R, Dominguez A. Enhancement of geothermal borehole heat exchangers performances by improvement of bentonite grouts conductivity. Appl Therm Eng. 2012;33–34(1):92–9.

Desmedt J, Van Bael J, Hoes H, Robeyn N. Experimental performance of borehole heat exchangers and grouting materials for ground source heat pumps. Int J Energy Res. 2012;36(13):1238–46.

DIN 18136:2003–2011. Soil—investigation and testing—Unconfined compression test. 2003–2011.

Eppelbaum L, Kutasov I, Pilchin A. Applied geothermics. Berlin: Springer-Verlag; 2014. p. 751.

Erol S, Francois B. Efficiency of various grouting materials for borehole heat exchangers. Appl Therm Eng. 2014;70(1):788–99.

Erol S, Francois B. Freeze damage of grouting materials for borehole exchangers: experimental and analytical evaluations. Geomech Energy Environ. 2015;5:29–41.

Fleuchaus P, Blum P. Damage event analysis of ground source heat pump systems in Germany. Geothermal Energy. 2017;5:10. https://doi.org/10.1186/s40517-017-0067-y.

Florides G, Kalogirou S. Ground heat exchangers—a review of systems, models and applications. Renew Energy. 2007;32(15):2461–78.

Fukai J, Kanou M, Kodama Y, Miyatake O. Thermal conductivity enhancement of energy storage media using carbon fibers. Energy Convers Manag. 2000;41:1543–56.

Indacoechea-Vega I, Pascual-Muñoz P, Castro-Fresno D, Calzada-Perez MA. Experimental characterization and performance evaluation of geothermal grouting materials subjected to heating-cooling cycles. Constr Build Mater. 2015;98:583–92.

Ingersoll LR, Zobel OJ, Ingersoll AC. Heat conduction: with engineering and geological applications. New York: McGraw-Hill Book Co; 1954.

Jobmann M, Buntebarth G. Influence of graphite and quartz addition on the thermo-physical properties of bentonite for sealing heat-generating radioactive waste. Appl Clay Sci. 2009;44(3–4):206–10.

Kalaiselvam S, Parameshwaran R. Thermal energy storage technologies for sustainability—systems design, assessment and applications. Elsevier: Academic Press; 2014. p. 430.

Karaipekli A, Sari A, Kaygusuz K. Thermal conductivity improvement of stearic acid using expanded graphite and carbon fiber for energy storage applications. Renew Energy. 2007;32:2201–10.

Kavanaugh SP, Rafferty K. Ground source heat pumps—design of geothermal systems for commercial and institutional buildings. ASHRAE Application Handbook. USA: American Society of Heating, Refrigerating and Air-Conditioning Engineers (ASHRAE); 1997.

Kim D, Kim G, Baek H. Relationship between thermal conductivity and soil-water characteristic curve of pure bentonite-based grout. Int J Heat Mass Transf. 2015;84:1049–55.

Kim D, Kim G, Park S, Baek H. Changes in the thermal conductivity of bentonite-based grouts with varying volumetric water content. Geosyst Eng. 2013;16(4):257–64.

Lee C, Lee K, Choi H, Choi HP. Characteristics of thermally-enhanced bentonite grouts for geothermal heat exchanger in South Korea. Sci China Ser E Technol Sci. 2010;53:123–8.

Lee C, Park M, Nguyen TB, Sohn B, Choi JM, Choi H. Performance evaluation of closed-loop vertical ground heat exchangers by conducting in situ thermal response tests. Renew Energy. 2012;42:77–83.

Remund CP. Borehole thermal resistance: laboratory and field studies. ASHRAE Trans. 1999;105:1.

Sari A. Thermal energy storage characteristics of bentonite-based composite PCMs with enhanced thermal conductivity as novel thermal storage building materials. Energy Convers Manag. 2016;117:132–41.

Smith MD, Perry RL. Borehole grouting: field studies and thermal performance testing. ASHRAE Transactions, Proceedings of the 1999 ASHRAE Winter Meeting. 1999, vol. 105, code 55431.

Tang AM, Cui YJ, Le TT. A study on the thermal conductivity of compacted bentonites. Appl Clay Sci. 2008;41:181–9.

UNI EN 12664. Thermal performance of building materials and products. Determination of thermal resistance by means of guarded hot plate and heat flow meter methods. Dry and moist products of medium and low thermal resistance. 2002.

UNI 11152. Aqueous suspensions for injections of hydraulic binders—characteristics and test methods (in Italian). 2005.

UNI EN 445. Grout for prestressing tendons—test methods. 2007.

7

Response surface method for assessing energy production from geopressured geothermal reservoirs

Esmail Ansari[*] [iD] and Richard Hughes

*Correspondence:
eansar2@lsu.edu
Louisiana State University,
70803 Baton Rouge, LA, USA

Abstract

Developing low-enthalpy geothermal resources along the US Gulf Coast is attractive for reducing global warming and providing clean energy. In this work, synthetic yet representative models for typical geopressured geothermal reservoirs located along the US Gulf Coast are considered. A Box–Behnken experimental design is used to select a small set of these models to perform detailed reservoir simulation runs. Full quadratic linear models are fit to the simulation results, and their sufficiency is confirmed by comparing them to kriging response surfaces. To achieve a higher degree of efficiency in using the response surfaces, Hammersley sequence sampling (HSS) method is used instead of traditional Monte Carlo sampling. HSS ensures that the factor space is sampled more uniformly and the response distribution is converged in less time. By evaluating these proxy models in the sampled factor space, the sensitivity and uncertainty of the response to the factors can be assessed. In this work, the sensitivity and uncertainty of engineered convection is assessed. For quantifying engineered convection, five uncertain reservoir attributes were selected. The response was defined as the net extracted enthalpy. In particular, two different designs for harvesting energy from geothermal reservoirs were compared using the response surfaces. In the modeled systems, results show that the regular design is more effective than the reverse design for extracting energy from geopressured geothermal reservoirs.

Keywords: Geothermal reservoir, Experimental design, Response surface, Sampling, Forced convection, Heat extraction

Background

Reducing greenhouse gases and providing the world's future energy require searching for clean alternative energy resources that can substitute for fossil fuel. Geopressured geothermal reservoirs along the US Gulf Coast are an alternative energy resource which have been considered as marginal and have not been developed extensively. The information available about these resources comes from well test data performed at the time of their development (John et al. 1998). Assessing the uncertainties associated with the commerciality of these reservoirs by simulating and history matching each case independently is an expensive and time-consuming process and should be reserved for the project design stage. One way for quickly assessing these assets is to study the sensitivity of produced energy to uncertain features using reservoir modeling.

Though computational speed and memory for solving problems is improving over time, detailed modeling for history matching of each reservoir or using Monte Carlo simulations is not feasible because running many models are numerically expensive. To overcome this limitation, there are three alternatives as follows: (1) procedures that efficiently create the history matched models; (2) surrogate reduced order models; and (3) statistical proxy models. The first approach uses efficient gradient-based or gradient-free algorithms for history matching production data and efficiently makes field development plans (Shirangi and Durlofsky 2015; Shirangi 2014). The second approach is to build surrogate reduced order models using piecewise linearization algorithms. These algorithms increase the efficiency of the Newton loop by creating the Jacobian matrices around previously simulated points instead of traditionally solving the flow equations (Ansari 2014; He and Durlofsky 2014; Cardoso and Durlofsky 2010). The third approach, which is used in this work, is to run the detailed model using specific combinations of factors sampled by experimental design and then fit a proxy response surface to the factor space (Ansari 2016). Experimental design and response models are popular and used widely (Fisher and Genetiker 1960; Mishra et al. 2015). Schuetter et al. (2014) compare the use of Box–Behnken sampling and quadratic polynomial regression with Latin Hypercube sampling, multivariate adaptive regression spline (MARS), and additivity variance stabilization (AVAS) techniques for geological CO_2 sequestration. They conclude that the model developed using Box–Behnken and quadratic polynomials performs the best. Following Schuetter et al. (2014), this work uses Box–Behnken experimental design. Experimental design and response surfaces have also been used in the context of geothermal reservoir engineering (Hoang et al. 2005; Quinao and Zarrouk 2014). Response surface models are fast and have adequate accuracy to represent the detailed model. Response surface models can be efficiently run thousands of times for uncertainty assessments. Traditionally, simple linear models are used to represent the actual model (Montgomery et al. 2012). For most of the cases, quadratic linear models (polynomials) are adequate. Once the proxy response surface model is constructed using a very small, yet statistically representative, set of detailed model runs, it can be used for sensitivity analysis and uncertainty assessments using sampling methods such as Monte Carlo (MC) or Hammersley sequences sampling (HSS).

This work compares different energy extraction designs for geopressured geothermal reservoirs and identifies the better technique. It further quantifies the uncertainty in the selected design using the developed proxy model and sampling methods.

This paper proceeds as follows: we first introduce design of experiments, response surface modeling, and sampling. Then, we apply these methods to compare two different heat extraction designs: regular design and reverse design. We select the best design for extracting energy by comparing their energy output. Regular design shows better performance than reverse design. For the regular design, the uncertainty of the factors is used to obtain the uncertainty in the cumulative energy produced.

Methods

Design of experiments (DOE)

Many factors influence the energy recovery and the focus should be on the factors that affect it the most. Experimental design is an efficient method for sampling the factor space and calculating the response with minimum number of runs (Montgomery 2008). Instead of changing the factors one-at-a-time, by which, factor interactions cannot be obtained and a large number of simulation runs is needed, factors are changed systematically in the experimental design to reveal effects and interactions using a smaller set of designed simulation runs. We used Box–Behnken design to fit a full quadratic response surface including all interaction terms to the simulation results (Box et al. 2005). This design needs fewer runs compared with other three-level counterparts (e.g., central composite designs, Fig. 1). Five geologic factors were chosen to test for their effects and uncertain attributes on the response. The reason for selecting these factors is that because of sparse well locations and measurements, these factors cannot be accurately measured and the uncertainty in their values always persists. One reason for selecting the Box–Behnken design is that it requires 41 runs for five factors, compared to 243 runs required for a full three-level factorial and 32 for a full two-level factorial design. We avoid full two-level factorial design because it cannot be used for modeling quadratic effects (second-degree curvature).

In reservoir engineering, factors can be categorized into three types: controllable, observable, and uncertain. The controllable factors may be engineered or selected, such as well location. The observable factors may be accurately measured such as the reservoir thickness at each well location. However, some uncertain factors, such as porosity far from wells, can neither be measured nor engineered. These factors are the important factors, on which the sensitivity analysis should be based.

Response surface methods

Once the results of runs suggested by the designs are obtained, response surfaces are used to determine the correlation between the factors and the response (Montgomery and Myers 1995). Two widely used formulations for the response surfaces are regression and kriging.

Regression

The method of least squares is conventionally used to estimate the regression coefficients and develop response surface models (Montgomery et al. 2012). The fitted linear or quadratic model to the sparse detailed runs can be used to estimate the effect of each parameter on the objective function (Eq. 1).

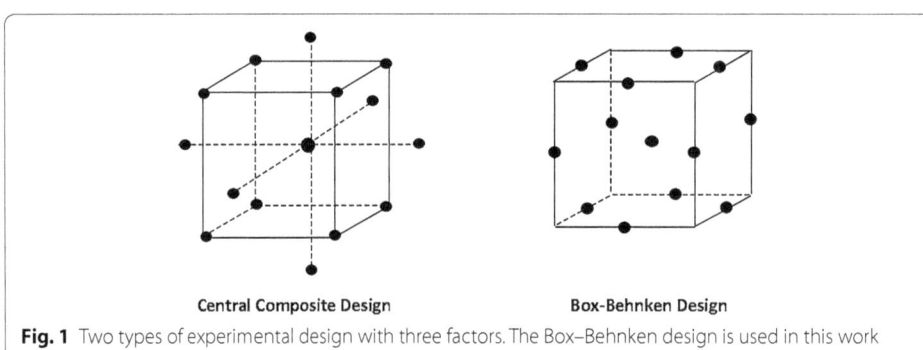

Central Composite Design   **Box-Behnken Design**

Fig. 1 Two types of experimental design with three factors. The Box–Behnken design is used in this work [from Kalla (2005)]

$$\hat{y} = \beta_0 + \sum_{i=1}^{k} \beta_i x_i + \sum_{i=1}^{k} \sum_{j>i}^{k} \beta_{ij} x_i x_j + \sum_{i=1}^{k} \beta_i i x_i^2 \tag{1}$$

In this equation, $\hat{y}$ stands for the predicted response, x stands for the variable factor of interest, and the β values are the regression coefficients. Equation 2 represents the least squares method for calculating the coefficient vector $\boldsymbol{\beta}$.

$$\boldsymbol{\beta} = (\mathbf{X}^T \mathbf{X})^{-1} \mathbf{X}^T \mathbf{y} \tag{2}$$

A large regression coefficient does not imply that the factor is significantly influencing the response and a small regression coefficient does not imply that the factor is not influencing the response. Thus, in order to eliminate the effect of factor units on the regression coefficient (for example using bar instead of pascal for pressure), it is necessary to scale the factors to have a range between −1 and 1 and make them scale-invariant. For accurately estimating the response and evaluating nonlinear effects, factor interaction and quadratic effects may also be needed. In these situations, a second-degree or reduced polynomial model can be used. A reduced model is a second-degree polynomial model in which the unimportant factors and interactions are removed. The p value statistical parameter is generally used for selecting the important factors.

Kriging

An alternative to using polynomials for producing response surfaces is ordinary kriging (Landa and Güyagüler 2003). This method linearly combines weighted observations (Eq. 3) and the weights depend on distances between the target point and the observations (Eq. 5). The distance is calculated in the k-dimensional factor space where k represents the number of factors (Eq. 6). In Eqs. 3, 4, and 5, $\mathbf{R}$ represents the correlation function and yields the relation between observations.

$$\hat{y} = \hat{\beta}_0 + \mathbf{r}(x)^T \mathbf{R}(\mathbf{y} - \hat{\beta}_0 \mathbf{1}) \tag{3}$$

$$\hat{\beta}_0 = (\mathbf{1}^T \mathbf{R}^{-1} \mathbf{1})^{-1} \mathbf{R}^{-1} y \tag{4}$$

$$\mathbf{r}^T(\mathbf{x}) = [\mathbf{R}(\mathbf{x}, \mathbf{x}_1), \mathbf{R}(\mathbf{x}, \mathbf{x}_2), \mathbf{R}(\mathbf{x}, \mathbf{x}_3), ..., \mathbf{R}(\mathbf{x}, \mathbf{x}_n)]^T \tag{5}$$

The correlation function can be modeled as a Gaussian, an exponential or any other positive definite function. Equation 6 shows a Gaussian representation for $\mathbf{R}$.

$$\mathbf{R}(\mathbf{x_i}, \mathbf{x_j}) = \exp \left(- \sum_{m=1}^{k} \theta_m |x_{m_i} - x_{m_j}|^2 \right) \tag{6}$$

In this equation, θ indicates a vector of parameters used to fit the model, n is the number of runs, k is the number of factors, and x_{m_i} and x_{m_j} are the mth factor levels of design runs. The distance between the target points and the observations is modeled using semivariogram models in the k-dimensional factor space. Covariance between points depends on the distance between them and decreases as the distance increases. Another useful feature of kriging is that it considers data redundancy and ensures that close points impose appropriate effect in predicting the target. These properties of kriging make it the best linear unbiased (BLUE) estimator for correlated data.

Sampling

Once the proxy models are constructed, a sampling method is needed to sample the factors and to translate the uncertainty from the input to the response. For doing this, a Monte Carlo or quasi Monte Carlo method, such as HSS, is generally used (Kroese et al. 2011).

Unlike straight Monte Carlo which samples n-dimensional space randomly, Hammersley sequence fills the space more uniformly (Fig. 2). This characteristic is known as low-discrepancy sequence sampling. In Hammersley sequences, the design point p (which is less than the total dimension n) is conditioned on the previous $p - 1$ points and the total dimension n, thus making the sample points dependent. The points generated in low-discrepancy sampling methods are highly ordered and exhibit much more regularity. The result of sampling using these sequences converges more efficiently than multidimensional Monte Carlo (Kroese et al. 2011). The only drawback of Hammersley sequences is that the number of points should be specified before simulation and if, due to lack of accuracy of the results, the number of points changes, the process needs to be repeated discarding previous results. The Hammersley sequences span the n-dimensional space with a small yet representative sample. The procedure of obtaining a Hammersley sequence is described below.

Any positive integer n can be expressed by a prime base p as follows:

$$n = \beta_0 + \beta_1 p + \beta_2 p^2 + \cdots + \beta_r p^r, \tag{7}$$

where every β_i is an integer number in the range $[0, p - 1]$. Now, a function ϕ_p of n can be defined as follows:

$$\phi_p(n) = \frac{\beta_0}{p} + \frac{\beta_1}{p^2} + \frac{\beta_2}{p^3} + \cdots + \frac{\beta_r}{p^{r+1}} \tag{8}$$

Hammersley points in the m-dimensional space can be given by

$$x_m(n) = \left(\frac{n}{N}, \phi_{R_1}(n), \phi_{R_2}(n), \ldots, \phi_{R_{m-1}}(n) \right)^{\mathrm{T}}, \tag{9}$$

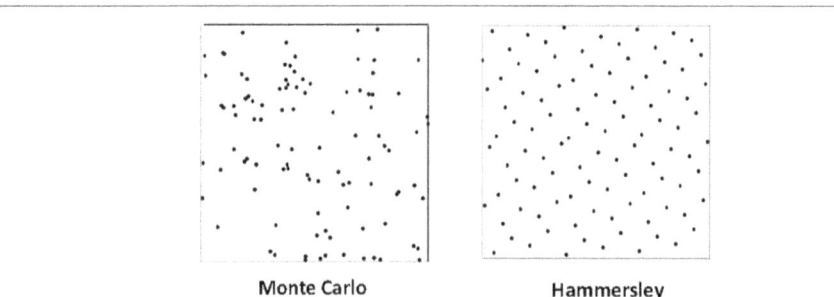

Fig. 2 Monte Carlo vs. Hammersley method for sampling a two-dimensional factor space. HSS fills the factor space uniformly (from Kalla (2005))

where $n = 1, 2, \ldots, N$ and $\boldsymbol{x_m}$ is the location of the point in the m-dimensional space. N is the total number of Hammersley sample points and R_1, R_2, R_{m-1} are $m - 1$ prime numbers for m dimensions.

Regular vs. reverse design

A hypothetical yet representative base model was developed from data obtained from the Camerina A sand zone located in the Gueydan Dome area, Vermillion Parish, Louisiana (Fig. 3). The model is a two-dimensional vertical cross section. This model has been proposed by Plaksina et al. (2011) and its characteristics are shown in Table 1. Similar models have been used in waterflooding (Shook et al. 1992) and CO_2 flooding

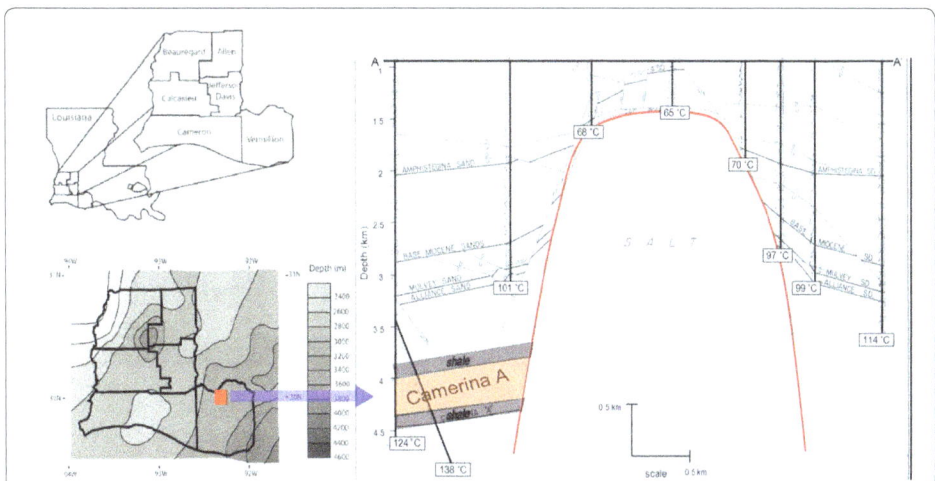

Fig. 3 Vertical cross section of the Gueydan dome (*right figure* is modified from Robinson (1967) and *left figure* is from Szalkowski and Hanor (2003). The Gueydan Dome, located in the Vermilion parish, LA, is shown by the *red dot*. The Gueydan salt dome and the Camerina A sand zone are shown schematically in the *right figure*

Table 1 Characteristics of the base hypothetical model [after Plaksina et al. (2011)]

Properties	Base value	Unit
Temperature of top cell	135	C
Matrix compressibility	2.0×10^{-5}	kPa^{-1}
Dip angle	*15*	*Degree*
Reservoir length	*2000*	*m*
Cross-section width	100	m
Reservoir thickness	*30*	*m*
Permeability	*300*	*md*
Porosity	*0.2*	*–*
Rock heat capacity	2.6×10^6	$J/(m^3 C)$
Rock bulk density	2.3	g/cm^3
Thermal heat conductivity	172,800	$J/(m\,day\,C)$
Water thermal expansion	8.8×10^{-4}	C^{-1}
Water compressibility	4.5×10^{-7}	kPa^{-1}
Water molecular weight	0.01802	kg/gmol
Water molar density	55,500	$gmol/m^3$
Water density	1.02	g/cm^3

Factors shown by italic are used in the experimental design

studies (Wood et al. 2008). The average temperature of the zone (Fig. 3) is calculated to be 142.5 °C with a range varying from 128 to 160 °C from top to bottom of the sand (Gray and Nunn 2010). A corner point grid system with 30 grids along the X-axis and 7 grids along the Z-axis was found sufficient and accurate enough for modeling this vertical cross section. The temperature of each grid block was calculated by setting the temperature gradient along the Z direction to 18 °C/km and the temperature of topmost grid block to 135 °C. Viscosity and density of the water depend on temperature and pressure. The geopressure zone extends from 4200 m to between 4650 and 4880 m depth; thus, the depth assigned to the topmost grid block is 4200 m. In the Gueydan Dome area, a shale sequence with a thickness ranging from 365 to 426 m overlies the Camerina A sand and a shale sequence of 150 m underlies it. Hence, the Camerina A structure represents a four-way closure for an area with one side enclosed by a salt dome (Ansari et al. 2014). The model does not consider the salt dome because Gray and Nunn (2010) found that the Gueydan salt dome does not have optimum burial depth to cause an increase in the temperature of the reservoir fluid.

The equilibrium state obtained from natural convection simulations (1000 years of simulation without injection or production) served as the initial condition for the forced convection. For the natural convection period, the temperature of the reservoir boundary is the same as its surrounding cap/base rock. As sediment cools down by cold water injection, it starts to gain conductive heat from the cap/base rock. A model, developed by Vinsome and Westerveld (1980), is used for peripheral boundary heat gain. The model is based on a semi-analytical impermeable heat conduction formulation which adequately describes the boundary condition at the interface. This model ensures adequate accuracy because in practice the thermal conduction coefficients between the reservoir and the cap/base rock are not precisely known.

Figure 4 compares three different boundary conditions for the base case considered in Table 1. The first case assumes that the reservoir is sealed and there is no heat conduction between the reservoir and cap/base rocks. In the second boundary condition, the temperature of cap/base rocks does not change as the reservoir cools down. The third

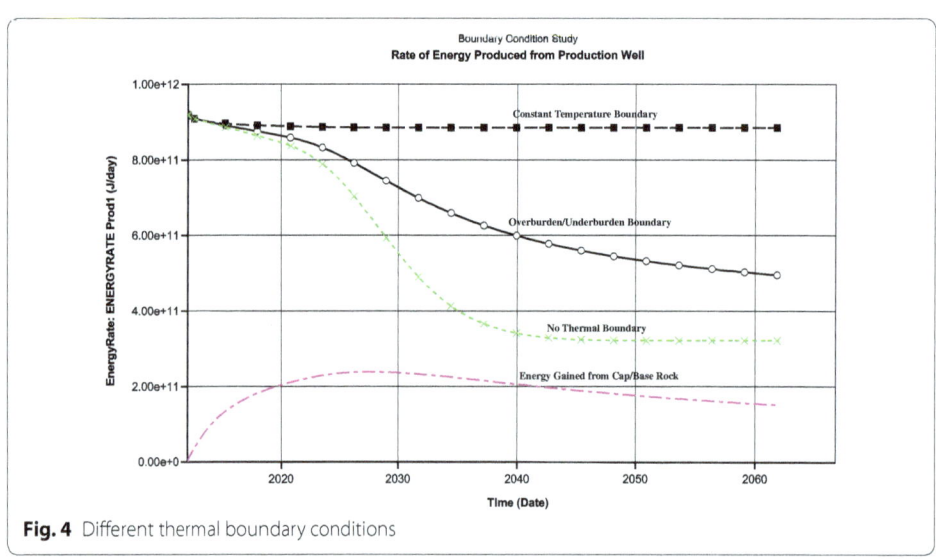

Fig. 4 Different thermal boundary conditions

case is the semi-analytical Vinsome and Westerveld's (1980) model. The heat-insulated boundary condition shows much lower produced energy, and the constant temperature boundary condition shows much higher produced energy than that shown by the more realistic boundary condition proposed by Vinsome and Westerveld (1980). Cumulative produced energy from the semi-analytical model is 38% more than the insulated reservoir and 22% less than constant temperature boundary condition after 30 years.

Figure 4 also implies that the value of the natural heat flux from the earth ($50\,\mathrm{mW/m^2}$), attributed to radioactive decay, has no perceptible influence on the reservoir once exploitation begins. The natural heat flux (i.e., 20,000 W for the considered vertical cross section with a length of 4000 m and a width of 100 m) is negligible compared with the amount of heat gained from the cap/base rock (order of 10^{11} W). Fig. 4 also shows that the rate of recharge from the cap/base rock to the reservoir initially increases and then decreases as the base/cap rock cools down.

Two designs for extracting energy are considered in this study: regular design and reverse design. Regular design places the production well at the bottom of the reservoir. The cool water is re-injected back into the reservoir at its top. Reverse design produced the hot geofluid from the updip reservoir and injects the cooled geofluid into the deeper sections. In the models, the wells are perforated only at the topmost and the bottommost grid blocks (Fig. 5). In all models, the produced geofluid is injected back into the reservoir. The production and injection rates are set to $2000\,\mathrm{m^3/day}$. The temperature of the injected water is set to 70 °C which is typical for low-enthalpy geothermal reservoirs. The models were simulated for 30 years. No salt concentration or chemical reactions are considered for the geofluid. The permeability and porosity are uniform and remain

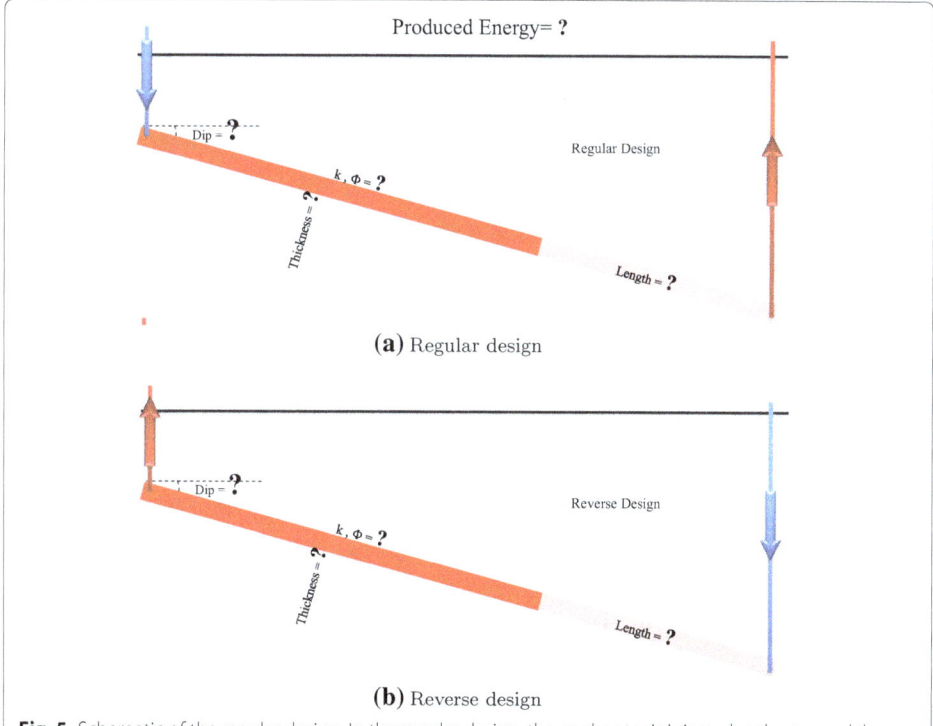

Fig. 5 Schematic of the regular design. In the regular design, the cool water is injected at the *top* and the hot water is produced from the *bottom*. In the reverse design, the hot water is produced from the top of the reservoir and is injected back at the *bottom*

constant during the production period. Five factors are more uncertain than the others which are colored as gray in Table 1. The range of possible change in these factors is summarized in Table 2. This study aims to recognize which heat extraction design is more effective and how the change in these factors affects the produced energy. A three-level Box–Behnken experimental design with 41 runs is used for developing the response surface models. Both polynomial and kriging types of the response surface models are investigated.

Results and discussion

The regular and reverse designs are compared for all the combinations of factors using Eq. 10

$$\epsilon_r = \frac{E_{reg} - E_{rev}}{E_{rev}} \times 100 \tag{10}$$

in which ϵ_r shows the percent difference between the energy that can be extracted from the regular and reverse designs.

In Fig. 6, the blue dots show the detailed model runs using the simulator. The contour lines show the response surface fitted to these model runs and projected onto the various subsets of factor space. These plots clearly demonstrate local gradients (i.e., local sensitivity) and average change in the response. The ϵ_r increases as the dip angle and length increase. At small dip angles, the effect of length on the ϵ_r is small and at large dip angles, the effect of length is great (sharp gradient in Fig. 6a). The ϵ_r is more sensitive to the dip angle than thickness (Fig. 6b). The increase in the thickness from 30 to 50 m increases ϵ_r 0.2%, while the increase in porosity increases ϵ_r 0.05% (Fig. 6c). Permeability has less effect on the ϵ_r compared with other factors (Fig. 6d).

The kriging response surface (Fig. 7) is comparable to the polynomial response surface (Fig. 6), and there are only subtle differences between them with the biggest difference being the permeability–porosity relationship. ϵ_r in all of these figures is positive indicating that the regular design is more effective than the reverse design in the modeled systems. For the modeled systems, the permeability range tested does not favor one energy extraction design over the other.

Regular design

The regular design was modeled in detail. A quadratic linear response surface model was fit to the simulation results. Our experience shows that using a second-degree polynomial instead of a first degree results in a better polynomial fit. The factors and interaction terms with p values less than 0.05 were selected as important (Table 3). Then, each factor was assigned a specific distribution. Both Monte Carlo and Hammersley sequence

Table 2 Levels of factors in the Box–Behnken design [Plaksina et al. (2011)]

Levels	Length (m)	Thickness (m)	Dip angle	Porosity	Permeability (md)
+1	2000	30	0	0.15	200
0	3000	40	15	0.20	500
−1	4000	50	30	0.25	800

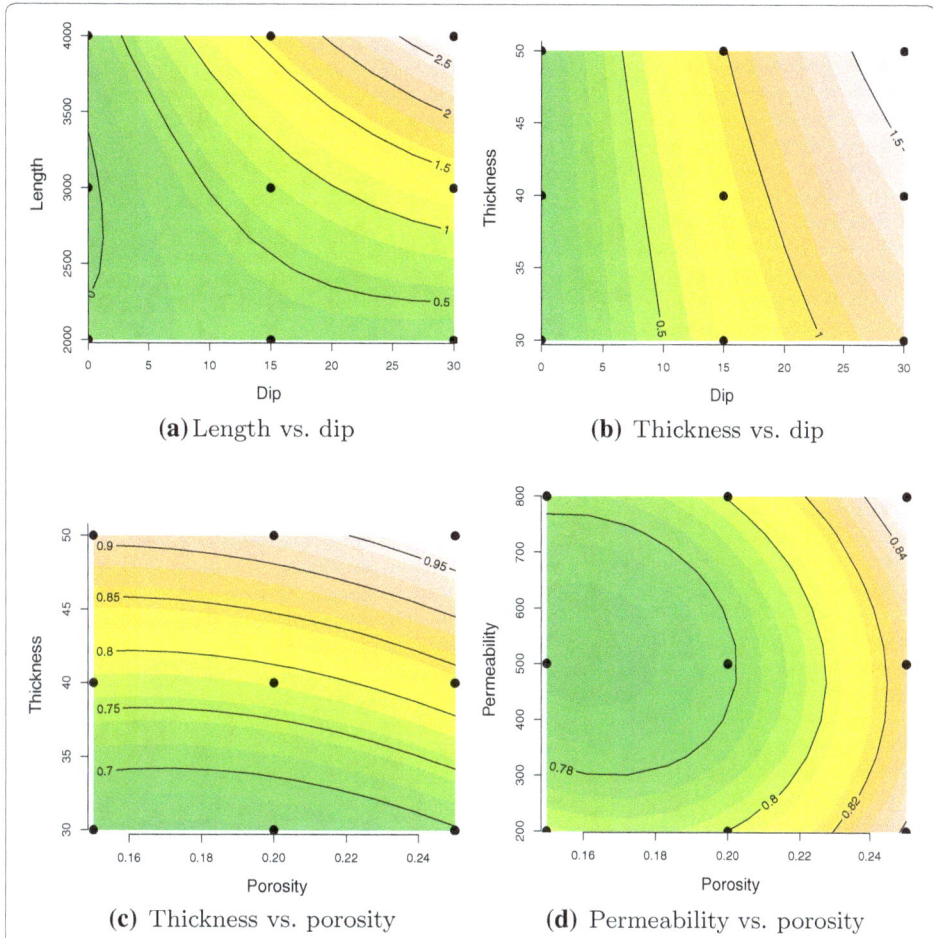

Fig. 6 Response surface results for the ϵ_r using the polynomial method. The *blue dots* show the detailed model runs using the simulator. The *contour lines* show the response surface fitted to these detailed model runs and projected into the various subspaces of factor space. The reservoir's length and the dip angle have the maximum effect on the ϵ_r. At high dip angles, the permeability has the least effect on the ϵ_r

sampling methods were used to sample these factors' distribution and translate the uncertainty from the factors to the response (net extracted energy).

The p value for all the factors except the permeability is less than 0.001 (Table 3) which means that all the factors have significant effect on the heat production except permeability. For the range of permeability considered for modeling, knowledge of the permeability map is less important for predicting the thermal energy recovery presumably due to the constant well rates assumption. This makes sense because the pressure of a geopressured geothermal reservoir is very high and this pressure constraint can provide the flow rate constraint imposed on the production well for the modeled range of permeability (Table 2).

A low p value and a positive coefficient for the porosity in Table 3 indicate that an increase in porosity would increase the produced energy. The fundamental idea in geothermal reservoirs is to extract the heat stored in the rock and use the fluid as the conduit. The increase in the fluid content of the system increases produced energy because the thermal capacity of the brine is more than the rock.

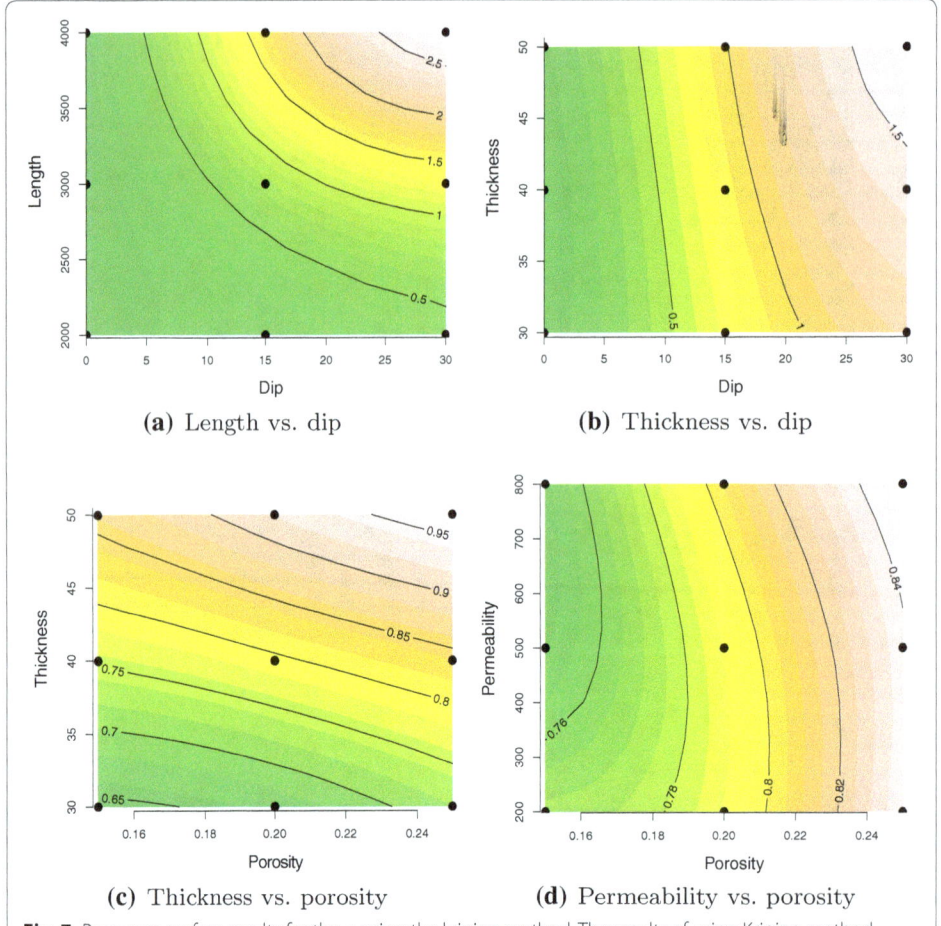

Fig. 7 Response surface results for the ϵ_r using the kriging method. The results of using Kriging method confirm the results of using polynomial response surface (Fig. 6). There are only subtle differences between these results

The dip angle, length, and thickness of the reservoir affect the heat production because all of these factors increase the temperature of the production well grid block. When the change in the response caused by the level change of one variable is not the same at all levels of another variable, the interaction term between the two variables is nonzero. For example, the impact of the reservoir's length on the produced energy for zero dip angle differs from when the dip angle is high. Thus, the interaction term between the length and the reservoir's dip angle is significant. Similarly, when the regular design was being compared with the reverse design (Figs. 6a, 7a), the interaction term between the length and the reservoir's dip angle was significant.

To test the model, the observation is plotted versus the model prediction (Fig. 8). The observation versus the model prediction falls on the 45° line which means the model is adequate. This model can be used for the Monte Carlo sampling of factor distributions. For quantifying the uncertainty in produced energy, a distribution was assigned to each factor. It is assumed that length, thickness, reservoir dip angle, and porosity, each follow a normal distribution (Fig. 9). A log-normal distribution is assigned to permeability.

Both MC and HSS methods give comparable results for the distribution of the response (Fig. 10). HSS was about 10 times more efficient than the MC simulation (MC

Table 3 Summary of the second-order linear regression for the regular design: representing the interaction between two factors

Factors	β value	Standard error	t value	Pr (> |t|)
(Intercept)	−0.027	0.009	−3.037	0.006
Thickness	0.218	0.003	83.748	0.000
Length	0.745	0.003	289.663	0.000
Permeability	0.000	0.003	0.162	0.873
Porosity	0.039	0.003	15.264	0.000
ReservoirDip	0.230	0.003	88.528	0.000
I (thickness^2)	−0.007	0.005	−1.277	0.214
I (length^2)	−0.050	0.005	−9.904	0.000
I (porosity^2)	−0.001	0.005	−0.245	0.809
I (permeability^2)	0.001	0.005	0.216	0.831
I (reservoirDip^2)	−0.035	0.005	−6.750	0.000
Thickness: length	0.041	0.005	7.929	0.000
Thickness: porosity	0.007	0.005	1.344	0.192
Thickness: permeability	0.001	0.005	0.220	0.828
Thickness: reservoirDip	0.020	0.005	3.927	0.001
Length: porosity	0.007	0.005	1.316	0.201
Length: permeability	0.002	0.005	0.353	0.727
Length: reservoirDip	0.138	0.005	26.580	0.000
Permeability: porosity	0.001	0.005	0.178	0.861
Porosity: reservoirDip	0.004	0.005	0.734	0.471
Permeability: reservoirDip	0.000	0.005	0.065	0.949

Estimate stands for the coefficient value of the regression model. Standard error, t value and p value for each coefficient is given. Gray color shows the important terms that should be retained in the reduced model. The p value is used for selecting the important predictors for the reduced model

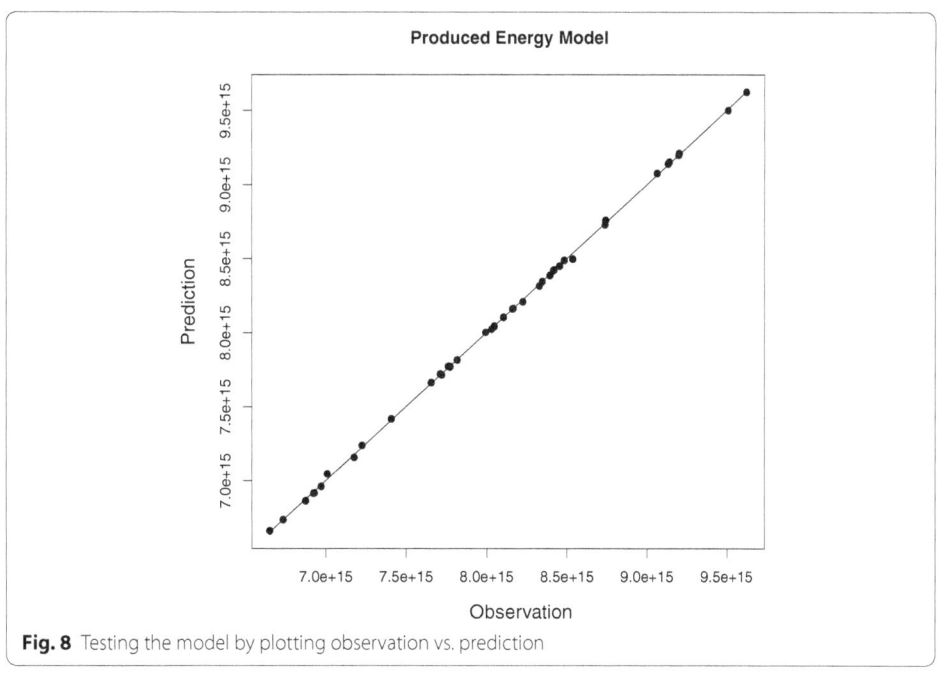

Fig. 8 Testing the model by plotting observation vs. prediction

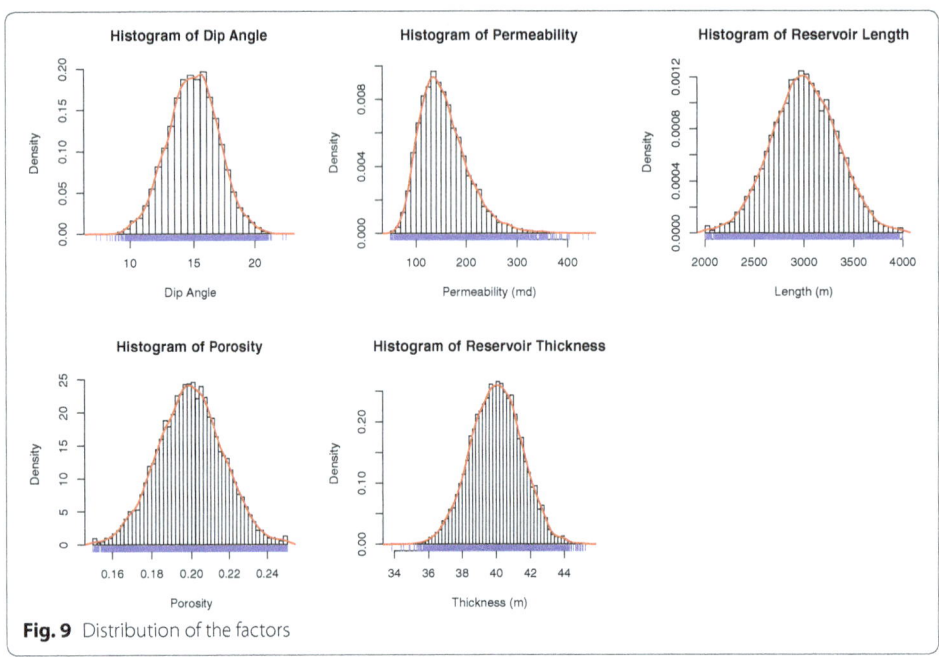

Fig. 9 Distribution of the factors

took 121.81 s while HS 12.73 s) due to its quasi-random and low-discrepancy property. The mean and median of the energy recovery distribution are around 8.1×10^{15} J for both. The obtained distribution for the extracted energy has a normal shape. The shape of this distribution makes sense because the response is sensitive to the length, dip, and thickness; all of which have a normal distribution. The extracted energy ranges between 6.8 and 9.2×10^{15} J and the box plot above the distribution indicates that 50% of the distribution lies between 7.85 and 8.35×10^{15} J.

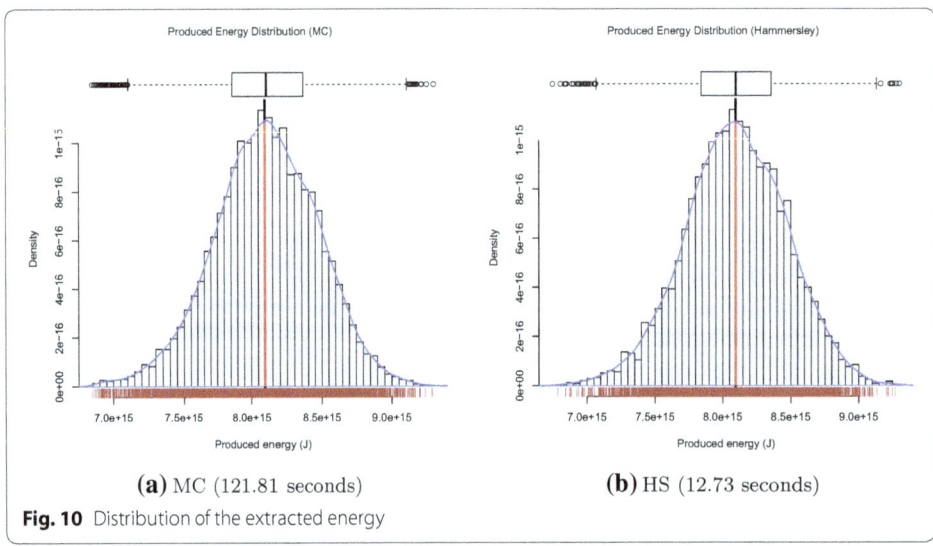

(a) MC (121.81 seconds)   **(b)** HS (12.73 seconds)

Fig. 10 Distribution of the extracted energy

Conclusion

Regular design outperforms reverse design for heat production in the modeled systems. This result is confirmed using polynomial and kriging response surfaces. The heat recovery from the regular design improves as the reservoir length, dip angle, or thickness increase. The results indicate that important criteria in evaluating a set of geothermal reservoirs with adequately high temperature is the size of the reservoir. For the reservoir models within the studied ranges, reservoir dip angle is less important that the reservoir size. The proxy models were efficiently used to construct produced energy distribution from the factor distributions. For having even more efficiency, HSS was used. HSS was about 10 times faster than the Monte Carlo simulation. For the considered problem, the produced energy was between 7×10^{15} and 9×10^{15} J. Future research should focus on testing the uncertainty in structural and isopach maps of a real reservoir model and compare the results with the results published in this work.

Authors' contributions
EA carried out the modeling, coding, and developing the results. EA also drafted the manuscript. The second author, RH supervised the research and guided the interpretation of results. RH also considerably edited and improved the drafts. Both authors read and approved the final manuscript.

Acknowledgements
The R source codes and the datasets for this study are available on request to the authors. The authors gratefully acknowledge financial support for this work from the US Department of Energy under grant DE-EE0005125. We thank Computer Modeling Group for providing reservoir simulation software. We also thank Christopher D. White and the members of the LSU Geothermal team for their comments, suggestions, and ideas supporting our efforts.

Competing interests
The authors declare that they have no competing interests.

References
Ansari E. Development of a surrogate simulator for two-phase subsurface flow simulation using trajectory piecewise linearization. J Pet Explor Prod Technol. 2014;4(3):315–25.
Ansari E. Mathematical scaling and statistical modeling of geopressured geothermal reservoirs. Baton Rouge: Louisiana State University; 2016.
Ansari E, Hughes R, White CD. Well placement optimization for maximum energy recovery from hot saline aquifers. In: 39th Workshop on Geothermal Reservoir Engineering, SGP-TR-202. Stanford: Stanford University; 2014.
Box GE, Hunter JS, Hunter WG. Statistics for experimenters: design, innovation, and discovery, vol. 2. Hoboken: Wiley Online Library; 2005.
Cardoso M, Durlofsky LJ. Linearized reduced-order models for subsurface flow simulation. J Comput Phys. 2010;229(3):681–700.
Fisher SRA, Genetiker S. The design of experiments. Edinburgh: Oliver and Boyd; 1960.
Gray T, Nunn J. Geothermal resource assessment of the Gueydan salt dome and the adjacent Southeast Gueydan field. Gulf Coast Assoc Geol Soc Trans. 2010;60:307–23.
He J, Durlofsky LJ. Reduced-order modeling for compositional simulation by use of trajectory piecewise linearization. SPE J. 2014;19(05):858–72.
Hoang V, Alamsyah O, Roberts J. Darajat geothermal field expansion performance-a probabilistic forecast. In: Proceedings world geothermal congress. 2005. p. 24–29.
John C, Maciasz G, Harder B. Resource description, program history, wells tested, university and company based research, site restoration. Gulf Coast geopressured-geothermal program summary report compilation. Baton Rouge: Tech. Rep., Louisiana State University, Basin Research Institution; 1998.
Kalla S. Use of orthogonal arrays, quasi-monte carlo sampling, and kriging response models for reservoir simulation with many varying factors. Master's thesis. Baton Rouge: Louisiana State University ; 2005.
Kroese DP, Taimre T, Botev ZI. Handbook of Monte Carlo methods. Hoboken: Wiley; 2011.
Landa JL, Güyagüler B. A methodology for history matching and the assessment of uncertainties associated with flow prediction. In: SPE annual technical conference and exhibition, society of petroleum engineers. 2003.
Mishra S, Ganesh PR, Schuetter J, He J, Jin Z, Durlofsky LJ. Developing and validating simplified predictive models for co_2 geologic sequestration. In: SPE annual technical conference and exhibition, society of petroleum engineers. 2015.
Montgomery DC. Design and analysis of experiments. New York: Wiley; 2008.

Montgomery DC, Myers RH. Response surface methodology: process and product optimization using designed experiments. New York: Wiley; 1995.

Montgomery DC, Peck EA, Vining GG. Introduction to linear regression analysis. New York: Wiley; 2012.

Plaksina T, White C, Nunn J, Gray T. Effects of coupled convection and CO_2 injection in stimulation of geopressured geothermal reservoirs. In: 36th Workshop on geothermal reservoir engineering. Stanford: Stanford University; 2011. p. 146–154.

Quinao JJ, Zarrouk SJ. Applications of experimental design and response surface method in probabilistic geothermal resource assessment–preliminary results. In: Proceedings, 39th Workshop on geothermal reservoir engineering. Stanford: Stanford University; 2014.

Robinson E. Acadia and vermilion parishes. Plano: The Pure Oil Company, Geomap Company; 1967.

Schuetter J, Ganesh PR, Mooney D. Building statistical proxy models for co_2 geologic sequestration. Energy Procedia. 2014;63:3702–14.

Shirangi MG. History matching production data and uncertainty assessment with an efficient tsvd parameterization algorithm. J Pet Sci Eng. 2014;113:54–71.

Shirangi MG, Durlofsky LJ. Closed-loop field development under uncertainty by use of optimization with sample validation. SPE Journal. 2015.

Shook M, Li D, Lake LW. Scaling immiscible flow through permeable media by inspectional analysis. In Situ. 1992;16(4):311.

Szalkowski DS, Hanor JS. Spatial variations in the salinity of produced waters from southwestern louisiana. Gulf Coast Assoc Geol Soc Trans. 2003;53:798–806.

Vinsome P, Westerveld J. A simple method for predicting cap and base rock heat losses in thermal reservoir simulators. J Can Pet Technol. 1980;19(3).

Wood DJ, Lake LW, Johns RT, Nunez V. A screening model for CO_2 flooding and storage in Gulf Coast reservoirs based on dimensionless groups. SPE Reserv Eval Eng. 2008;11(03):513–20.

8

Flow-through experiments on the interaction of sandstone with Ba-rich fluids at geothermal conditions

Pia Orywall[1,2]* [iD], Kirsten Drüppel[3], Dietmar Kuhn[1], Thomas Kohl[4], Michael Zimmermann[5] and Elisabeth Eiche[6]

*Correspondence: pia.orywall@
kit.edu; p.orywall@rbs-wave.de
[1] Institute of Nuclear and Energy
Technology, Karlsruhe
Institute of Technology,
Herrmann-von Helmholtz-Platz 1,
76344 Eggenstein-Leopoldshafen,
Germany
Full list of author information is
available at the end of the
article

Abstract

It is commonly known that heat extraction and decompression can lead to mineral precipitation and reservoir clogging in geothermal systems. In the Upper Rhine Graben, the precipitating minerals are mainly barite and calcite. This study focuses on clogging processes due to mineral precipitation in porous reservoir rocks, i.e., sandstone. The goal is to develop, build, and put into operation the HydRA apparatus, a facility for performing experiments on forced precipitation of barite in the pore spaces of sandstone under geothermally relevant pressure and temperature conditions. Barite precipitation during the flow-through is provoked by using barite-supersaturated solutions with a saturation index (SI) of 1.75. Scanning electron microscopy (SEM) investigations are used to detect barite crystal agglomerations and clogging of the pore spaces by overgrowths on these agglomerates. Following this, different crystal shapes are observed. The results are confirmed by permeability analyses before and after the flow-through experiments. Comparison of the major and trace element compositions of the original and reacted sandstones indicates element mobility due to water–rock interaction, even during the short-time experimental runs (max. 24 h).

Keywords: Flow-through experiments, Percolation, Barite, Precipitation, Dissolution, Water–rock interactions, Geothermal energy, Permeability, Pore space clogging

Background

When heat is extracted from geothermal fluids, the chemical equilibrium in the geothermal system is changed and thus some mineral phases become supersaturated and precipitate. Depending on the origin of the fluid and the degree of cooling, the main precipitating mineral phases observed are carbonates ($CaCO_3$), sulfates ($CaSO_4$, $BaSO_4$, $SrSO_4$), silica, and sulfides (FeS, PbS, CuS) (Stober and Bucher 2012). During a geothermal cycle, this modified fluid is re-injected into the reservoir and thus may change the mineralogical composition of the rock drastically. Dissolution of minerals may alter the rock structure, whereas precipitation of mineral phases from the fluid may have an effect on the permeability.

Flow-through experiments with sandstones were performed in numerous studies with regard to geothermal energy use for heat storage, energy extraction, or CO_2-storage. In all these studies, a fluid is forced to flow through reservoir rocks, like limestone, crystalline rocks, or sandstones. Investigations on formation damages due to particle redistribution in sandstone reservoirs were performed and the existence of a critical flow rate was confirmed by Ochi and Vernoux (1998). Rosenbrand et al. (2015) attributed the reduction of the permeability of reacted sandstones to the migration of fine particles, depending on the salinity of the percolating fluid and temperature. Another study shows that the decrease in permeability of kaolinite containing sandstones is related to the dissolution and re-precipitation of the kaolinite (Rosenbrand et al. 2014).

By performing experiments with oxidized water at the geothermal site of Neustadt-Glewe, Kühn et al. (1998) showed that a decrease of the permeability is caused by the precipitation of Fe-hydroxides and/or particle redistribution. Research work on water–rock interaction of granite was performed by Savage et al. (1992), implying that temperature, flow rate, and fluid composition are crucial parameters which determine the progress of the chemical reactions and hydraulic changes of the rock. Banks et al. (2014) present an experimental design for predicting the scaling risks concerning barite mineralization in basin-hosted enhanced geothermal systems.

In geothermal exploration and the following energy production, barite supersaturation and precipitation is caused by extracting the heat from a reservoir fluid, which is supersaturated or slightly undersaturated with regard to barite (Pauwels et al. 1993). Either barite particles nucleate from the fluid and clog the pore gussets or precipitation occurs in the available pore space by overgrowing matrix particles. One good example is Soultz-sous-Forêts (France), where geothermal heat extraction leads to massive barite deposits in the tube on the reinjection side (Scheiber et al. 2014). The influence of rising temperature is studied by investigations on barite in a closed system. Compared to ambient temperature, barite precipitation rates are increasing with decreasing temperatures (Blount 1977).

Christy and Putnis (1993) studied barite precipitation and dissolution in barite-supersaturated NaCl brines at temperatures of up to 85 °C. They state that the growth and dissolution of barite is not sensitive to the pH value and NaCl concentration and that the barite precipitation follows second-order kinetics. In a more recent study, however, the influence of the pH of the solution was demonstrated in nanoscale experiments. Under alkaline pH-conditions, barite growth stopped during progressive precipitation suggesting a distortion of the barite structure, which may be caused by the OH^-/CO_3^{2-} ions of the alkaline solution. At high pH values, a smaller particle size of barite was observed (Ruiz-Agudo et al. 2015).

In the present work, the barite precipitation in porous sandstone is studied. A special experimental setup is designed to simulate realistic geothermal reservoir conditions. A flow-through apparatus is used for the experiments, in which an artificial geothermal fluid is forced through a common sandstone of the Upper Rhine Graben. The experiments are carried out at elevated temperatures of up to 150 °C and pressure ranges of up to 300 bar with a fixed flow rate of 2 cm^3/min. These experimental parameters are closely aligned to the operational parameters of geothermal sites in the Upper Rhine Graben, as for instance, Soultz-sous-Forêts (France) and Bruchsal (Germany) (Herzberger et al. 2010; Genter et al. 2010). In the experiments, an artificial fluid with a well-defined supersaturation of barite (SI 1.75) is used. This composition is similar to the composition of the natural fluid of Soultz-sous-Forêts (Sanjuan et al. 2010).

The results of this study provide information on the processes and the risks of pore space clogging related to barite precipitation at geothermal sites and help to specify the effects of the reinjection of a chemically modified geothermal fluid into the reservoir, particularly with respect to the permeability change of the reservoir formation.

Methods

Experimental apparatus and procedure

All experiments were performed in HydRA (hydrothermal reaction apparatus), in which rock samples are percolated at a fixed flow rate under geothermally relevant temperature and pressure conditions by a barite-supersaturated fluid to induce barite precipitation. HydRA is designed and constructed exclusively for this kind of experiments by the Department of Energy and Process Engineering (Institute of Nuclear and Energy Technologies) at the Karlsruhe Institute of Technology (KIT).

The centerpiece of HydRA (Fig. 1) is the autoclave in which the rock sample is mounted. It is suitable for cylindrical samples with the dimensions of 50.8 mm × 25.4 mm. Temperatures of up to 150 °C and pressures of up to 350 bar can be used as boundary conditions in the experiments. Flow rates can be adjusted in a range of 2–20 cm^3/min.

Further main components of HydRA are two reservoir tanks for the hydrochemical solutions, three pumps (LEWA membrane pumps), the hydrochemical sampling point, a scale, a heater, and two pressure and two temperature sensors. All components are connected by steel pipes (Herfurth and Orywall 2015). The pipes, the autoclave, and the reservoir tanks are made from non-corrosive austenitic steel (DIN EN 10088-3 2014) 1.4571 with the following composition of the main elements: Fe 66.7 wt %, Cr 16.7 wt %, and Ni 10.72 wt %. The steel can be used up to temperatures of 550 °C.

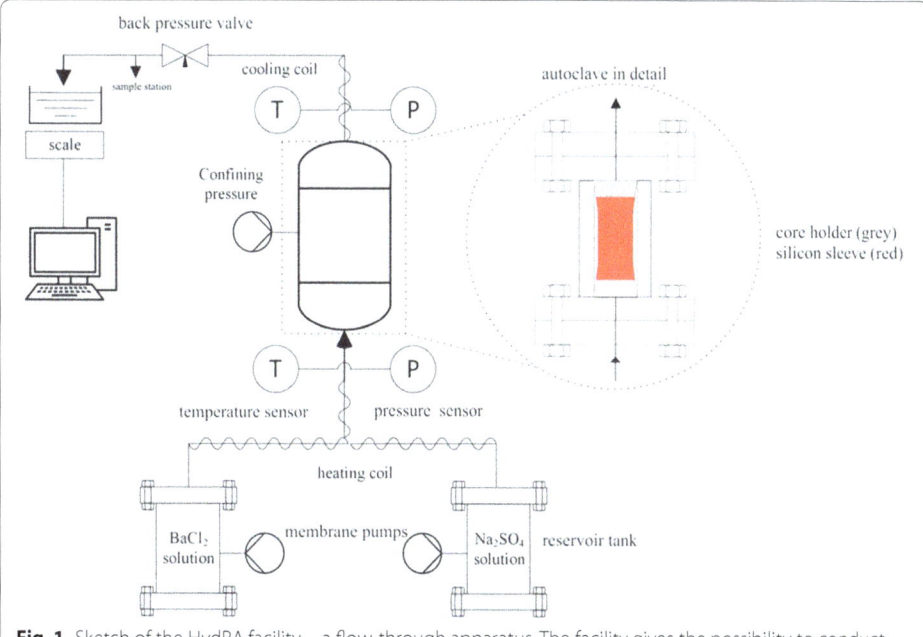

Fig. 1 Sketch of the HydRA facility—a flow-through apparatus. The facility gives the possibility to conduct experiments under geothermal in situ pressure and temperature conditions, thereby the solution, the flow rate, and the kind of rock are variable

The rock sample is fixed in a sample holder (Fig. 1) and tightly enclosed by a silicon tube (Fig. 1, red marked) with a wall thickness of 3 mm. To avoid a flow between this silicon tube and the sample, a confining pressure, which is always 10% higher than the pressure in the percolation system, is applied, using a separate pump and a closed water cycle with distilled water.

To fit into the autoclave, the cylindrical rock sample has to have a length of 50.8 mm and a diameter of 25.4 mm. The length of the sample has to be precisely adhered to fit into the core holder of the autoclave. The samples used were homogeneous, without any visible layering or damages at the edges of the rock body. The sandstones were dried under ambient conditions in a hood (protected in a glass bowl) until their weight was constant.

The fluid, which is pumped through the sample, circulates in a separate cycle. Reservoir tank 1 mainly contains the solution of a background electrolyte with barium chloride ($BaCl_2$), whereas reservoir tank 2 contains a solution of the background electrolyte with potassium sulfate (Na_2SO_4). The choice of the background electrolyte is based on two considerations: (1) the composition of the solution should be as close as possible to the natural fluid of the Upper Rhine Graben and (2) the supersaturated mineral phases should not precipitate immediately after mixing (Canic et al. 2015; Kaufmann-Knoke 1992). The two solutions are mixed at the entry of the rock sample. The flow rate of the solutions is adjusted via two pumps, one for each reservoir tank. Both pumps have an individual water cycle which is filled with distilled water. This design separates the technical equipment of the pumps from the saline solution and thus avoids corrosion.

After leaving the rock sample, the solution is conducted through a heat exchanger that cools down the hot percolated fluid. Behind this, a back pressure valve is installed to control the pressure in the system. At the exit of the apparatus, the flow quantity of the solution is measured using a scale and fluid samples are collected (Fig. 1).

All system data (temperatures, pressures, and mass flows) are recorded digitally with the software OPAL (Daubner and Krieger 2010). The apparatus is controlled using a Siemens programmable logic control (PLC) in which also safety functions are implemented. If, for example, the temperature of the heating system or the differential pressure exceeds the permitted range (all values are monitored by the PLC), HydRA turns off automatically and an error message is displayed.

Experimental parameters

The experiments are conducted at a fixed flow rate of approx. 1 cm^3/min for both pumps (i.e., reservoir tanks) which amounts to a total volume flow rate through the rock sample of 2 cm^3/min. The temperature range is chosen according to the conditions at the natural injection wells and is set to 20, 60, and 150 °C. The pressure is set to 20 bar (operational pressure), 300, and 350 bar (reservoir pressure).

The parameters recorded during the experiments are mass flow, temperature, upstream pressure, downstream pressure (resulting in differential pressure over the rock sample), and the confining pressure. In Table 1, the selected experiments and their respective conditions are listed.

Table 1 Experimental parameters of selected experiments

Experimental run	Temperature (°C)	Pressure (bar)	Barite concentration (mg/L)	Flow rate (cm³/min)	Duration of the experiments (h)	Sample name
PV10	20	20	112.3	2	24	12.1
PV09	20	300	111.0	2	24	8.2
PV07	60	20	216.7	2	~ 14	7.2
PV14	60	300	219.0	2	~ 18	17.1
PV06	60	350	219.0	2	~ 19	7.1
PV13	150	300	358.0	2	~ 11	16.1

Rock sample

This study focuses on porous media and thus the experiments are conducted with rocks of the Buntsandstein Group, which is, inter alia, one reservoir rock of the Bruchsal location (Herzberger et al. 2010). The rock samples, used for the experiments, belong to the Middle Buntsandstein that consists of the Eck-formation Horizon, the Bausandstein-formation, and the general conglomerate (Geyer and Gwinner 2011).

The rock was collected from a quarry close to Lahr in the Black Forest, at the eastern border of the Upper Rhine Graben. Blocks of around 300–400 mm length and width and a height of 150 mm were cut out. In the laboratory, the cores were brought to the correct dimensions of 25.4 mm in diameter and 50.8 mm in length.

Fluid-numerical modeling

In order to model the solution, which was used for the percolating of the rock sample, the software PHREEQC (Parkhurst and Appelo 2013) is used. In a first step, the salt concentration of the solution (NaCl, CaCl$_2$) was calculated. In the second step, these values were used as input parameters to model the required barite and sulfate concentrations with a saturation index (SI) of 1.75 according to

$$SI = \log \frac{IAP}{LP}, \tag{1}$$

where IAP is the ion activity product and LP the solubility product (Tutolo et al. 2015; Merkel and Planer-Friedrich 2008).

The calculations were made using the geochemical software PHREEQC Version 3 and the llnl database (Parkhurst and Appelo 2013).

Fluid composition

An artificial fluid that reproduces the major hydrochemistry of Soultz-sous-Forêts was prepared according to the geothermal fluid composition described in Sanjuan et al. (2010). The pH value of the solution is in the range of 5.3–5.5. This geothermal fluid contains Na$^+$, Ca^{2+}, and Cl$^-$ and therefore the artificial fluid was prepared using the salts NaCl and CaCl$_2$ * H$_2$O. To gain the desired concentrations for Na$^+$ (1.27 mol/L) and Ca^{2+} (0.17 mol/L), 74.4 g/L NaCl and 25.3 g/L CaCl$_2$ * H$_2$O were used. The resulting concentration of Cl$^-$ is 1.62 mol/L.

During an experiment, one reservoir tank contains the artificial fluid (consisting of NaCl and $CaCl_2$) together with a modeled concentration of $BaCl_2$, while the other reservoir tank is filled with a similar artificial fluid (NaCl and $CaCl_2$) and a modeled concentration of Na_2SO_4. At the inlet of the rock sample (Fig. 1), the two fluids are mixed and the resulting solution shows a well-defined supersaturation with respect to barite [see formula (2)]

$$BaCl_2 + Na_2SO_4 \rightarrow BaSO_4 \downarrow + 2Na^+ + 2Cl^-. \tag{2}$$

Hydraulic parameters

Permeability

To monitor the permeability changes during the experiment and to control the values of the permeability before and after the experiment, the differential pressure was determined by measuring the fluid pressures up- and downstream of the sample. The differential pressure was adjusted, depending on the flow rate used. These descriptive values were taken before and after the percolation experiment. By using Darcy's law, the calculation is as follows:

$$k(t) = \frac{\eta Q l}{A \Delta p(t)} \tag{3}$$

$k(t)$ is intrinsic permeability (m^2) ($1D = 9.87 \times 10^{-13} m^2$); η is dynamic viscosity of the fluid [Pa s]; l is length of the sample (m); Q is flow rate (m^3/s)]; A is cross section (m^2); and Δp is differential pressure [Pa] (1 bar = 100 kPa).

The permeability measurements were conducted with the HydRA facility by using distilled water with a well-known viscosity. Further known parameters are the dimensions of the sample and the differential pressure, which was recorded by the pressure load cells (Fig. 1). So the requirements for a standardized permeability calculation according to DIN 18130-1 (1998) are fulfilled for each experiment and thus the complete saturation of the sample with distilled water and the measurement should be performed in the range of linear flow to ensure a Darcy flow (Soni et al. 1978).

Effective porosity

To gain more information on the inherent properties of HydRA and to characterize the flow-through behavior of the mounted rock sample, tracer tests using different salt solutions were carried out. With K and Li as tracer cations, it is possible to determine the effective porosity of the rock samples according to

$$n_{eff} = \left(\frac{Q * t_{0,5}}{V} \right), \tag{4}$$

where Q is the flow volume (cm^3/min); V is the volume of the sample (cm^3); $t_{0,5}$ time to breakthrough (min) (Klotz et al. 1982).

The composition of the tracer solution is the same as for the background electrolyte solution described above, with the inert tracer being added.

By measuring the initial porosity of some samples, before and after the experiments, it was noted that the changes of the porosity values were in the range of measurement accuracy.

Analytical methods

Rock–mineralogical analysis

Scanning electron microscopy (SEM)

Scanning electron microscopy was used to analyze the mineralogy of the sandstone rock samples before and after the experiment. Thin sections were prepared as longitudinal sections of this cylindrical rock samples with a thickness of 30 µm at the Mineralogical and Geochemical Lab, Institute of Applied Geoscience (AGW), KIT.

The thin sections were sputtered with an 8-mm-thick layer of Au/Pd (80/20), prior to the analysis, using a Cressington Sputter Coater 208. SEM analyses were performed with a LEO Gemini 982 from Zeiss. The determination of the chemical composition of micro areas, including line scans, was performed using an Oxford INCA Penta FETx3 EDX-System. Both instruments are housed at the IKFT (Institute of Catalysis Research and Technology) at KIT. One image of the initial mineralogical composition was made at the KIT-LEM (Laboratory of Electron Microscopy), where the sample was sputtered with a layer of carbon.

X-ray computer-assisted tomography (CT)

One CT analysis was performed for an unpercolated rock sample that was reworked to a diameter of 40 mm to fit into the core holder of the CT. Tomographic 3D-datasets were recorded with the CT scanner ProCon X-Ray. The measurements were performed at an acceleration voltage of 130 kV, a current of 180 µA, and an exposure time of 180 ms. The datasets were reconstructed with a Volex reconstruction engine (Fraunhofer-Allianz Vision 2012). The reconstruction algorithm is based on a Radon transform by convolution and back filter (Feldkamp et al. 1984). Corresponding voxel size is 21.87 µm. After the 3D volume reconstruction, the images were processed using the software package Avizo 9.1 (ZIB 2016).

X-ray fluorescence (XRF)

The initial chemical rock composition was analyzed by X-ray fluorescence. Analyses were performed with a wave length dispersive XRF (S4 Pioneer, Bruker AXS). For the analysis, the rock samples were crushed and grinded to powder. Afterwards a fusion bead was synthesized and the measurement was performed against a matrix-matched calibration at 60 °C and 300 bar.

Inductively coupled plasma mass spectrometry (ICP-MS)

The major and trace element geochemistry of the rock samples (7.2, 8.2, 15.1, and 16.1) before and after the experiment was determined by ICP-MS (X-series 2, Thermo Fisher Scientific) after HNO_3–HF–$HClO_4$ acid digestions of the powdered material (100 mg). To assure a complete silicate decomposition, 40% HF (Suprapur), 70% $HClO_4$ (Normapur), and the pre-oxidized sample (65% HNO_3, sub-boiled) were heated in a closed vessel for 16 h at 120 °C. After evaporating the acids to incipient dryness, the residue was re-dissolved in 65% HNO_3 and evaporated again (three times) for purification purposes. The final residue was dissolved in 50 mL of ultrapure water. To assure the quality of the whole procedure, three blanks and two certified reference materials [GS-N, SY-2; Govindaraju (1994)] were included into the digestion process (accuracy: ± 10%). The reproducibility ($\pm$ 5% for most elements) was checked by digesting one sample in triplicate. The quality assurance for the ICP-MS measurement was done by including the certified reference material CRM-TMDW-A (High-Purity standards, Inc.) into the protocol (accuracy: $\pm$ 7% for most elements).

Fluid analysis

A total of 25 fluid samples was collected for each experiment at the sampling point (Fig. 1) and analyzed on their Ba concentration using inductively coupled plasma optical emission spectrometry (ICP-OES). Hereby, samples 1 and 2 are the starting solutions [background electrolyte with $BaCl_2$ (1) and with Na_2SO_4 (2)], while samples 3 and 4 were taken from the mixture of the solutions (background electrolyte with $BaSO_4$) to determine the real inlet concentration. To prevent further reactions of Ba^{2+} and SO_4^{2-}, the final solutions of the flow-through experiment had to be diluted immediately after sampling.

The sampling began 30 min after starting the pumps, since the effluent needs 30–35 min to reach the sampling station. Samples were taken after 30, 32, 36, 38, 40, 50, 60, 90, 160, 210, and 270 min (calculated from the start of the run). Each sample was collected manually with a 100 µL pipette and transferred into a bottle containing HNO_3 (0.3 mL, conc., sub-boiled) with the internal standard Yttrium and distilled water (9.6 mL) to gain the predefined dilution of 1:100. During the night, five samples were taken using an automatic sampling carrousel and on the second day sampling was done every 60 min until the test duration of 24 h was over.

All samples were analyzed, processed, and evaluated with an ICP-OES Optima 4300 DV (PerkinElmer Instruments) and the implemented software. The calibration of the ICP-OES was performed with the background electrolyte. For the measurements, the solutions had to be diluted by a factor of 1:100. The used specific wave lengths were 455.403 and 493.408 nm with a detection limit of 1 µg Ba/L.

Results
Hydraulic parameters
Permeability

To detect permeability changes during the experiments, two pressure cells (one upstream and one downstream of the rock sample) were used. Measurements were done with distilled water before and after the experimental runs. For a measurement, the rock sample was fully saturated with distilled water and then installed to the core holder before it was mounted into the autoclave. Then temperature and pressure were adjusted. During the measurement, it was necessary to adjust the stable laminar Darcian flow.

Figure 2 illustrates the intrinsic permeability of the rock samples of the analyzed experimental runs. For all rock samples, the permeability is higher before the flow-through experiment than afterward (Fig. 2).

A considerable difference in the rock permeability can be found for sample 17.1 with an eightfold decrease from 23 to ~ 2.9 mD within 1100 min. For sample 12.1 only a 3.6-fold decrease is detectable with values of 20 mD before and ~ 5.5 mD after the experiment. With 59 mD the sample 8.2 has the highest initial permeability of all samples, which dropped to 17 mD after 24 h. Sample 7.2 is, apart from sample 16.1, the sandstone with

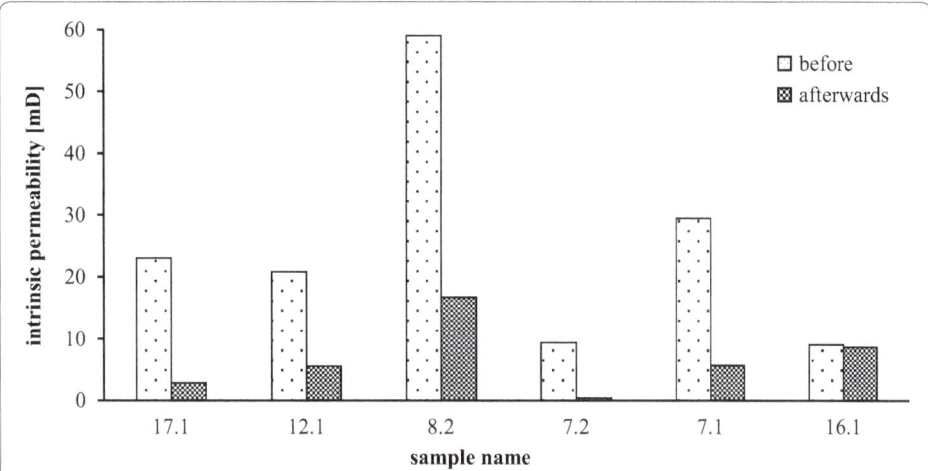

Fig. 2 Intrinsic permeability of the rock samples before and after the experimental runs. On the *y* axis the intrinsic permeability (mD) is shown. The *x* axis displays the used samples (identified by their number/name). White bars with few dots illustrate the permeability of the unpercolated rock samples whereas the strongly dotted bars show the permeability of the percolated rock samples after the experiment

the lowest initial permeability of ∼ 9 mD, which drops to 0.4 mD and thereby shows a tenfold decrease. A fivefold decrease in permeability can be determined for sample 7.1 with a drop from 29.5 to 5.7 mD (duration of the run: 1155 min). Almost no permeability drop is observed for sample 16.1, which shows an initial permeability of 9 mD that declines to 8.6 mD. No particle clogging is observed at the inlet of the rock sample, while in some cases a minor amount of sand grains are observed between the sample and the tube.

Effective porosity

Tracer breakthrough   Figures 3 and 4 show the Ba, Li, and K concentrations as evolution over the experiment duration in minutes. For the experiments, a barite saturation index (SI) of 1.5 is chosen. The temperature was set to 60 °C, the pressure was 300 bar, and the flow rate was adjusted to 2 cm^3/min. For the two tracer tests (test A and test B), two different rock samples were used.

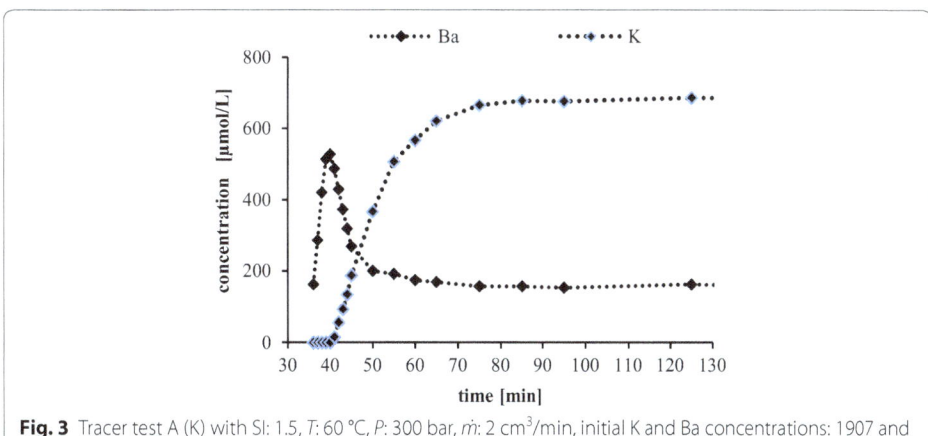

Fig. 3 Tracer test A (K) with SI: 1.5, *T*: 60 °C, *P*: 300 bar, $\dot{m}$: 2 cm^3/min, initial K and Ba concentrations: 1907 and 1054 µmol/L

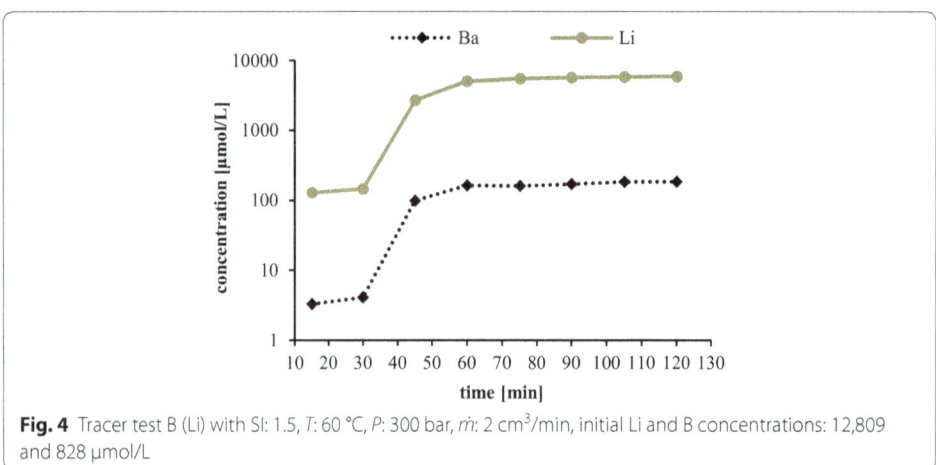

Fig. 4 Tracer test B (Li) with SI: 1.5, *T*: 60 °C, *P*: 300 bar, *ṁ*: 2 cm³/min, initial Li and B concentrations: 12,809 and 828 µmol/L

In test A (Fig. 3, 19 hydrochemical analyses) K was the tracer. The initial concentrations of K and Ba were 1907 and 1054 µmol/L, respectively. According to Fig. 3, the Ba concentration reaches a sharp maximum concentration of 528 µmol/L after 40 min before it approaches an almost stable concentration of 156 µmol/L. The increase in the K concentration starts after 41 min from 15 µmol/L to reach an almost stable value of 653 µmol/L after 65 min. The loss of K is ~ 65% (1907–653 µmol/L).

For test B (Fig. 4, seven hydrochemical analyses), the concentrations are plotted logarithmically to improve clarity. The initial concentrations of Li and Ba are 12,809 and 828 µmol/L, respectively. According to Fig. 4, Li is detected after 15 min with a concentration of 138 µmol/L. Later, the concentration increases to a value of approx. 2680 µmol/L before it reaches an almost stable value of 5688 µmol/L after 75 min. After a test duration of 15 min, Ba is detected with a concentration of 3 µmol/L. Within 30 min the concentration of Ba increases to a value of 99 µmol/L and after a test duration of 90 min an almost stable concentration 180 µmol/L is reached. The overall loss of Li is ~ 56% (12,809–5688 µmol/L).

The effective porosity is calculated with the following input parameters: $Q = 2$ cm³/min $t_{0.5}$ (K) = 365 min, $t_{0.5}$ (Li) = 127 min, and the volume for the samples is 25.54 cm³. For the Tracer test A (with K) the effective porosity of the rock sample n_{eff} is around 28.6, while for the second Tracer test B (with Li) n_{eff} is around 10.

Fluid–Ba concentration

In Table 1, selected experimental runs are listed with their sample names and the respective temperature, pressure, and flow conditions. For all experiments, the saturation index of barite was calculated with PHREEQC (Parkhurst and Appelo 2013) and had a value of 1.75.

Prior to an experiment, the pipes and the rock samples were percolated with distilled water. With the start of the experiment, the distilled water in the two reservoir tanks was replaced by the artificial fluid.

In Figs. 5 and 6, the dissolved Ba concentrations in µmol/L (logarithmic) are plotted over the duration of the experiment in minutes. All curves are characterized by an initial increase of the Ba concentration. This feature can be interpreted as a result of an initial

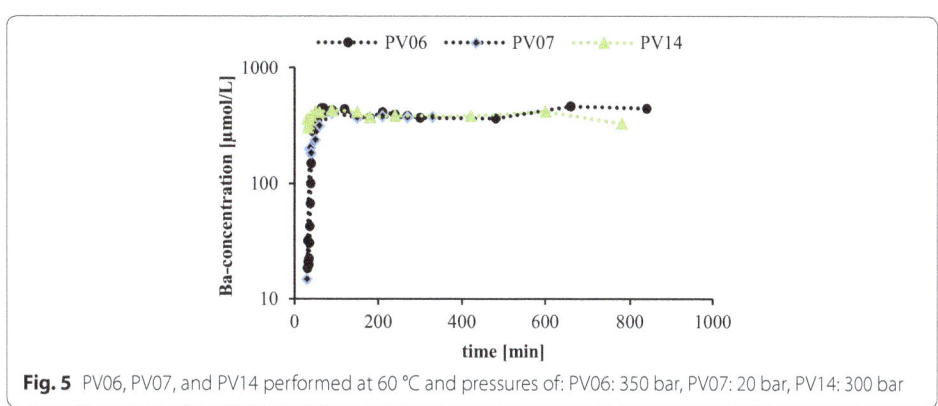

Fig. 5 PV06, PV07, and PV14 performed at 60 °C and pressures of: PV06: 350 bar, PV07: 20 bar, PV14: 300 bar

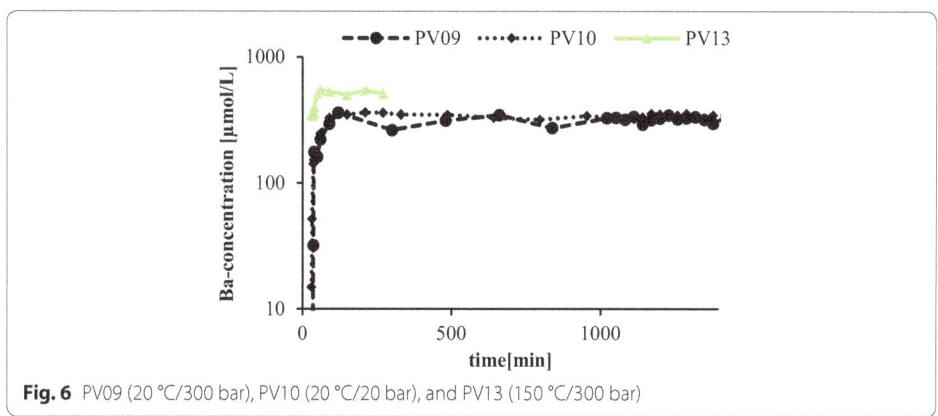

Fig. 6 PV09 (20 °C/300 bar), PV10 (20 °C/20 bar), and PV13 (150 °C/300 bar)

dilution by the residual distilled water in the sample before the whole system is flooded by the undiluted, Ba-containing artificial fluid. All Ba concentrations increase up to a maximum value and remain more or less constant afterward.

Figure 5 shows the Ba concentration curves for the experimental runs PV06 (60 °C/350 bar), PV07 (60 °C/20 bar), and PV14 (60 °C/300 bar). The initial Ba concentration for the three runs is 1487 μmol/L. The Ba concentration in PV07 shows a distinct peak of 406 μmol/L after 90 min before it approaches an average concentration of 375 μmol/L, resulting in a total Ba concentration loss of approx. 75%. PV14 also shows a distinct peak in the Ba concentration after 90 min with a value of 431 μmol/L, before it decreases to an average concentration of 378 μmol/L (total Ba concentration loss of approx. 74%). In PV06, the distinct peak with a Ba concentration of 445 μmol/L is observed already after 70 min. Later the curve approaches an average concentration of 381 μmol/L resulting in a total Ba concentration loss of approx. 74%.

Figure 6 illustrates the evolution of Ba concentrations for the experimental runs PV09 (20 °C/300 bar), PV10 (20 °C/20 bar), and PV13 (150 °C/300 bar) using a logarithmic scale. The initial Ba concentration in PV09 is 763 μmol/L. A distinct Ba peak with a maximum value of 361 μmol/L is observed after 120 min. Later an average concentration of 315 μmol/L is approached, implying a total Ba concentration loss of approx. 58%. PV10 (Fig. 6) has an initial Ba concentration of 775 μmol/L and a distinct peak with a Ba concentration of 362 μmol/L that develops after 210 min. Afterward, the Ba concentration decreases to an almost stable value of 340 µmol/L, resulting in a total Ba concentration loss of approx. 56%. In PV13, the initial Ba concentration is 2613 µmol/L which drops very fast to a value of approx. 360 µmol/L. Then the concentration increases up to a maximum value of 547 µmol/L before it decreases again to 503 µmol/L after 150 min. The final resilient concentration value of this experimental run is 547 µmol/L Ba after a test duration of 210 min. The total concentration loss of Ba is approx. 78%.

Mineralogical composition

Initial rock

The used sandstone of the stratigraphic unit Bausandstein–Geröllsandstein (su–sm) contains high amounts of SiO_2, while Al_2O_3 and K_2O are present in a low percentage range, mainly resulting from minor feldspar content (Table 2). The values of all other major elements are very low. In addition, the chemical composition of the sandstone given by Hirsch (2008) is given as a reference. A comparison of the values shows that the composition of the sandstone has a poor variance. Therefore the rock samples can be considered as representative for the stratigraphic unit.

The sandstone is red colored and medium-grained. It is quartz-rich with a clay matrix. The quartz and K-feldspar grains are rounded to subrounded (Fig. 7). Locally, clay enrichments (called Tongallen) are observed but test pieces with these textures were avoided as also pieces with prominent layering. Some clay minerals are found as space filling (Fig. 8). The mica content of less than 1 vol % is very low, the content of hematite/Fe-hydroxide is ≤ 1 vol % and the modal content of matrix reaches values of up to 5 vol %.

Table 2 Major element composition of the initial sandstone compared to the reference values of Hirsch (2008)

Mass %	SiO_2	Al_2O_3	K_2O	TiO_2	Fe_2O_3	Na_2O	MgO	CaO
IAM-AWP (2016)[a]	93.70	3.55	2.26	0.33	0.28	0.10	0.07	0.07
Hirsch (2008)	94.54	2.89	1.93	0.03	0.25	0.09	0.04	0.06

[a] Performed at 60 °C and 300 bar (IAM-AWP—Institute for Applied Materials, 2016)

Fig. 7 Texture of the original sandstone of the stratigraphic unit Bausandstein–Geröllsandstein (BSE image, SEM; LEM)

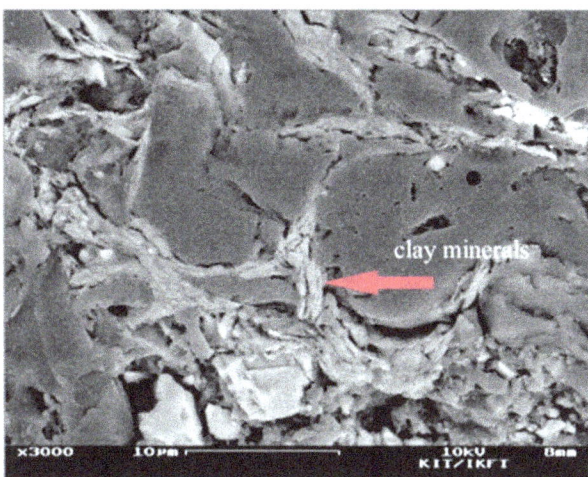

Fig. 8 Local pore space filling by clay minerals of the original sandstone (SE image, SEM; IKFT)

Similar results can also be found by CT scans (Fig. 9). The images are segmented into binary images of pores (blue) and grains (red) by applying a watershed algorithm on the unfiltered gray value images (Beucher and Lantuéjoul 1979). Following this procedure, the pore space of the reference sample is 5.6 vol % and the amount of quartz and feldspar is calculated to approx. 94 vol %.

Hirsch (2008), among others, analyzed the same rock during an investigation of the different sandstones used to build up the Freiburger Münster. For this purpose she examined the rocks regarding their mineralogical composition, density, porosity, water absorption coefficient, and the origin of the sandstone. She examined similar mineralogical and geochemical compositions as well as values for porosity like those observed in the samples of this study.

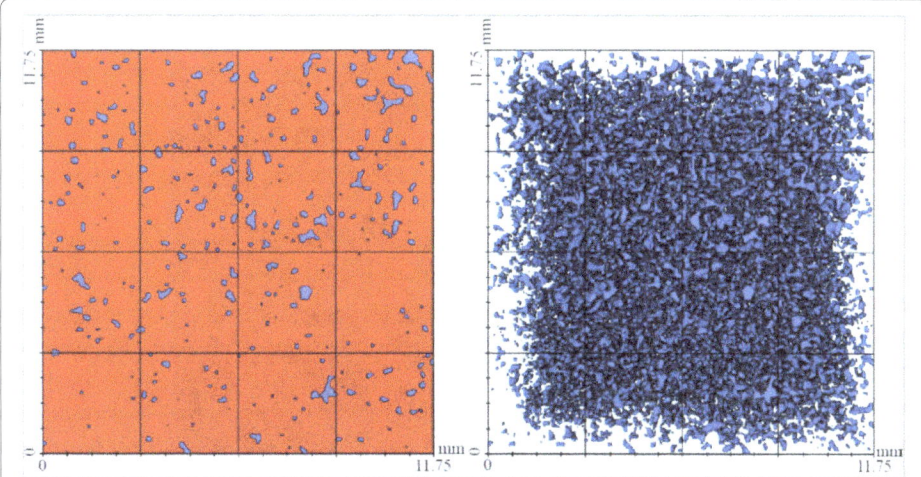

Fig. 9 CT scan of the rock sample: pore spaces are blue while quartz and feldspar are colored red. Left: the grains are visible as red area with blue spots for the pores. Right: visualization of the pores, the matrix is omitted in this representation

Flow-through rock

In situ precipitations of barite crystals   Barite precipitations are provoked by using a barite-supersaturated solution. After several pre-experiments, the barite crystallization was successfully provoked in the pore spaces of the sandstones and not in the pipe or at the inlet area. The barite crystals are visible as white crystals in BSE images of the SEM (Figs. 10, 11, 12, 13, 14, 15, 16). Supersaturation of more than 1.75 led to precipitations of barite in the pipes and at the inlet of the sample while supersaturation from 1 to 1.75 led to poor amounts of barite precipitation, which were not detected by the analytical methods used. Hence a supersaturation of 1.75 was chosen and worked reliably in all experiments.

Detailed investigation with SEM shows that barite occurs in only some of the pore spaces of an individual rock samples, whereas others remain almost unchanged. The growth features observed for barite imply a crystallization sequence in the experiments.

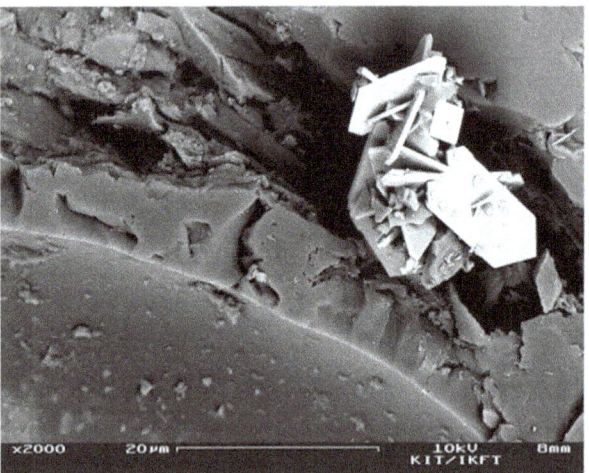

Fig. 10 Growth of euhedral, tabular barite crystals sample 12.1 (PV10; 20 °C/20 bar)

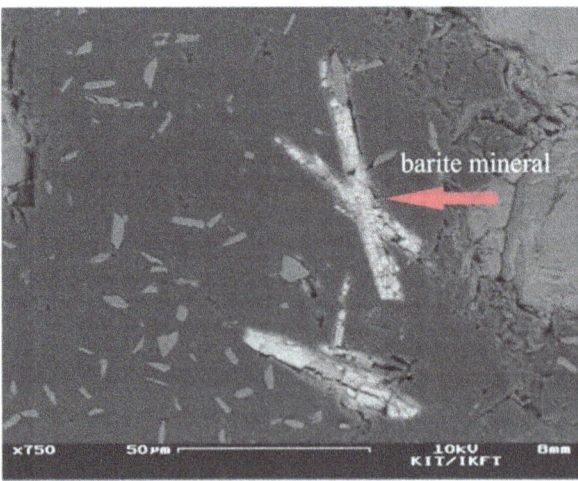

Fig. 11 Intergrowth of barite, sample 7.2 (PV07; 60 °C/20 bar)

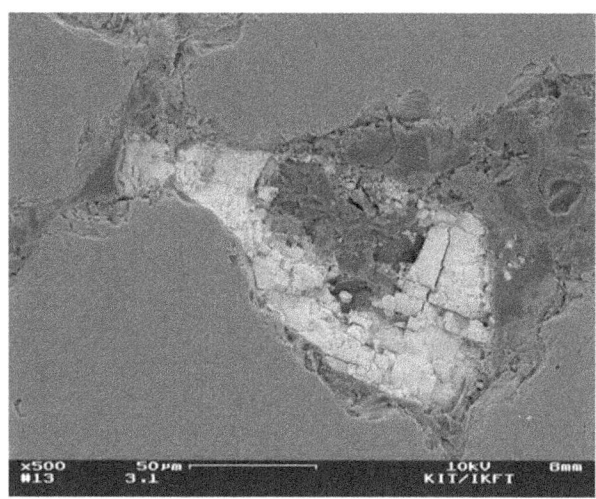

Fig. 12 overgrowth of illite by barite sample 3.1

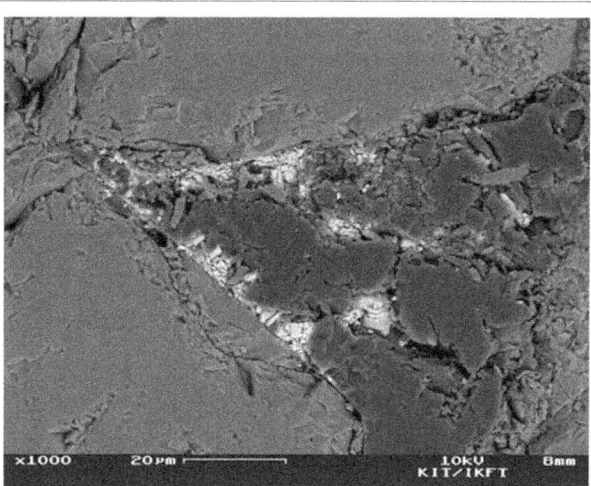

Fig. 13 Preferential accumulation of the barite crystals in the tips of the pore spaces, sample 14.2 (180 °C/350 bar, the experiment is not part of this study, since HydRA stopped the experiment automatically after 9 h)

In the beginning, the growth of the barite crystals is tabular (Fig. 10) with a size of approx. 15 μm in the larger pore spaces. Once a barite crystal is formed, it acts as a seed and is overgrown by further barite precipitations forming a crystal agglomerate (Fig. 11). The crystal shape of the agglomerate is mostly platy with a well-defined cleavage. Crystal agglomerates can be found in a part of the pore spaces of all sandstone samples after experiment, which were either unfilled before the experiment or partially filled with clay minerals and hematite. In the latter case, the barite overgrows the illite along the margins of the pore (Fig. 12). Quartz grains have smooth surfaces with small holes and often show dissolution edges (Fig. 18), whereas such features are absent in the feldspar. Especially in the smaller pores and cracks, subhedral baryte crystals grow together forming crystal aggregates. With increasing growth or narrowing of the pore spaces, the crystals

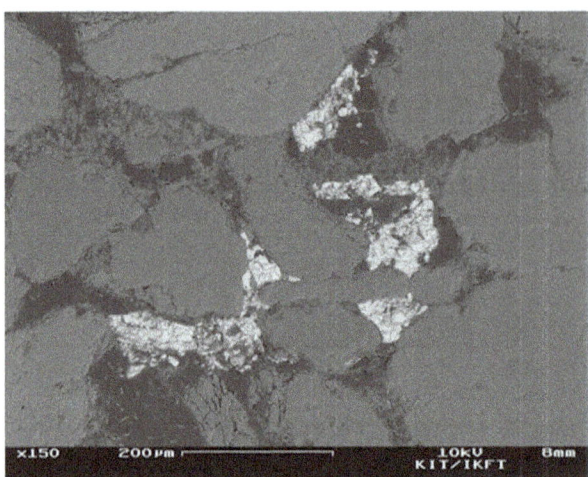

Fig. 14 Growth of barite crystals in some of the pore spaces, sample 7.2 (PV07; 60 °C/20 bar)

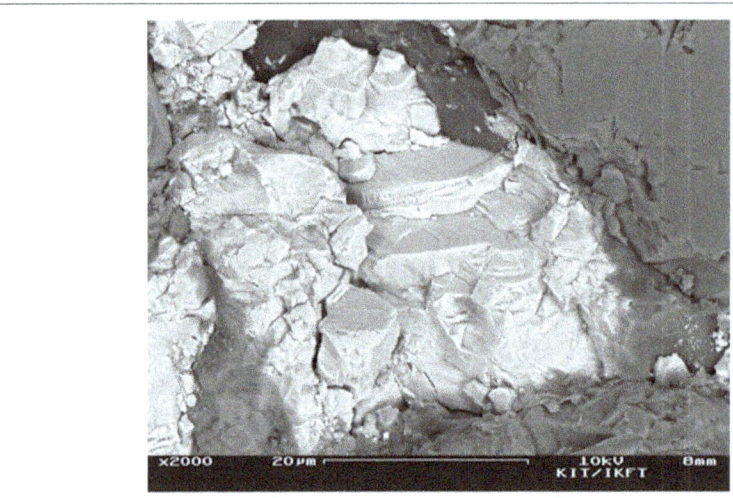

Fig. 15 Crystal agglomeration of barite, sample 7.2 (PV07; 60 °C/20 bar)

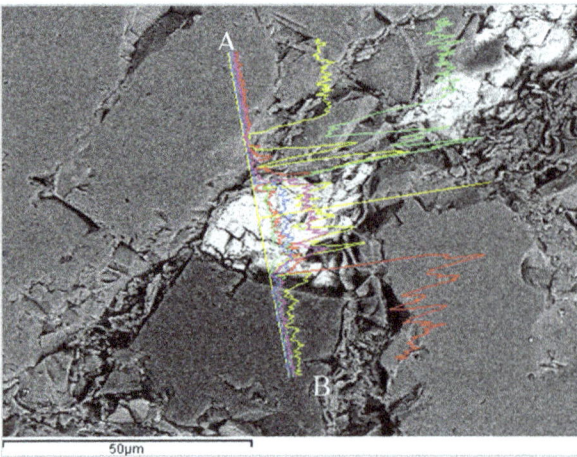

Fig. 16 Position and results (for details see Fig. 17) of a line scan over a barite crystal to analyze its chemical composition

preferentially accumulate at the bottlenecks of the pore spaces (Fig. 13). This can lead to a clogging of the whole interspaces. Finally, the pore spaces and cracks get completely filled with barite crystals (Figs. 14 and 15).

Most of the crystals are weakly fractured. This fragmentation is probably caused by the permeability measurement after the experimental runs and the subsequent sample preparation steps, i.e., sawing and preparation of the thin sections.

The line scan (Figs. 16 and 17) shows the intensities of the characteristic x-ray signals of Si (Kα), O (Kα), C (Kα), S (Kα), and Ba (Lα) over the distance from the starting point of the scan.

Precipitations of secondary mineral phases   During investigation with SEM, some spherical structures were observed, that only occur in some of the reacted samples (Figs. 18 and 19) as small spheres with diameters of < 1 μm. They grew on the surfaces of sandstone minerals, also including the newly formed barite in the pore spaces. Accordingly, the appearance of these phases is not related to replacement of a special mineral phase, or a specific experimental condition.

Energy dispersive XRF measurements with the SEM reveal that these phases consist of O (53.27 wt %), Si (30.12 wt %), and C (15.57 wt %). The residue to 100% can be attributed to the sputtering materials Au and Pd.

Trace elements

The geochemistry of the rock samples was further analyzed with regard to trace element composition after full acid digestion. The geochemistry of the sandstone was characterized before and after the experiments and thus allows the investigation of chemical changes during the experiments due to water–rock interaction processes.

After the experiments, the samples 7.2, 8.2, 15.1, and 16.1 (see Table 1 for experimental conditions) show a decrease in the concentrations of Al, K, Na, Ti, Rb, Sn, and Tl, when compared to the corresponding initial samples. Apart from Na, these elements

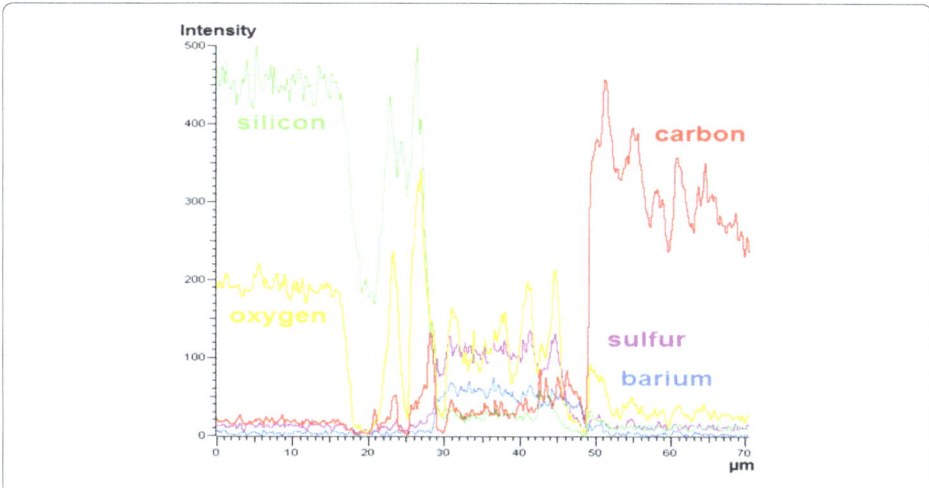

Fig. 17 Result of the line scan with concentration curves of the measured elements. For position of the line, see Fig. 16. The colored lines show the intensity of the elements (yellow-oxygen, green-silicon, red-carbon, purple-sulfur, blue-barium) over the length of the line

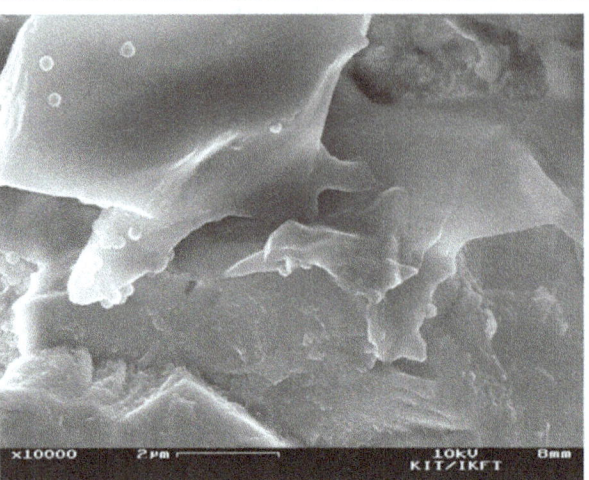

Fig. 18 SEM (SE image) of barite covered with spherical phases, sample 10.1 (20 °C/200 bar)

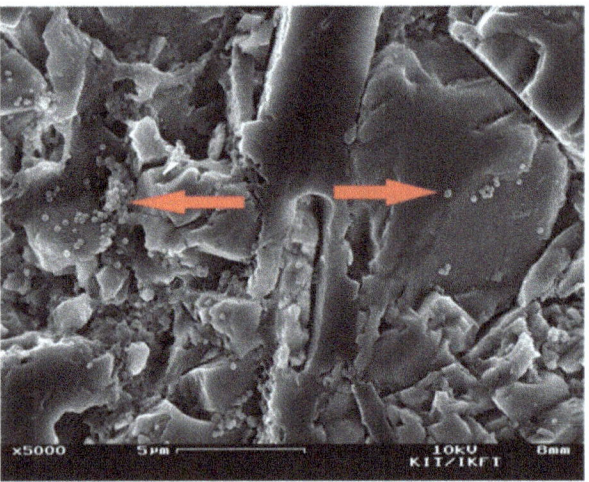

Fig. 19 SEM (SE image) of barite with numerous spherical phases formed on its crystal phases, sample 7.2 (PV07; 60 °C/20 bar)

were not present in the initial artificial fluid, but they were found in the fluid after percolating the sandstone. Consequently, they were released from the sandstone due to fluid–rock interactions. In all rock samples, higher Ni and Ba concentrations can be detected after the experiments. The additional Ni likely originates from dissolution and transport of the pipe material to the sandstone, while the increase in the Ba content is due to the barite precipitation.

In Figs. 20 and 21 the results of the analyses of the rock samples 7 and 7.2 are compared.

Decreasing concentrations are observed for the elements Fe, Ca, P, Li, Cr, Co, Cs, Cu, U, Sb, and Mo. The most significant decreases were detected for Cr (5.39–4.12 mg/kg, decrease of approx. 23%), Co (1.36–0.43 mg/kg, approx. 68%), Cs (3.33–2.53 mg/kg, 24%), and Mo (0.10–0.07 mg/kg, 32%).

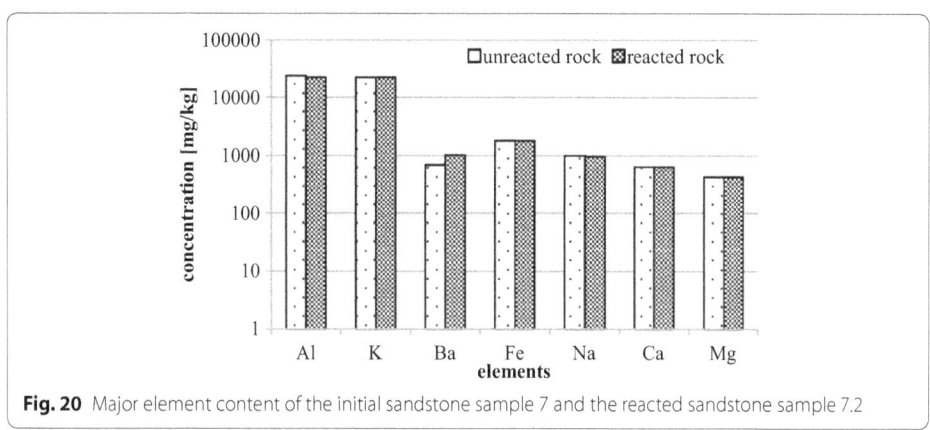

Fig. 20 Major element content of the initial sandstone sample 7 and the reacted sandstone sample 7.2

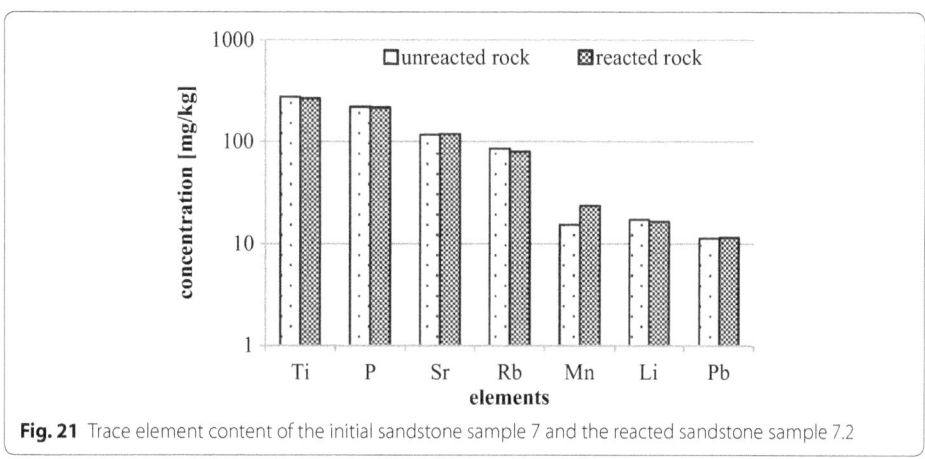

Fig. 21 Trace element content of the initial sandstone sample 7 and the reacted sandstone sample 7.2

Increasing concentrations are measured for Ba, Mg, Sr, Mn, Pb, Y, V, Zn, Ni, As, and Cd with most significant increases for Ba (692–1024 mg/kg, approx. 32%), Mn (15.45–23.6 mg/kg, approx. 34%), Zn (3.61 up to 7.23 mg/kg, 50%), and Ni (2.95–46.5 mg/kg, approx. 94%).

Since the initial and reacted sandstone samples for geochemical analysis were taken from different regions of the cylindrical sample, differences in the composition are possible that this is not related to the experimental procedure and originate from natural heterogeneities (Additional file 1).

Discussion
Effect of temperature
Fresh test solution is supplied with a constant flow rate during the complete experiment duration. Therefore the reaction time remains the same resulting in a constant concentration and thus a constant precipitation of barite. According to the experimental results, the final Ba concentration in the fluid strongly depends on the temperature while pressure changes seem to be of minor importance. Ba losses of 56–63%, 74–75%, and 78% were found at temperatures of 20, 60, and 150 °C, respectively, indicating progressive Ba precipitation in the porous sandstone with increasing temperature.

However, it has to be taken into account, to get an SI of barite of 1.75, the input concentration of the $NaSO_4$ and $BaCl_2$ is much higher at 150 °C compared to the runs at 20 and 60 °C. This is reasoned by the solubility of $BaSO_4$ because this undergoes a maximum close to 150 °C (Blount 1977).

Some elements like K, Na, Rb, Al, P, or Sr show a higher concentration decrease in low temperature experiments (20 °C) compared to higher temperatures (60, 150 °C) indicating an increased dissolution of primary sandstone minerals at lower temperatures. This does not go in line with the known low solubility and general solubility with increasing temperature for most minerals like K-feldspar, mica, clay minerals, or apatite, that are present in sandstone. Rather it can be expected that dissolution and re-precipitation are dominant processes in the high-tempered experiments compared to washout effects of pre-experimental weathering at low temperatures. This fact is shown by Schmidt et al. (2017). He observed dissolution of quartz, illite, kaolinite, and feldspar and precipitation of analcime, chlorite, and albite at high temperatures (200 and 260 °C, experiment duration 45–55 days).

The elements Mn, P, Cr, Ni, Mo, and Ti, which show a concentration increase in the sandstones after the experiments, might originate from the pipe material of the apparatus.

Change of hydraulic parameters

Effective porosity   In the tracer tests, more than half of the input quantity of the used tracer cations K and Li was lost during the percolation of the rock sample (Li: 65%, K: 58%). The shape of the K concentration curves, which was assumed to show a conservative behavior (Fig. 3), leads to the assumption that K accumulates in the rock sample through adsorption on negatively charged surface sites of clay minerals or Fe oxyhydroxides, while Ba passes the sample more or less unreacted. Thus a high peak can be observed at the beginning of the Ba concentration curve that rapidly decreases afterwards. In the test with Li (Fig. 4), the Ba concentration behaves completely different. In this experiment, the concentration curves of Ba and Li have the same shape and proceed almost parallel. Both curves increase until they reach an almost stable concentration after approx. 60 min.

The results of both tracer tests lead to the assumption that the used tracer cations K and Li are not suitable for the determination of the effective porosity of the present rock sample. The ions do not behave conservatively and therefore the calculations using formula (4) produce inaccurate results. The calculated porosity value for the test with Li is 10, while it is 28.6 for the test with K. This large difference in the porosity values is a further hint on the infeasibility.

Permeability changes   All experimental runs show a decrease of the permeability of the reacted sandstones (Fig. 2) that is in accordance with the observed Ba precipitation using SEM analysis. The experimental runs at a temperature of 20 °C show a decrease of the permeability of approx. 68% for PV09 and approx. 75% for PV10. At a temperature of 60 °C the experimental runs PV 07, PV14, and PV 06 show higher decreases of 95, 90, and 80%, respectively. But in test PV13 at the highest used temperature investigated (150 °C), the decrease of the permeability is only 6%. This contradictory effect—lowest permeability at highest temperature with largest loss in Ba concentration—may be due to the solubility maximum of barite around 150 °C.

The determination of the barite-precipitation mass was not feasible, because during the removal of the sample (out of the silicon tube) some rock material got lost and falsified the values. Additionally, the mass of the precipitated barite was extremely low and it was not possible to quantify it exactly. Therefore the relation between the permeability loss and the mass increase of barite cannot be given here.

Reaction sequence

Formation of barite

Due to the constant supply of barite-supersaturated solution, barite can grow continuously once precipitated, eventually until the pore spaces are completely filled. Barite is found as a part of the pore spaces of all the sandstone samples after the experiment. Smaller pores and cracks are completely filled with barite after experiments, whereas larger pores preserve open spaces. In larger pores, homogeneous nucleation and growth of barite is observed initially. As a result, large, subhedral to euhedral barite plates were formed in the beginning, which apparently grew inside the pores, away from a mineral surface. Subsequently grown barite preferentially overgrows already existing barite grains, indicating a change of the nucleation style from homogeneous to heterogeneous, i.e., on pre-existing crystal faces. In the smaller pores, heterogeneous nucleation and growth dominate. In this case, already the initial nucleation of barite preferentially occurs on the mineral faces surrounding the pores and these primary barite grains are then overgrown by the continuously precipitating barite during the experiment. Nucleation and growth mainly occurs in the tips of the pores and small cracks, finally leading to clogging of the open spaces. As a result the permeability of the sandstone decreases in most experiments.

Formation of secondary mineral phases

The observed microspheres are no residues of the abrasive, which was used to produce the thin sections. They are always well rounded and smaller than 1 μm in diameter; larger spheres are not found. Their morphology and O–Si–C-rich composition resembles that of polycarbosilane phases (Chen et al. 2011). Their small grain size and their formation on the crystal faces of the sandstone phases, also including the newly formed barite suggest that they form late, possibly during cooling of the sample. Their occurrence implies elevated Si concentrations of the solution after reaction and may be related to the observed dissolution of quartz during the experiment (Fig. 18).

Implications for geothermal operation

The most relevant information for geothermal energy usage that can be concluded from the experimental runs of this study is the drastic change of the permeability in the reservoir rock. This observation is definitely decisive for the profitability of a geothermal power plant. If the permeability of the reservoir rock is high, high flow rates can be expected. At the discharge wells, which are considered in this study, the capacity of the reservoir rocks is of major relevance for the return of the used geothermal fluid to the underground. If the permeability is decreasing with time, further investments for cost-intensive compensation measures like active pumping (to press the fluid back into the underground) and/or chemical enhancing methods might become necessary.

Conclusions

In the present work, HydRA, an apparatus for performing experiments under geothermally relevant conditions, was developed, built, and put into operation. Using the HydRA facility it was also possible to fulfill the second aim of this work, the experimental investigation of the effects of barite precipitation in reservoir rocks under similar temperature and pressure conditions like those observed in the Upper Rhine Graben.

For this purpose also a method for inducing barite precipitation in the rock sample during percolation of a barite-supersaturated solution was developed and tested. The method, in which two solutions with different compositions are mixed at the entry of the sample, offers the opportunity of studying the influence of in situ precipitation during forced flow-through.

The experimental part of this study focused on possible changes of the permeability of reservoir rocks. A strong decrease of the permeability of the present reservoir rocks due to barite precipitation was observed in most experimental runs and different stages of pore clogging due to the barite precipitation were identified in the samples and visualized via SEM. This is of special importance for the operators of geothermal power plants, since the permeability of reservoir rocks is decisive for the profitability of a geothermal power plant. If the permeability at a geothermal site decreases, cost-intensive compensation measures have to be initiated.

Additionally, mineral dissolution in the sandstone was observed as a result of water–rock interaction during flow-through even in the short experimental residence times. In geothermal power plants, the temperature and pressure reduced fluids (by that the saturation degree of some mineral phases are raised up) would be injected into the reservoir rocks where they could alter the composition of the reservoir rocks drastically.

In forthcoming experiments, the residence time of the fluid in the sample should be elongated. This can be realized by reducing the flow rate. In the future, the HydRA facility can also be used to investigate other rock compositions, according to existing geothermal reservoirs. Moreover, the experimental setup can be adapted to simulate other geochemically relevant scenarios, as for example, a calcite supersaturation of the fluid. Even natural geothermal water can be used for the experiments. So HydRA offers many opportunities for performing experiments under geothermally relevant conditions.

Authors' contributions

PO performed the experiments and was the leader of the working group concerning HydRA. DK has contributed the financial support and the general idea how to realize such a flow-through apparatus. EE conducted the analyzing of the rock samples regarding the trace elements. She gave technical and analytical support to measure as good as possible the rock samples and contributed text. MZ performed all SEM and SEM–EDX measurements and contributed some text blocks. KD had read the manuscript several times and has given some constructive criticism. TK gave the initial idea to do this kind of research. All authors read and approved the final manuscript.

Author details

[1] Institute of Nuclear and Energy Technology, Karlsruhe Institute of Technology, Herrmann-von Helmholtz-Platz 1, 76344 Eggenstein-Leopoldshafen, Germany. [2] RBS wave GmbH, Postfach 311508, 70475 Stuttgart, Germany. [3] Institute of Applied Geoscience-Division of Mineralogy and Petrology, Karlsruhe Institute of Technology, Kaiserstrasse 12, 76131 Karlsruhe, Germany. [4] Institute of Applied Geoscience-Division Geothermics, Karlsruhe Institute of Technology, Kaiserstrasse 12, 76131 Karlsruhe, Germany. [5] Institute of Catalysis Research and Technology, Karlsruhe Institute of Technology, Herrmann-von Helmholtz-Platz 1, 76344 Eggenstein-Leopoldshafen, Germany. [6] Institute of Applied Geoscience-Aquatic Geochemistry Division, Karlsruhe Institute of Technology, Kaiserstrasse 12, 76131 Karlsruhe, Germany.

Acknowledgements

We wish to thank David Hillesheimer and Michael Scholz for technical input and thoughts. A heartfelt thank you goes to Dominik Mayer and Klaus Thomauske for their technical support regarding HydRA. I would also point out the very conscientious work and great support of Sabine Baur. She gave me enormous assistance in the chemical laboratory and had always time for fruitful discussions. I am grateful to Sarah Herfurth and Andreas Friedrich, for reading the manuscript precisely.

For producing the thin sections, special thanks goes to Kristian Nikoloski from the Institute of Applied Geosciences (AGW). Additional ICP-MS analyses were performed at this Institute (Division of Aquatic Geochemistry), KIT), special thanks for that big support.

For the supply regarding the sandstone material, we thank Dr. Steffen Klumbach from the University of Bayreuth. To Mathias Nehler from the International Geothermal Centre of Bochum goes best wishes for contributing the CT scan of the reference rock. A big thank you goes to the Institute for Applied Materials, Department for Applied Materials Physics at the KIT for performing the XRF and the fluid analysis for this research.

Competing interests

The authors declare that they have no competing interests.

References

Banks J, Regensburg S, Milsch H. Experimental method for determining mixed-phase precipitation kinetics from synthetic geothermal brine. Appl Geochem. 2014;47:74–84.

Beucher S, Lantuéjoul C. Use of watersheds in contour detection. Fontainbleau: Centre de Geostatique et de Morphologie Mathematique; 1979.

Blount CW. Barite solubilities and thermodynamic quantities up to 300 °C and 1400 bars. Am Miner. 1977;62:942–57.

Canic T, Baur S, Bergfeldt T, Kuhn D. Influences on the barite precipitation from geothermal brines. In: Proceedings World Geothermal Congress 2015, Australia, Melbourne, 19–25 April 2015. 2015.

Chen Y, Li S, Luo Y, Xu C. Fabrication of polycarbosilane and silicon oxycarbide microspheres with hierarchical morphology. Solid State Sci. 2011;13:1664–7.

Christy AG, Putnis A. The kinetics of barite dissolution and precipitation in water and sodium chloride brines at 44–85 °C. Geochim Cosmochim Acta. 1993;57:2161–8.

Daubner M, Krieger V. Betriebsmessdatenvisualisierung und -erfassung mit OPAL (OPC-Panel Livegraph), KIT Scientific Reports 7558, Karlsruhe. 2010.

DIN 18130-1. 05 Bestimmung des Wasserdurchlässigkeitswertes Teil 1 Laborversuche, Berlin. 1998.

DIN EN 10088-3. Stainless steels—part 3: technical delivery conditions for semi-finished products, bars, rods, wire, sections and bright products of corrosion resisting steels for general purposes. Berlin. 2014.

Feldkamp LA, Davis LC, Kress JW. Practical cone-beam algorithm. J Opt Soc Am. 1984;1(6):612–61. doi:10.1364/JOSAA.1.000612.

Fraunhofer-Allianz Vision. Volex. Volume explorer software. Version 6.3., Fürth. 2012.

Genter A, Evans K, Cuenot N, Fritsch D, Sanjuan B. Contribution of the exploration of deep crystalline fractured reservoir of Soultz to the knowledge of enhanced geothermal systems (EGS). C R Geosci. 2010;342:502–16.

Geyer FO, Gwinner MP, Geyer M, Nitsch E, Simon T. Geologie von Baden-Württemberg, 5. Auflage, 627 Seiten, 185 Abbildungen, 4 Tabellen. ISBN: 978-3-510-65267-9. Schweizerbart. 2011.

Govindaraju K. Compilation of working values and sample description for 383 geostandards. Geostand Newsl. 1994;18:1–158.

Herfurth S, Orywall P. Annual report 2015 of the Institute of Nuclear and Energy Technology. Karlsruhe: Scientific Publishing KIT; 2015. doi:10.5445/KSP/1000057984.

Herzberger P, Münch W, Kölbel T, Bruchmann U, Schlagermann P, Hötzl H, Wolf L, Rettenmaier D, Steger H, Zorn R, Seibt P, Möllmann GU, Sauter M, Ghergut J, Ptak T. The geothermal power plant Bruchsal. In: Proceedings world geothermal congress 2010, Indonesia, Bali. 25–29 April 2010. 2010.

Hirsch A. Gesteinsuntersuchungen und Bausteinkartierung am Turmhelm des Freiburger Münsters. Diploma Thesis. Albert-Ludwigs-University Freiburg. 2008.

Kaufmann-Knoke R. Zur Problematik von Mineralausfällungen insbesondere von (Ba,Sr) SO4-Mischkristallen bei der Erdölförderung, Berichte Geolog.-Paläont. Institut, Christian-Albrechts-Universität Kiel. 1992.

Konrad-Zuse Zentrum für Informationstechnik Berlin (ZIB). Avizo 9, Avizo users guide, FEI. 2016. p. 850.

Klotz D, Lang H, Moser H, Behrens H. Ausbreitung von Radionukliden der Elemente I, Sr, Cs und Ce in oberflächennahen Lockergesteinen, GSF Bericht R 289, Gesellschaft für Strahlen- und Umweltforschung mbH, 84 Seiten, München. 1982.

Kühn M, Vernoux J-F, Kellner T, Isenbeck-Schröter M, Schulz HD. Onsite experimental simulation of brine injection into a clastic reservoir applied to geothermal exploitation in Germany. Appl Geochem. 1998;13(4):477–90.

Merkel BJ, Planer-Friedrich B. Grundwasserchemie: Praxisorientierter Leitfaden zur numerischen Modellierung von Beschaffenheit, Kontamination und Sanierung aquatischer Systeme. Berlin: Springer Verlag; 2008.

Ochi J, Vernoux J-F. Permeability decrease in sandstone reservoirs by fluid injection hydrodynamic and chemical effects. J Hydrogeol. 1998;208:237–48.

Parkhurst DL, Appelo CAJ. Description of input and examples for PHREEQC version 3: a computer program for speciation, batch-reaction, one-dimensional transport, and inverse geochemical calculations. U.S. geological survey techniques and Methods, book 6, chapter A 43: Denver. 2013.

Pauwels H, Fouillac C, Fouillac A-M. Chemistry and isotopes of deep geothermal saline fluids in the Upper Rhine Graben: origin of compounds and water-rock interactions. Geochim Cosmochim Acta. 1993;57:2737–49.

Rosenbrand E, Kjoller C, Riis JF, Kets F, Fabricius IL. Different effects of temperature and salinity on permeability reduction by fines migration in Berea sandstone. Geothermics. 2015;53:225–35.

Rosenbrand E, Haugwitz C, Munch Jacobsen PS, Kjoller C, Fabricius IL. The effect of hot water injection on sandstone permeability. Geothermics. 2014;50:155–66.

Ruiz-Agudo C, Putnis CV, Ruiz-Agudo E, Putnis A. The influence of pH on barite nucleation and growth. Chem Geol. 2015;391:7–18.

Sanjuan B, Millot R, Dezayes C, Brach M. Main characteristics of the deep geothermal brine (5km) at Soultz-sous-Forêts (France) determined using geochemical and tracer data. C R Geosci. 2010;342:546–59.

Savage D, Bateman K, Richards HG. Granite-water interactions in a flow-through experimental system with applications to the hot dry rock geothermal system at Rosemanowes, Cornwall, U.K. Appl Geochem. 1992;7:223–41.

Scheiber J, Seibt A, Birner J, Cuenot N, Genter A, Moeckes W. Barite scale control at the Soultz-sous-Forêts (France) EGS site. In: Proceedings, 38 workshop on geothermal reservoir engineering, Stanford University, Stanford, California. 24–26 February 2014. 2014.

Schmidt RB, Bucher K, Drüppel K, Stober I. Experimental interaction of hydrothermal Na-Cl solution with fracture surfaces of geothermal reservoir sandstone of the Upper Rhine Graben. Appl Geochem. 2017;81:36–52.

Soni JP, Islam N, Basak P. An experimental evaluation of non-Darcian flow in porous media. J Hydrogeol. 1978;38:231–41.

Stober I, Bucher K. Geothermie. Heidelberg: Springer; 2012.

Tutolo BM, Luhmann AJ, Kong X, Saar MO, Seyfried WE. CO2 sequestration in feldspar-rich sandstone: coupled evolution of fluid chemistry, mineral reaction rates, and hydrogeochemical properties. Geochim Cosmochim Acta. 2015;160:132–54.

9

Matrix permeability of reservoir rocks, Ngatamariki geothermal field, Taupo Volcanic Zone, New Zealand

J. L. Cant, P. A. Siratovich, J. W. Cole, M. C. Villeneuve[*] and B. M. Kennedy

*Correspondence: marlene.villeneuve@canterbury.ac.nz
Department of Geological Sciences, University of Canterbury, Private Bag 4800, Christchurch 8140, New Zealand

Abstract

The Taupo Volcanic Zone (TVZ) hosts 23 geothermal fields, seven of which are currently utilised for power generation. Ngatamariki geothermal field (NGF) is one of the latest geothermal power generation developments in New Zealand (commissioned in 2013), located approximately 15 km north of Taupo. Samples of reservoir rocks were taken from the Tahorakuri Formation and Ngatamariki Intrusive Complex, from five wells at the NGF at depths ranging from 1354 to 3284 m. The samples were categorised according to whether their microstructure was pore or microfracture dominated. Image analysis of thin sections impregnated with an epoxy fluorescent dye was used to characterise and quantify the porosity structures and their physical properties were measured in the laboratory. Our results show that the physical properties of the samples correspond to the relative dominance of microfractures compared to pores. Microfracture-dominated samples have low connected porosity and permeability, and the permeability decreases sharply in response to increasing confining pressure. The pore-dominated samples have high connected porosity and permeability, and lower permeability decrease in response to increasing confining pressure. Samples with both microfractures and pores have a wide range of porosity and relatively high permeability that is moderately sensitive to confining pressure. A general trend of decreasing connected porosity and permeability associated with increasing dry bulk density and sonic velocity occurs with depth; however, variations in these parameters are more closely related to changes in lithology and processes such as dissolution and secondary veining and re-crystallisation. This study provides the first broad matrix permeability characterisation of rocks from depth at Ngatamariki, providing inputs for modelling of the geothermal system. We conclude that the complex response of permeability to confining pressure is in part due to the intricate dissolution, veining, and recrystallization textures of many of these rocks that lead to a wide variety of pore shapes and sizes. While the laboratory results are relevant only to similar rocks in the Taupo Volcanic Zone, the relationships they highlight are applicable to other geothermal fields, as well as rock mechanic applications to, for example, aspects of volcanology, landslide stabilisation, mining, and tunnelling at depth.

Keywords: Pores, Volcaniclastic, Confining pressure, Microfractures, Connected porosity

Background

New Zealand relies on geothermal energy to generate approximately 16.5% of its electricity (MBIE 2017). The generation portfolio was increased in 2013 with the addition of 82 MWe generation capacity at the Ngatamariki geothermal field (NGF) (Fig. 1). Understanding the nature and behaviour of the geothermal reservoir at Ngatamariki is of upmost importance for the efficiency and longevity of the geothermal resource. Two key properties that control reservoir behaviour are porosity and permeability of the reservoir rocks (Jafari and Babadagli 2011). Porosity is the measure of void volume (empty space) and permeability indicates how easily a fluid can pass through a medium (Guéguen and Palciauskas 1994). Because porosity does not indicate the shape, size, and distribution of the voids, it provides limited information about the ability for a fluid to flow through the rock. Porosity includes both connected porosity and unconnected porosity. Unconnected porosity refers to void spaces that are not interconnected with the rest of the void network and, therefore, cannot be accessed by fluids. Connected (or effective) porosity refers to void spaces that are interconnected and can, therefore, contribute to permeability. This study focuses only on connected porosity. Many studies have described the control of the primary rock textures on permeability, with large differences between intrusive, volcanic, and sedimentary rocks (e.g., Géraud 1994; Heap et al. 2015, 2017b; Ruddy et al. 1989; Farrell and Healy 2017). Alteration, dissolution, and mineralisation associated with hydrothermal fluids also affect connected porosity (e.g., Wyering et al. 2014). Permeability as defined by Henry Darcy in the mid-1800s applies to non-turbulent (Darcian) flow (Glassley 2010). It is scale dependent with distinct differences between macro- (metre scale fractures) and micro (matrix)-scale permeability

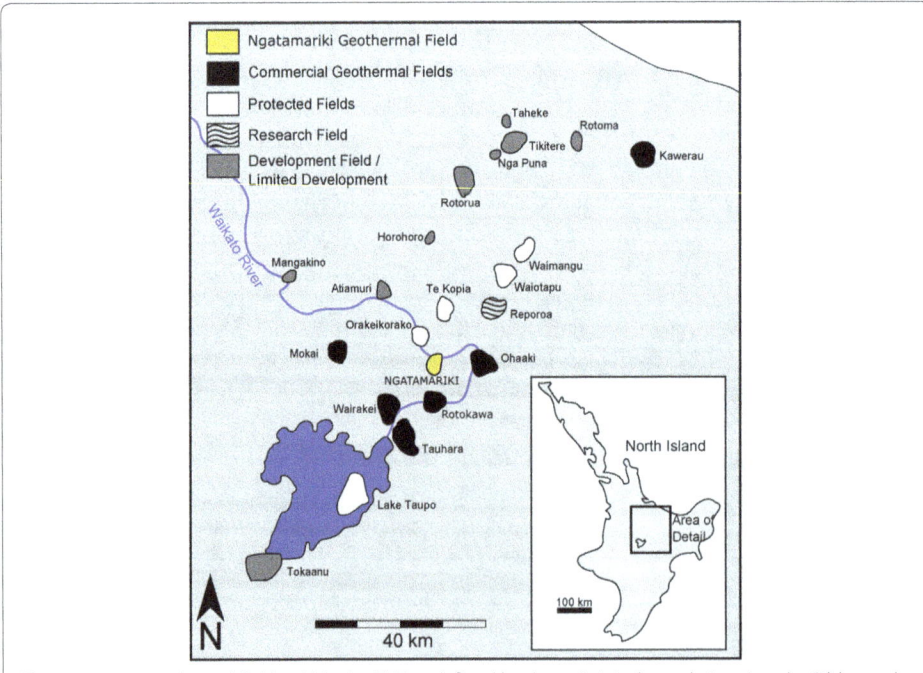

Fig. 1 Known geothermal fields within the TVZ as defined by the resistivity boundaries given by Bibby et al. (1995) (Modified from Boseley et al. 2012)

(Heap and Kennedy 2016) and can be partially attributed to the random distribution of void structures throughout a rock mass (Glassley 2010).

A common approach to modelling a geothermal system is to assume dual porosity/permeability, where two interactive continua, macro-fracture, and matrix permeability are assumed to have their own unique properties (Jafari and Babadagli 2011). Natural macro-fractures within a geothermal system, resulting from unconformities, cooling joints, and tectonic stress discontinuities, strongly control fluid flow due to their high permeability (Murphy et al. 2004), and generally control the permeability in geothermal systems (Jafari and Babadagli 2011). Testing of macro-fracture permeability is usually done in situ with injection flow rate tests used to identify areas of high permeability associated with fractured zones (Watson 2013). In this study, we focus on sample scale (i.e., matrix) properties of recovered drill core. This provides the second component of rock mass permeability, which, when combined with in-situ testing, forms the basis for permeability inputs for reservoir modelling.

In this paper, we present laboratory tests of rocks from the Tahorakuri Formation and Ngatamariki Intrusive Complex (NIC), from five wells at depths of 1354–3284 m. We made measurements of porosity, dry bulk density, ultrasonic velocity (saturated and dry), and permeability (at a range of confining pressures). We also present an assessment of the sensitivity of the permeability of the tested lithologies to confining stress. In particular, we focus on the type and shape of the pores and microfractures (Siratovich 2014; Lamur et al. 2017) and how they affect the fluid flow properties of rocks from NGF. Finally, we explore the relationships between burial diagenesis, hydrothermal alteration, and physical properties. An understanding of the reservoir rock's physical properties can help with field exploration and operation, as well as provide guidance to numerical models that can guide future field optimisation.

Geological setting

The Taupo Volcanic Zone (TVZ) is a rifted arc in the centre of the North Island of New Zealand, related to the subduction of the Pacific plate below the Australian plate at the Hikurangi margin. The geology, volcanology, and structure of the TVZ have been thoroughly described elsewhere by authors such as Wilson et al. (1995) and more recently with a geothermal perspective by authors such as Wilson and Rowland (2016). NGF is situated in the central part of the TVZ (Fig. 1) and is operated as a geothermal power generation site by Mercury NZ Limited. Twelve deep boreholes have been drilled between 1980 and 2016 with the most recent (NM12) in 2014. There are currently four production and five injection wells along with 34 monitoring wells, used to observe shallow, intermediate, and deep aquifer fluid and pressure trends. A small number of these wells also provide monitoring of the separation between NGF and the nearby protected Orakei-Korako geothermal area.

The subsurface stratigraphy encountered at NGF (Fig. 2; Table 1) has been described by Bignall (2009), Boseley et al. (2012) and Chambefort et al. (2014). The Tahorakuri Formation is the unit of primary interest in this study for several reasons. It is one of the significant reservoir rocks at NGF (e.g., Coutts 2013), and spot cores recovered during drilling cover a wide range of depths within this unit, allowing investigation of the changes in physical properties with depth. The Tahorakuri Formation is a sequence

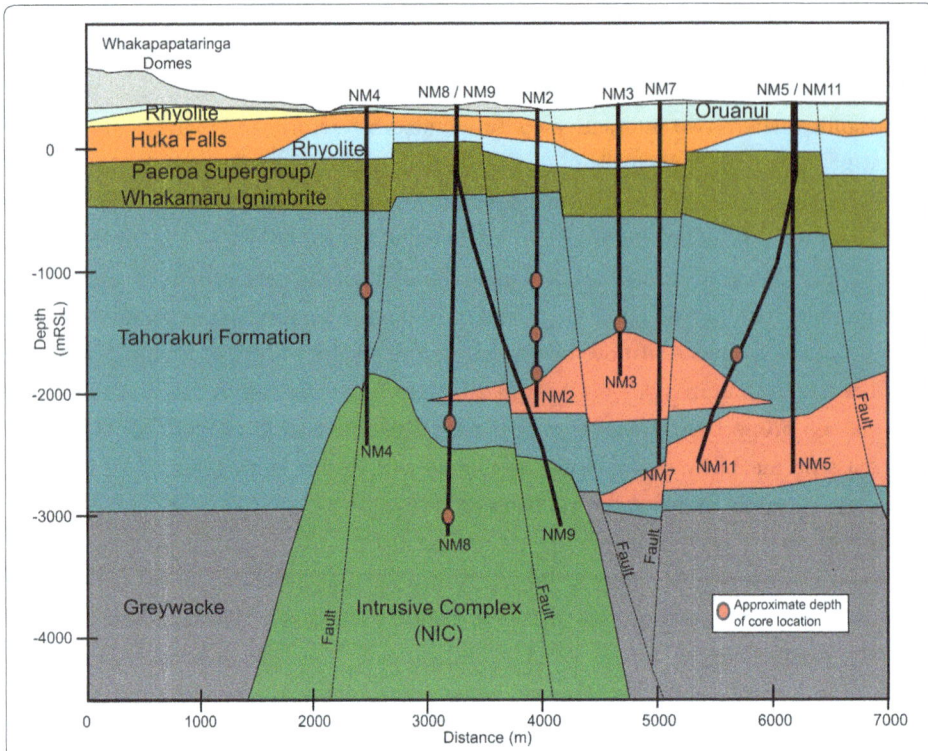

Fig. 2 Geological cross section of the Ngatamariki geothermal field from the NW to the SE with boreholes NM1–NM11 projected (after Bignall 2009; Boseley et al. 2012; Chambefort et al. 2014 and Mercury NZ Limited internal document)

of deposits between the Whakamaru Group ignimbrites and the Torlesse greywacke basement. At NGF, the Tahorakuri Formation comprises a thick pyroclastic sequence (~ 1000–1500 m thick) of primary tuffs, volcaniclastic (reworked) tuffs, and welded ignimbrites, overlain by sediments and tuffs in the northern and central part of the field (Chambefort et al. 2014). In the north to northwest of the field a quartz-phyric tonalite intrusion was encountered in three boreholes. This study also incorporates measurements of samples from this intrusion and the Tahorakuri volcaniclastic tuffs and ignimbrites. Dating of the Tahorakuri Formation (Eastwood 2013; Chambefort et al. 2014) indicates the unit was deposited over 1.22 Ma. Figure 2 shows locations of cores used in this study.

Methods

All rock preparation, measurement, and analysis were carried out in the rock mechanics laboratories, at the Department of Geological Sciences, University of Canterbury (Cant 2015). Samples were taken from core supplied by Mercury NZ Limited and Tauhara North No. 2 Trust. A drill press was used to extract 25–20 mm diameter cylinders from the core using a diamond tipped coring bit. The cylinders were all oriented parallel to the long axis of the core samples, making them approximately vertical within the stratigraphic column. A small piece of each cylinder was removed for thin section preparation for petrophysical analysis and void structure investigation. The cylinders were cut for a length-to-diameter ratio between 1:1.8 and 1:2.2, and ground flat for parallel ends as

Table 1 Subsurface lithology of Ngatamariki from wells NM1-7, as described by Bignall (2009)

Ngatamariki stratigraphy		
Formation name	Thickness (m)	Lithological description
Orakonui Fm	0–10	Pumice breccia, with common volcanic lithics, quartz and minor feldspar
Orunanui Fm	15–85	Cream to pinkish vitric–lithic tuff, with vesicular pumice and lava lithics, plus quartz, feldspar and rare pyroxene crystal fragments
Huka falls Fm	> 70–285	Coarse to medium grained sandstone, minor gravel (laminated lacustrine sediments)
Waiora Fm	0–10	An upper level interval of Waiora Formation, comprises pumice-rich vitric tuff, with volcanic lithics, quartz, rare biotite and pyroxene crystals
Rhyolite lava	115–315	Glassy rhyolite lava, with perlitic textures, phenocrysts are quartz, feldspar, pyroxene and magnetite
Waiora Fm	0–240	A lower interval of Waiora Formation, comprising pumice-rich vitreous tuff, intercalated with crystal tuff, tuffaceous coarse sandstone and tuffaceous siltstone
Wairakei ignimbrite	100–200	Crystal–lithic tuff/breccia, with abundant quartz, minor feldspar, rare biotite and pyroxene, minor volcanic lithics and pumice, in a fine ash
Rhyolite lava	0–285	Hard porphyritic quartz-rich rhyolite lava with phenocrysts of quartz, minor feldspar, and minor ferromagnesian minerals
Tahorakuri Fm	460–700	White to pale grey lithic tuff/breccia intercalated with fine sediments. In NM6 it is intercalated with 310 m of andesite lavas and breccias (tuffs and sediments)
Tahorakuri Fm	> 200–840	Pale grey, lithic tuff/breccia containing dark grey/brown lava, rhyolite pumice greywacke–argillite and sandstone clasts in a silty matrix (Akaterewa ignimbrite)
Tahorakuri Fm	> 830	Pale grey porphyritic feldspar, pyroxene and amphibole andesite lava and breccia (andesite lava, breccia)
Torlesse greywacke	Undefined	Pale grey to grey, massive meta-sandstone which lack obvious bedding, quartz veins

recommended by Ulusay and Hudson (2007) to allow for future unconfined compressive strength (UCS) testing. After coring and grinding the cylinders, they were placed in an ultrasonic bath with distilled water to clean and remove loose fractured material or clays formed during core drilling and grinding, then oven dried.

Thin section analysis

From the 21 samples collected, 17 were prepared for thin sections. Fluorescent epoxy was impregnated under vacuum into the sample, which was then polished. The microstructure was characterised using a Nikon Eclipse 80i epifluorescence microscope. The epifluorescent microscope uses a high-pressure mercury lamp that radiates ultraviolet light, which interacts with the fluorescent epoxy resin impregnated within the sample. Areas where the resin has accumulated (pores, vugs, fractures, etc.), glow under the light emitted by the mercury bulb (as in Heap et al. 2014). The advantage of this method of impregnation is the fluorescent dye only accumulates in the connected void spaces. The Nikon Eclipse 80i also has a standard microscope bulb, so features can be compared in fluorescent light and plane-polarised light. This allows areas that have been identified as void spaces in fluorescent light to be confirmed using plane-polarised light.

The thin sections were petrographically assessed to identify the primary and secondary mineralogy and textures in the samples to identify rock type and microstructure.

Photomicrograph maps of the fluorescing thin sections were taken for analysis of the two-dimensional microstructure. The computer program Autostitch was used to stitch 1620 individual photographs of the thin sections into one large image. Open source software ImageJ was then used to identify and isolate areas in which the fluorescent dye had aggregated (as in Heap et al. 2014).

To analyse the microstructure in the rocks of NGF, binary photomicrograph maps were created for each sample to identify areas of connected porosity. The resultant image (Fig. 3) has completely isolated the connected void spaces from the groundmass. From this image, quantitative analysis can be performed on the connected porosity. Using the analytical functions in ImageJ, the connected porosity was calculated on all binary thin section images as a percentage of the total area of the binary image that was black (void space).

Two types of micro-porosity were observed in the binary images: microfractures and pores. To differentiate the two forms of porosity, the definition applied by Heap et al. (2014) was used, where microfractures have a length-to-width ratio (aspect ratio) typically above 1:100 and pores typically range from 1:1 (perfectly circular) to 1:10 (oval).

Pore analysis was performed in ImageJ to ascertain the aspect ratio, circularity and roundness of the pores. The circularity is a measure how smooth the edges of the pore are, whereas the roundness is a measure of how close to the shape of a circle the pore is. Roundness is the reciprocal of aspect ratio, with additional scaling factors. A minimum pore area of 0.0002 mm^2 was used during all pore analysis. This precision is controlled

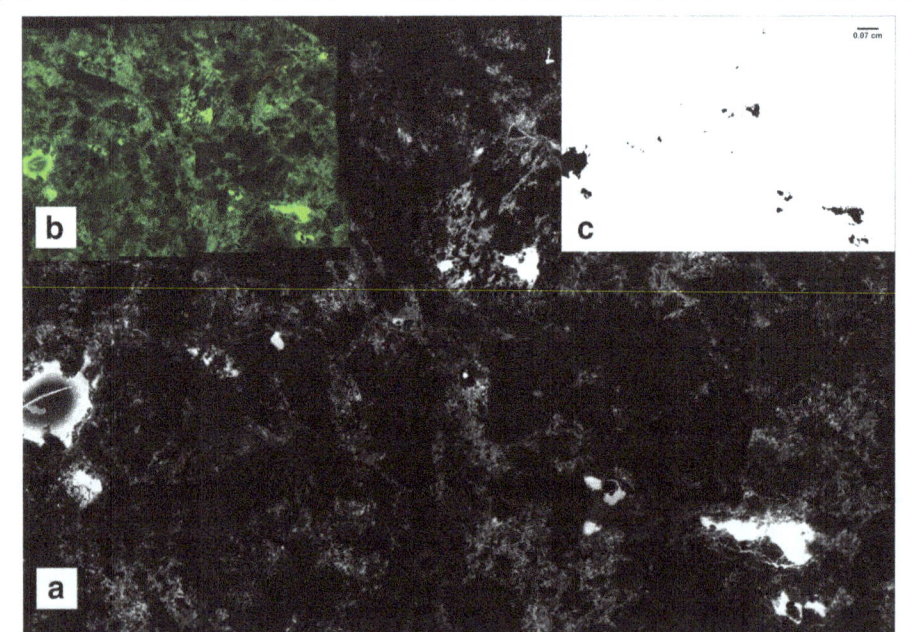

Fig. 3 a Thin section of sample with pore-dominated porosity and only one minor fracture visible; **b** thin section under fluorescent light; **c** TS3 binary image with the connected voids in black and unconnected voids and minerals in white. In the bottom right of the image, there are visibly fewer fractures surrounding the vug in **c** than in **a** due to the white intensity selected for the image. If the intensity was set to allow for the microfractures surrounding the vug in **c** to be visible, several falsely identified "void spaces" in the background would also be identified. While there are small areas of voids that are not identified in the final image, in **c,** there is a high degree of confidence that all identified void spaces are true pore spaces

by the quality of the images used. Below this value, the void space resolution becomes poor and no longer provides an adequate representation of the voids they characterise. To analyse the pores, first, they were converted into best fit ellipses to allow ImageJ to perform the quantitative analysis. These ellipses have the same area, orientation, and centroid as the pore they represent. ImageJ then measures the major and minor axis lengths and angles. Henceforth, all reference to the quantitative pore analysis will refer to the measurements performed on the best fit ellipses. The pore parameters were automatically calculated by the ImageJ software using Eqs. 1–4:

$$\text{Aspect ratio} = \frac{\text{Long axis}}{\text{Short axis}}. \tag{1}$$

$$\text{Area} = (\text{Major axis})(\text{Minor axis}). \tag{2}$$

$$\text{Circularity} = 4\pi \left(\frac{\text{Area}}{\text{Perimeter}^2} \right). \tag{3}$$

$$\text{Roundness} = 4 \left(\frac{\text{Area}}{\pi \left(\text{Major axis} \right)^2} \right). \tag{4}$$

Microfracture surface area was measured using classical stereological techniques outlined by Underwood (1969) and further described by Wu et al. (2000) and Heap et al. (2014). Using the binary images created in ImageJ the number of microfractures intersecting a grid of parallel and perpendicular lines spaced at 0.1 mm is recorded. The microfracture density per millimeter in each plane is then calculated from the known length and width of the image giving values for $P \parallel$ (microfractures intersecting parallel lines per millimeter) and $P\perp$ (microfractures intersecting perpendicular lines per millimeter). This allows the calculation of microfracture surface area per unit volume using Eq. 5 (Underwood 1969; Wu et al. 2000):

$$s_\text{v} = \frac{\pi}{2}P\perp + \left(2 - \frac{\pi}{2} \right)P \parallel, \tag{5}$$

where s_v = surface area per unit volume, mm^2/mm^3; $P\perp$ = microfracture density for intercepts perpendicular to orientation axis, mm^{-1}; $P \parallel$ = microfracture density for intercepts parallel to orientation axis, mm^{-1}.

Anisotropy of the microfracture intersection distribution was calculated using Eq. 6 (Underwood 1969; Wu et al. 2000):

$$\Omega_{2,3} = \frac{P\perp - P \parallel}{P\perp} + \left(\frac{4}{\pi} - 1 \right)P \parallel, \tag{6}$$

where $\Omega_{2,3}$ = anisotropy of microfracture distribution.

Connected porosity, dry bulk density, and ultrasonic wave velocity

Connected porosity and dry bulk density were determined using the saturation and buoyancy method (Ulusay and Hudson 2007). Axial P (compressional) and S (shear) wave velocities were measured using a GCTS (Geotechnical Consulting and Testing Systems) Computer-Aided Ultrasonic Velocity Testing System (CATS ULT-100) device. Piezoelectric transducers within the device are used to measure the arrival time of the compressional and shear waves from which the velocity can be calculated. A loading frame applied a load of 2.7 kN (5 MPa axial stress) to the samples to ensure solid contact between the sample and the Piezoelectric transducers. This results in consistent waveforms for all velocity measurements. The applied stress of 5 MPa was selected, such that it did not cause plastic deformation of the extensively altered rock mass, based on the likely strengths of the rocks given in Wyering et al. (2014).

Ultrasonic wave velocity tests were performed on the samples twice, once when the samples were oven dried and again when the samples had been saturated in distilled water under a vacuum. Dry samples were oven dried and stored in a desiccator before testing, while the saturated samples were stored submerged before testing. A total of 144 waveforms were captured for both the dried and saturated samples. First, 72 waveforms were captured in one axial direction and the samples were flipped to capture waveforms in the other direction. The values were then compared and averaged to obtain a representative value for the sample and account for any anisotropy of wave propagation associated with directivity. The waveform velocities were used to calculate dynamic Poisson's ratio and Young's Modulus using Eqs. 7 and 8 (Guéguen and Palciauskas 1994):

$$v = \frac{V_P^2 - 2V_S^2}{2\left(V_P^2 - V_S^2\right)}, \tag{7}$$

$$E = \frac{\rho V_S^2 \left(3V_P^2 - 4V_S^2\right)}{\left(V_P^2 - V_S^2\right)}, \tag{8}$$

where v = Poisson's ratio; V_p = compressional P-wave velocity (m/s); V_s = shear S-wave velocity (m/s); E = Young's modulus (Pa); ρ = dry bulk density (kg/m^3).

Permeability

Permeability measurements were conducted using a pulse decay permeameter (Corelab PDP-200) with confining pressure (Fig. 4). The sample was placed inside a Viton tube inside the core holder (testing cell) and a confining pressure ranging from 5 to 65 MPa

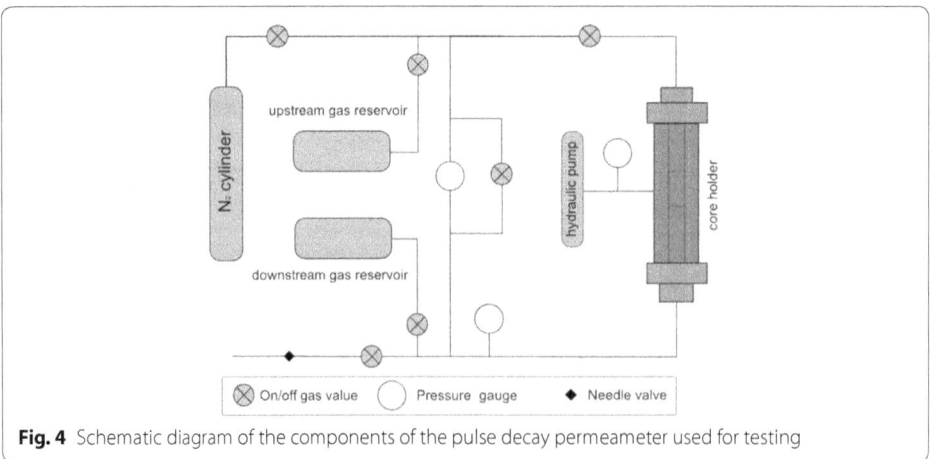

Fig. 4 Schematic diagram of the components of the pulse decay permeameter used for testing

was applied by a manual hydraulic pump. Pressurised nitrogen was applied to the sample and left to "soak" to allow the sample to equalise to the test pore pressure and temperature. The pore pressure was selected based on the expected permeability of the sample, which was inferred from the connected porosity. The higher the expected permeability (based on porosity measurements), the lower the pore pressure was used to ensure laminar gas flow. The gas valves were shut and the nitrogen gas bled from the downstream side of the core holder until a desired pressure differential was achieved. The bypass was then closed and the pressure differential across the sample was monitored as the pressure equilibrated by the gas traveling through the sample. The gas differential across the sample decays in logarithmic fashion that is recorded by the PDP 200's software.

The calculation for gas permeability according to a modified version of Darcy's law (Brace et al. 1968) is as follows:

$$k_{gas} = \left(\frac{2\eta L}{A} \right) \left(\frac{V_{up}}{P_{up}^2 - P_{down}^2} \right) \left(\frac{\Delta P_{up}}{\Delta t} \right), \tag{9}$$

where k_{gas} = gas permeability; η = viscosity of the pore fluid; L = length of the sample; A = cross-sectional area of the sample; V_{up} = volume of upstream pore pressure circuit; P_{up} = upstream pore pressure; P_{down} = downstream pore pressure; t = time.

Equation 9 is used by the PDP 200's software to calculate the gas permeability of the differential pressure decay curve and results in gas apparent permeability measurements. To determine the true permeability, a Klinkenberg correction is required (Klinkenberg 1941) to account for gas slippage within the sample:

$$k_{true} = k_{gas} \left(1 + \frac{b}{P_{mean}} \right), \tag{10}$$

where k_{true} = true permeability; k_{gas} = gas permeability at a particular pore pressure; b = Klinkenberg slip factor; P_{mean} = mean pore pressure.

Conducting a Klinkenberg correction requires the gas permeability test to be performed at several different pore pressures. By plotting gas permeability versus $1/P_{mean}$, the true permeability can be taken as the trend line intercept on the permeability axis.

A testing procedure was followed for each sample to achieve an accurate and repeatable result. The test started at the lowest possible confining pressure (5 MPa), where three-to-five apparent gas permeability tests would be measured. The confining pressure would then be increased by 10 MPa and the sample would be left to "soak" for the appropriate amount of time before testing the permeability using the method described above. The soak time varied depending on the permeability of the sample and ranged from 5 min to up to 24 h. Soak times were established through trial and error. If the sample had not fully equalised the test results showed a non-logarithmic decay, upon which the test was repeated with a longer soak time. This procedure was repeated until the confining pressure reached 65 MPa, except for two volcaniclastic samples that were only tested from to 55 MPa due to their low strength as determined by Wyering et al. (2014).

Lithostatic stress model

When investigating the effects of burial diagenesis on physical properties, it is important to conduct tests at the conditions from which the samples were taken. This was not possible for the thin section analysis, connected porosity, dry bulk density, and ultrasonic wave velocity, but possible for permeability using the PDP 200. To achieve this, a lithostatic stress model was compiled from the cross section and lithologies, as shown in Fig. 2. Due to the extensional nature of TVZ, σ_1 was assumed to be vertical (Hurst et al. 2002); this allowed the true lithostatic stress to be calculated using Eq. 11:

$$\sigma' = \sigma_{\text{bulk}} - \sigma_{\text{hydro}}, \tag{11}$$

where $\sigma' =$ true (effective) lithostatic stress; $\sigma_{\text{bulk}} =$ bulk lithostatic stress; $\sigma_{\text{hydro}} =$ hydrostatic stress.

The bulk lithostatic stress is the combined stress of each overburden unit as applied to each sample, which varies from sample to sample due to differing burial depths and/or different overlying lithological units. The hydrostatic stress is the total stress applied by the groundwater. The hydrostatic stress is assumed to be equal in all directions and results in a stress that acts against the lithostatic stress. This stress is experienced at pore and fracture boundaries within the rock mass (e.g., Peacock et al. 2017). Due to limited published data on the hydrology of the field, a very simple hydrostatic model was used that assumed a connected water column throughout the field of cold water (to maintain a constant density for calculation). The bulk lithostatic stress applied to any sample by the overlying intact rock is calculated by summing the bulk lithostatic stress for each overlying layer using Eq. 12:

$$\sigma_{\text{bulk}} = \rho g h, \tag{12}$$

where $\sigma_{\text{bulk}} =$ stress (MPa); $\rho =$ intact rock dry bulk density (kg/m^3); $g =$ gravitational force (m/s^2); $h =$ layer thickness (m).

The hydrostatic stress is calculated using Eq. 12, where ρ is the density of water at 20 °C (1000 kg/m^3), and h is the depth from which the sample was recovered. The effective lithostatic stress is then the lithostatic stress minus the hydrostatic stress.

Results

Lithological units

Rock types in the reservoir can be divided into two broad groups: volcaniclastic and intrusive. The Tahorakuri Formation comprises mainly volcaniclastic rocks, while the Ngatamariki Igneous Complex comprises predominantly tonalite.

The volcaniclastic rocks are largely silicic lapilli tuffs with recrystallised pumice clasts and occasional lithics (dominantly 2–5 mm) in an ash matrix containing small crystals of alkali feldspar, plagioclase, and quartz (dominantly 0.1–2 mm), although all glass has ubiquitously recrystallised (Fig. 5). Shallow samples (Fig. 5a, b) contain recrystallised pumice clasts of quartz and feldspar that in some samples appear rounded, with large pores partially infilled with zeolites, clay minerals, pyrite, and rare epidote. Some of the larger pores are rectangular relict feldspar, which are also variably infilled. In these shallow samples, the matrix is fine-grained quartz and feldspar or locally clay. Intermediate depth (Fig. 5c, d) samples have common epidote and calcite filling large pores locally

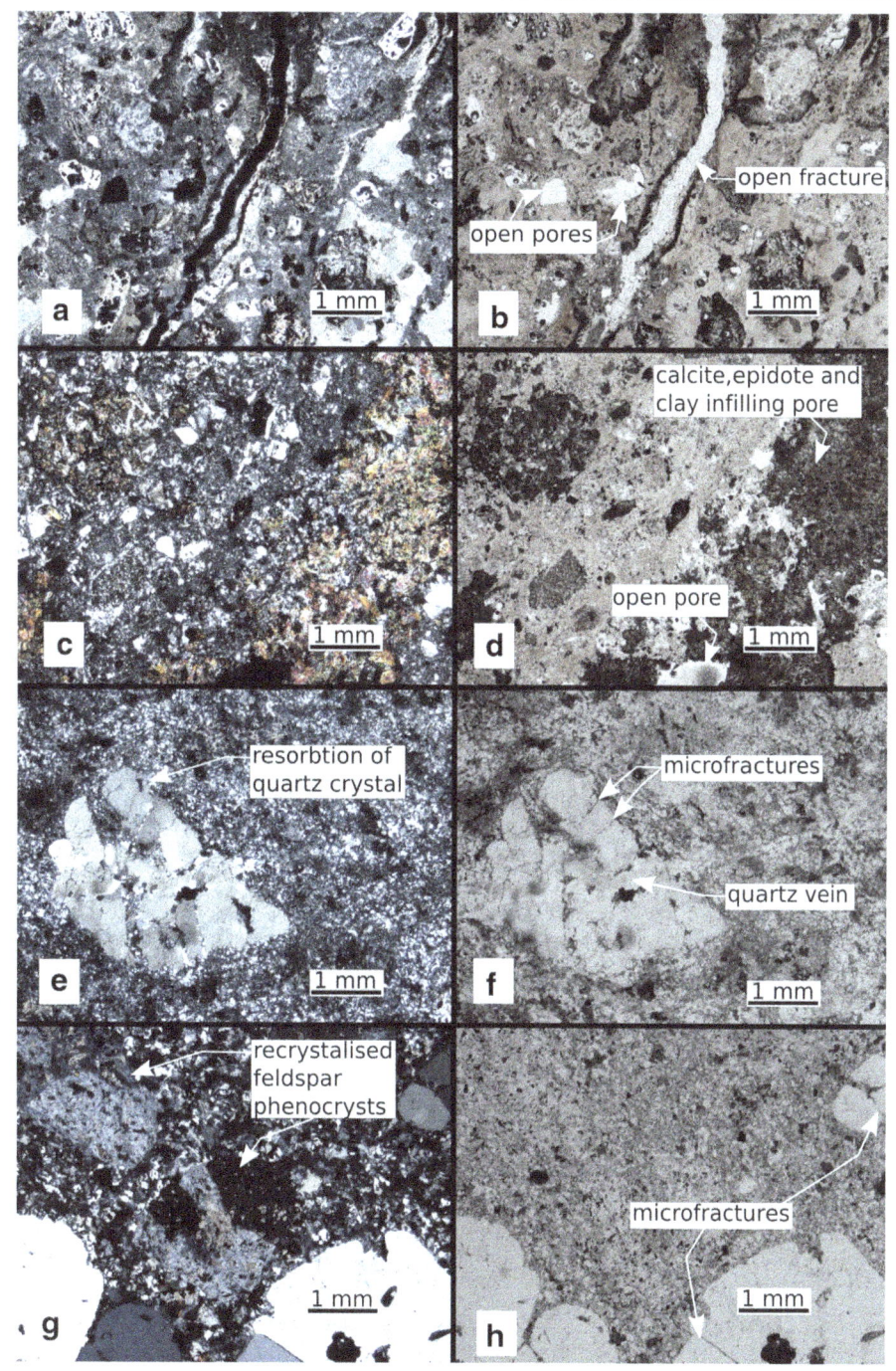

Fig. 5 Cross-polarised (left) and plane-polarised (right) photomicrographs of **a**, **b** shallow-, mixed pore-, and microfracture-dominated lapilli tuff with recrystallised pumice clasts, and partially precipitated microfractures with altered margins (thin section from NM2_1350B_01 Wyering et al. 2014); **c**, **d** intermediate depth, pore-dominated lapilli tuff with large pores infilled with epidote and clay, most original textures are no longer visible due to re-crystallisation (thin section from NM11 1.2_01 Wyering et al. 2014); **e**, **f** deep, microfracture-dominated lapilli tuff showing recrystallization textures, quartz vein, and resorption of quartz crystal in pumice clast (thin section NM8A C1_4_02 from Wyering et al. 2014); **g**, **h** deep, microfracture-dominated tonalite with quartz and feldspar phenocrysts showing resorption (quartz) and recrystallization (feldspar) (thin section NM8A C2_1_1 from Wyering et al. 2014)

filled with clay and little evidence of primary volcanic textures except the presence of fragmental and whole feldspar phenocrysts. In deeper (Fig. 5e, f) samples, relict pumice containing larger phenocrysts of quartz have a re-crystallisation fabric reminiscent of fiamme which could suggest welding in an ignimbrite.

The tonalite (Fig. 5g, h) is porphyritic with large quartz (~ 40%) and feldspar (~ 60%). Fractured quartz phenocrysts (≥ 5 mm) are sub rounded and embayed due to resorption. Plagioclase phenocrysts are generally more euhedral, but largely recrystallised into finer grained feldspar and clay. The fine-grained groundmass (~ 0.1 mm) is dominantly quartz and feldspar with minor chlorite and epidote.

Pore and microfracture analysis

Three distinct styles of microstructure are present in the thin sections: (1) shallow, relatively high connected porosity (~ 10–21%) samples contain variable microstructure with occasional large dissolution pores and partially precipitated microcracks with altered margins (Fig. 5a, b). (2) intermediate depth, with intermediate connected porosity (13–15%) consisting of large irregular pores, dissolution pores partially filled with secondary crystals with few/no visible microfractures (Fig. 5c, d). Some pores show evidence of pore collapse (crushed pore in bottom right of Fig. 3a, c). (3) deep, relatively low connected porosity samples (~ 3–7%) dominated by microfractures in and around phenocrysts (Fig. 5e, f). The microfractured volcaniclastic samples have no large pores due to complete re-crystallisation and microfractures are frequently filled with secondary mineralization (quartz vein in Fig. 5f). The tonalite is dominated by microfractures associated with large fractured phenocrysts (Fig. 5g, h).

For the pore-dominated samples, aspect ratio, pore area, circularity, roundness, and vug porosity were determined (Table 2). The pore-dominated thin sections are from NM11 within a depth range of 2083–2087 m. The permeability at the lowest confining pressure (5 MPa) was used to compare to the pore analysis, which was conducted at atmospheric pressure. Figure 6 indicates that only circularity correlates with permeability.

For the microfractured samples, microfracture densities were determined and used to calculate the microfracture area per unit volume and anisotropy (Table 3). Microfracture density (area per unit volume) ranged from 2.28 to 31.77 mm^2/mm^3 with anisotropy

Table 2 Quantitative analysis of thin sections with vug pore space

Sample ID	Thin section #	Vug porosity (%)	Average circularity	Average aspect ratio	Maximum aspect ratio	Average roundness
NM11 2083 B	TS1	1.2	0.41	1.99	3.94	0.56
NM11 2087.4 C	TS3	1.6	0.46	2.05	6.96	0.57
NM2 2254.7 A	TS4	9.5	0.32	2.10	3.78	0.53
NM11 2083 C	TS5	3.1	0.34	2.02	5.53	0.56
NM11 2083.34 A	TS6	0.1	0.47	1.73	2.32	0.63
NM11 2087.4 D	TS7	0.4	0.37	1.89	3.24	0.60
NM11 2087.4 A	TS8	4.4	0.33	2.12	7.35	0.55
NM11 2083 A	TS10	1.6	0.37	2.45	7.89	0.49
NM11 2087.4 B	TS11	1.5	0.37	2.45	7.89	0.49
NM8a 2525.5 C	TS12	2.4	0.39	2.32	9.78	0.51

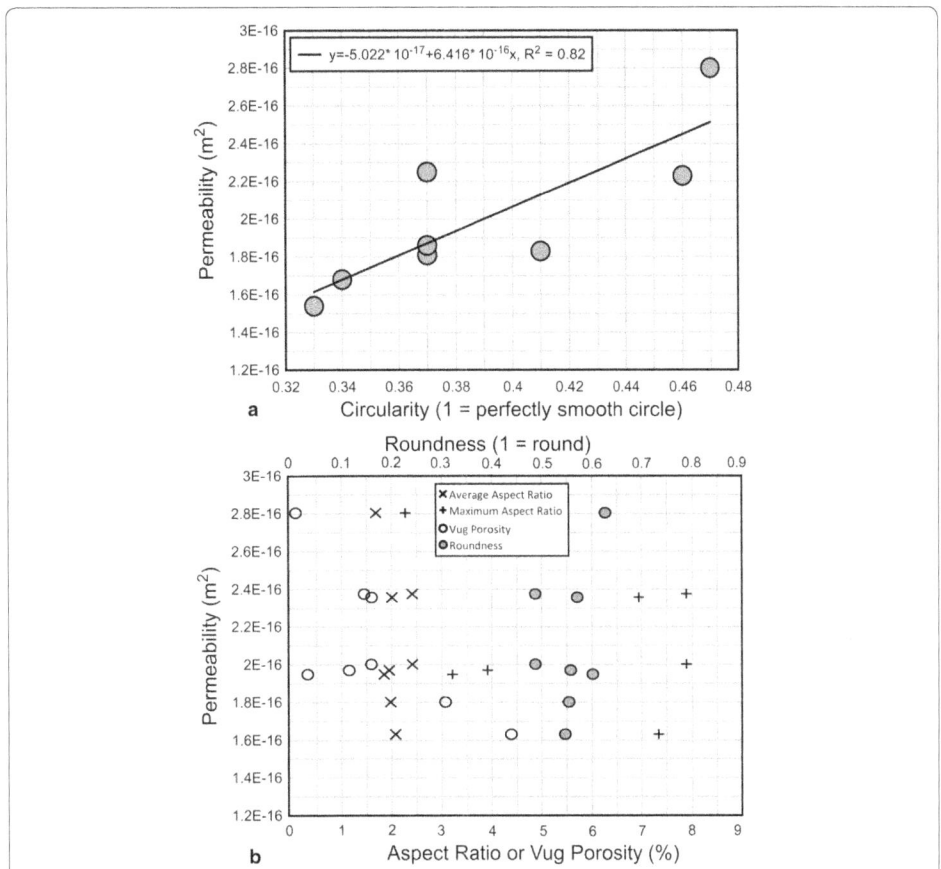

Fig. 6 Pore geometry measurements versus permeability at 5 MPa confining pressure for pore-dominated samples from NM 11 at 2083–2087.4 m depth. **a** circularity correlates with permeability; **b** aspect ratio, roundness, and vug porosity do not correlate with permeability

Table 3 Quantitative analysis of thin sections with microfractures

Sample ID	Thin section #	Parallel microfracture density per mm ($P \parallel$)	Perpendicular microfracture density per mm ($P \perp$)	Microfracture area per unit volume (S_v) (mm²/mm³)	Anisotropy ($\Omega_{2,3}$)
NM8a 2525.5 B	TS2	2.66	0.77	4.51	0.66
NM4 1477.2 A	TS13	4.67	2.00	8.19	0.51
NM8a 3284.7 C	TS14	16.59	13.30	31.77	0.16
NM8a 3280 C	TS15	10.74	25.94	28.00	0.85
NM2 1354.2 B	TS17	10.24	9.06	19.97	0.09
NM3 1743 A	TS19	4.67	4.64	9.32	0.00
NM3 1743 C	TS20	1.10	1.29	2.28	0.13

factors ranging from 0.00 (isotropic, equal number of microfracture intercepts on predetermined x/y planes) to 0.85 (fairly anisotropic, significantly more microfracture intercepts on one plane). As with the pore-dominated samples, the permeability at the lowest confining pressure (5 MPa) was used to compare to the microfracture analysis, which was conducted at atmospheric pressure. There was no clear correlation between microfracture density and connected porosity (Fig. 7a) or permeability (Fig. 7b).

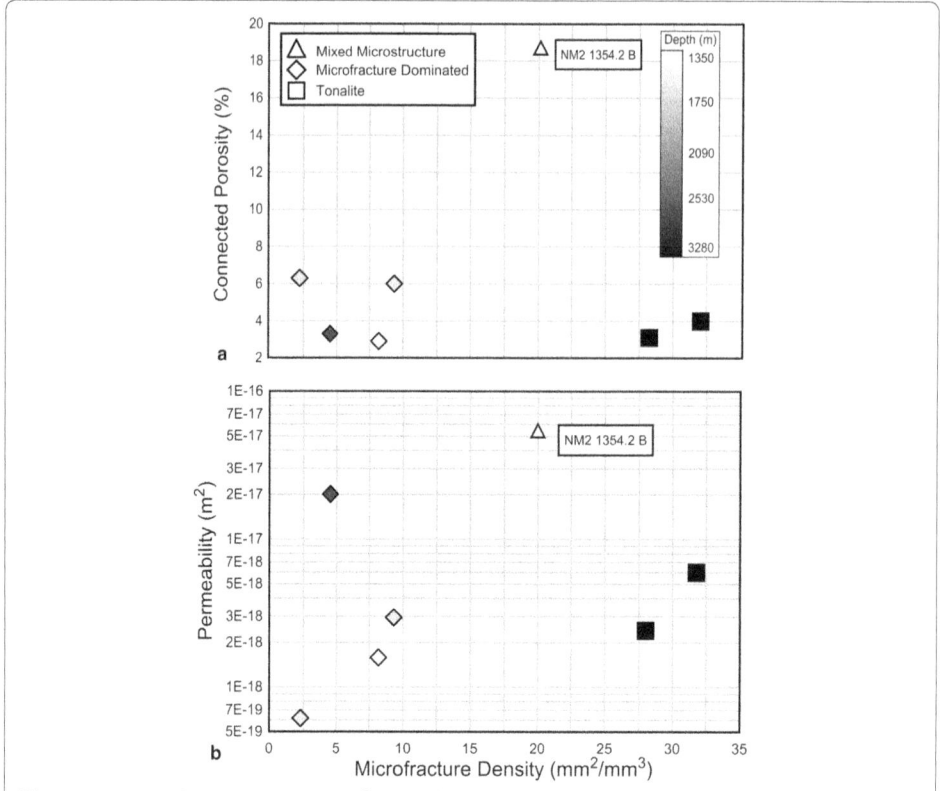

Fig. 7 a connected porosity versus microfracture density. NM2 1354.2 B appears as an outlier with a distinctively higher connected porosity; **b** permeability at 5 MPa versus microfracture density, using a log scale. NM2 1354.2 B appears as an outlier with a distinctively higher permeability. Tone darkens with depth of sample, as shown in legend

The mixed microfracture and pore samples are not plotted in Figs. 6 or 7, except for NM2 1354.2 B, which plots as an outlier in Fig. 7, because its mixed microstructure made it difficult to define either the circularity or the microfracture density to the same extent as the single microstructure samples.

Physical properties

There is a clear relationship between connected porosity and dry bulk density (Table 4, Fig. 8), as well as P-wave velocity (Fig. 9). Ultrasonic wave velocity for both saturated and oven dried samples and the calculated dynamic elastic moduli are given in Appendix. Water-saturated samples have a faster P-wave velocity by an average of 147 m/s and a slower S-wave velocity by an average of 35 m/s. Appendix contains the Klinkenberg corrected permeability for all samples tested in this study.

Figure 10 shows the permeability of each sample as a function of confining pressure. Figure 11 shows the connected porosity–permeability relationship according to the microstructure type with each sample tested at the lowest confining pressure (5 MPa) and at the highest confining pressure (55 MPa). The volcaniclastic samples have a wide range of both connected porosity and permeability. The shallow samples with mixed microstructure (Fig. 5a, b) have relatively high permeability and a wide range of (20–21%) connected porosity. The intermediate connected porosity (12–15%) samples from intermediate depths (2083–2087 m) and dominated by pore porosity (Fig. 5c, d) have the

Table 4 Sample cores and their associated sample ID [consisting of well number (NM#), depth and sample #], formation name, microstructure type, dry bulk density, and connected porosity

Sample ID (well #, depth, sample)	Formation name	Microstructure type	Dry bulk density (kg/m^3)	Connected porosity (%)
NM2 1354.2 A	Tahorakuri	Mixed pores and microfractures	2070	20.3
NM2 1354.2 B	Tahorakuri	Mixed pores and microfractures	2100	18.6
NM2 1354.4 A	Tahorakuri	Mixed pores and microfractures	2160	19.3
NM2 1788 A	Tahorakuri	Mixed pores and microfractures	2470	10.0
NM2 2254.7 A	Tahorakuri	Microfracture dominated	2570	4.9
NM3 1743 A	Tahorakuri	Microfracture dominated	2540	6.0
NM3 1743 C	Tahorakuri	Microfracture dominated	2510	6.3
NM4 1477.2 A	Tahorakuri	Microfracture dominated	2670	2.9
NM8a 2525.5 B	Tahorakuri	Microfracture dominated	2600	2.5
NM8a 2525.5 C	Tahorakuri	Microfracture dominated	2580	3.3
NM8a 3280 C	NIC Tonalite	Microfracture dominated	2510	3.1
NM8a 3284.7 C	NIC Tonalite	Microfracture dominated	2490	4.0
NM11 2083 A	Tahorakuri	Pore dominated	2290	14.3
NM11 2083 B	Tahorakuri	Pore dominated	2270	15.3
NM11 2083 C	Tahorakuri	Pore dominated	2290	14.7
NM11 2083.34 A	Tahorakuri	Pore dominated	2300	14.0
NM11 2083.34 B	Tahorakuri	Pore dominated	2280	14.9
NM11 2087.4 A	Tahorakuri	Pore dominated	2350	12.9
NM11 2087.4 B	Tahorakuri	Pore dominated	2310	14.4
NM11 2087.4 C	Tahorakuri	Pore dominated	2320	14.2
NM11 2087.4 D	Tahorakuri	Pore dominated	2340	13.4

NIC, Ngatamariki Intrusive Complex

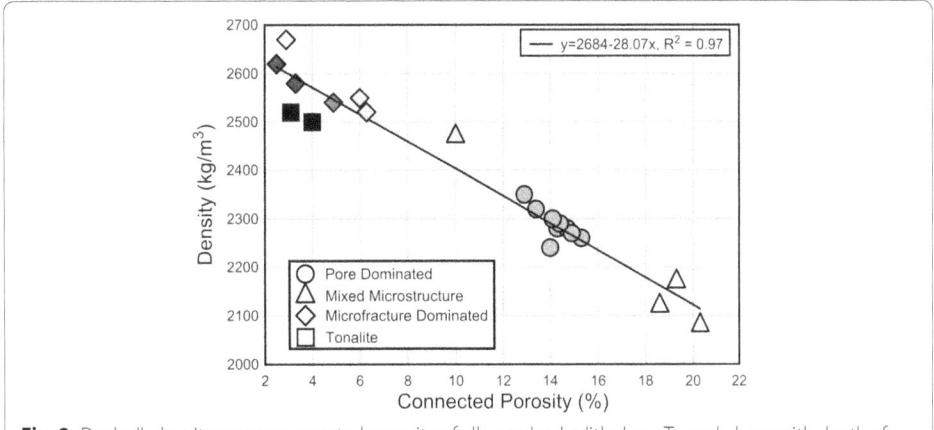

Fig. 8 Dry bulk density versus connected porosity of all samples, by lithology. Tone darkens with depth of sample, as shown in legend in Fig. 7a

highest permeability. The volcaniclastic samples with lower connected porosity (< 8%) and permeability correspond to a microfractured pore structure (Fig. 5e, f), but were sourced from a range of depths. The two samples of tonalite have low connected porosity (< 5%) and variable permeability; both samples are from great depth and have porosity dominated by microfractures (Fig. 5g, h).

Figure 10 also highlights that the four sample types behave in three different ways consistent with their respective microstructure. The volcaniclastic and tonalite samples with

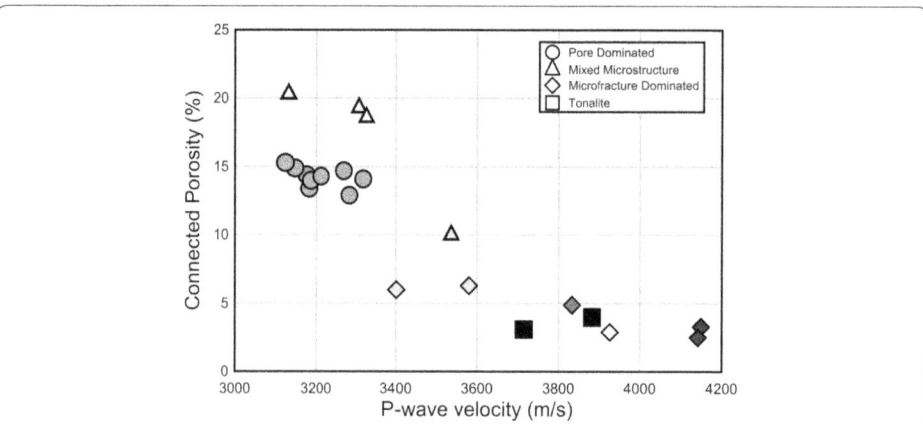

Fig. 9 Dry P-wave velocity versus connected porosity. Tone darkens with depth of sample, as shown in legend in Fig. 7a

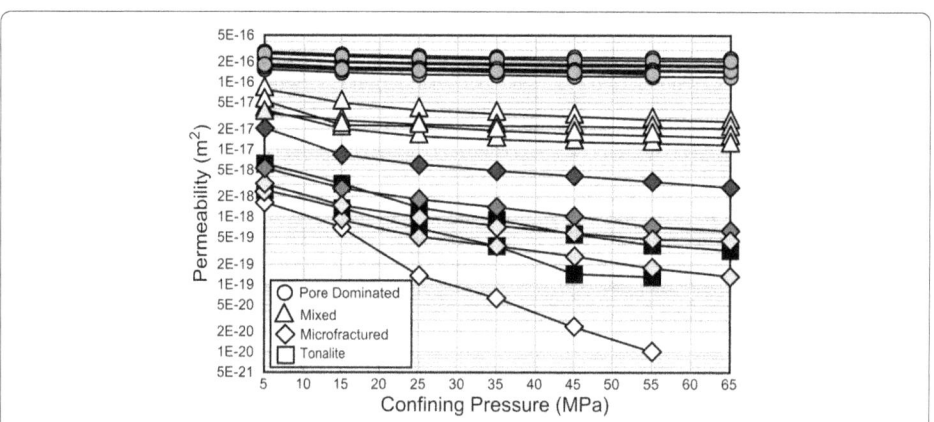

Fig. 10 Permeability versus confining pressure. Tone darkens with depth of sample, as shown in legend in Fig. 7a

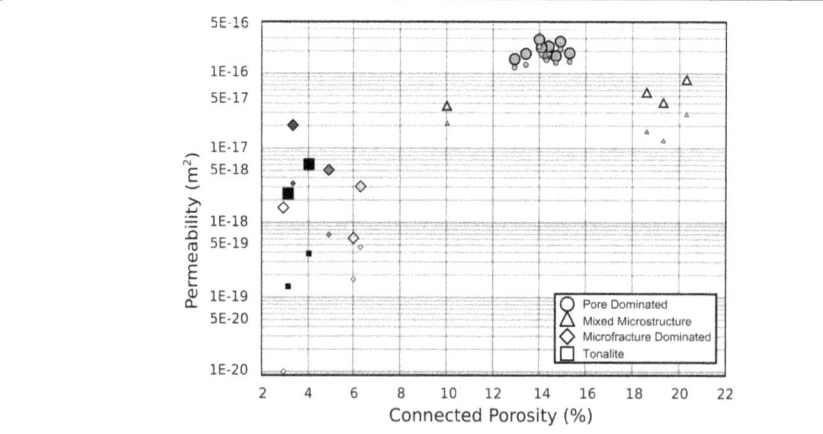

Fig. 11 Permeability versus connected porosity, showing permeability results from both 5 MPa (large symbols) and 55 MPa (small symbols) confining pressures. Tone darkens with depth of sample, as shown in legend in Fig. 7a

low, microfracture dominated, porosity experience a relatively large decrease in per-meability with increasing confining pressure. The samples with relatively high perme-ability and with mixed pore and microfracture-dominated microstructure show a small decrease in permeability with increased confining pressure. The samples with pores have a reduced response to increasing confining pressure.

Lithostatic stress model

Figure 2 shows formation thickness variability across the field. For the Tahorakuri pyro-clastic succession, this ranges from 740 m in NM3 to 1655 m in NM8a (Chambefort et al. 2014). Table 6 in the Appendix contains the stresses applied by the various for-mations of the NGF. Table 5 contains the tested samples, their sampling depth, and the corresponding effective lithostatic stress. The in-situ permeability was selected as the permeability at the tested confining stress nearest to the calculated effective lithostatic stress.

Discussion

Microfracture and pore analysis

The morphology of primary pores and fractures in an igneous rock is controlled by magma viscosity and gas content, emplacement processes, depositional and tectonic environments (e.g., Lewis and McConchie 1994; Shea et al. 2010; Davidson 2014; Heap et al. 2015; Colombier et al. 2017). The micropore structure, which controls fluid flow (Farquharson et al. 2015; Heap et al. 2017a), can be modified by post-depositional pro-cesses. Intrusive rocks have little initial porosity due to their holocrystalline matrix, but

Table 5 Effective lithostatic stress (confining pressure) for each sample with the corre-sponding permeability test pressure used to calculate the in-situ permeability

Sample ID	Effective lithostatic stress (MPa)	Tested confining stress (MPa)	In-situ permeability (m^2)
NM2 1788 A	18	15	2.70E−17
NM2 1354.2 A	12	15	4.94E−17
NM2 1354.2 B	12	15	2.29E−17
NM2 1354.4 A	12	15	2.04E−17
NM2 2254.7	24	25	1.76E−18
NM3 1743 A	17	15	8.62E−19
NM3 1743 C	17	15	8.62E−19
NM4 1477.2 A	15	15	6.76E−19
NM8a 2525.5 C	28	25	5.80E−18
NM8a 3280 C	34	35	3.63E−19
NM8a 3284.7 C	35	35	8.74E−19
NM11 2083 A	21	25	1.60E−16
NM11 2083 B	21	25	1.55E−16
NM11 2083 C	21	25	1.49E−16
NM11 2083.34 A	21	25	2.43E−16
NM11 2083.34 B	21	25	2.26E−16
NM11 2087.4 A	21	25	1.29E−16
NM11 2087.4 B	21	25	1.92E−16
NM11 2087.4 C	21	25	1.85E−16
NM11 2087.4 D	21	25	1.48E−16

their post-cooling porosity and permeability develops as a result of tectonic and thermal stresses that manifest as macroscopic and microscopic fractures (Géraud 1994; Lane and Gilbert 2008). Volcanic rocks have a wide range of porosities due to variables such as cooling time, transport, gas content, and weathering (e.g., Mueller et al. 2011; Olalla et al. 2010), while porosity in volcaniclastic and sedimentary rocks is generally controlled by the size and distribution of particles (e.g., Heap et al. 2017a; Bai et al. 2016).

In this study, we attempted to find a correlation between microstructure displayed in 2D thin sections and physical properties of the sample set. In the pore-dominated samples, only circularity appears to correlate with permeability (Fig. 6). This shows that smooth circular pores conduct fluid better than irregular pores, because for the same pore area, the circular pore will have a smaller amount of pore surface (pore perimeter) at which fluid velocity is 0. It is interesting that aspect ratio and roundness do not provide a good fit, which suggests that embayments along the pore perimeter do not contribute to fluid flow. This is analogous to the low permeability of some samples with micro-porosity characterised by high surface area in Farquharson et al. (2015), where many micropores did not contribute to fluid flow. We also attempted to determine if a correlation between increased microfracture density and increased connected porosity was present, as has been observed in reservoir rocks of the nearby Rotokawa Andesite (Siratovich et al. 2014). As shown in Fig. 7a, there is no clear correlation between microfracture density and connected porosity in samples with dominant microfractured microstructure. This may be the result of a difference in the microstructure between the two andesites. It could also be that the method used to measure both pore shape and microfracture density only observes these features on a predefined x and y plane (the thin section plane), and does not take into account the depth of pores or the width or length fractures in the third dimension.

To illustrate how small area 2D investigations may yield inaccurate results, we point to sample NM2 1354.2 B which displays a connected porosity much higher than would be expected for a sample, whose thin section indicates it only contains microfractures (Fig. 7a). Sample NM2 1354.2 A was extracted from the same piece of core and its thin section does not contain any visible microfractures, but contains several pores. Thin sections from this depth from Wyering (2014), reanalysed for this study, clearly show pores and microfractures (Fig. 5a, b). This illustrates that as shown in Kennedy et al. (2016) when considering the porosity of a system, there is a scale dependence on the structures that control a complicated fluid flow network, and sufficient care is required to ensure statistically relevant volumes.

Ultrasonic wave velocity

P-wave velocity increases with dry bulk density (Table 4) and decreasing connected porosity (Fig. 9), as also seen in Barton (2007), Vasconcelos et al. (2008), Khandelwal (2013) and Wyering et al. (2014). This shows that the ultrasonic velocity of samples from the Tahorakuri Formation and the Ngatamariki Intrusive Complex are primarily controlled by the connected porosity, and hence the microstructure. The sonic wave results show that the saturated samples have a noticeable P-wave velocity increase (average 4%) in wave velocity compared to the dry samples, while the saturated and dry S-wave velocities are generally smaller (average − 2%). This phenomenon has been observed in other studies (e.g., Heap et al. 2013, 2014). The microfracture-dominated samples tended to have a higher dry P- and S-wave velocities. This is accompanied by greater P-wave velocity increase and smaller S-wave decrease when saturated than the mixed or pore dominated, despite the porosity in pore-dominated samples being much higher. Our data set is too small to provide any statistical quantification of these differences; however, these results demonstrate that sonic wave velocity reduction is less severe in samples with pore dominated compared to microfracture microstructure.

Connected porosity–permeability relationship

The trend of decreasing permeability with decreasing connected porosity in the NGF rocks shown in Fig. 11 has also been observed in other studies (Heard and Page 1982; Klug and Cashman 1996; Saar and Manga 1999; Rust and Cashman 2004; Stimac et al. 2004; Mueller et al. 2008; Wright et al. 2009; Heap et al. 2014; Farquharson et al. 2015; Wadsworth et al. 2016; Heap and Kennedy 2016). There is scatter in the relationship, which results from the geometry and connectivity of the pores (as noted in Mueller et al. 2008 and others), amongst other factors. The complex geometry of the altered mineralogy, interclast pore spaces, and microfractures, creates highly tortuous flow paths. This is accentuated in the rocks with low connected porosity which, as a result, can be sensitive to confining pressure. Our data set does not contain sufficiently high porosity samples to identify a changepoint in the permeability–porosity relationship similar to the changepoint proposed by Farquharson et al. (2015); however, the critical porosity range of 12–15% from Heap et al. (2014) and 14–16% form Farquharson et al. (2015) to some extent captures the distinction between microfracture-dominated and mixed/pore-dominated microstructure we observe at porosity > 10–12%.

Porosity–permeability relationships and sensitivity to confining pressure

In this study, our textural analysis shows microfractures and irregular-shaped pores (Fig. 5). As with the previous studies (e.g., Guéguen and Palciauskas 1994; Lamur et al. 2017), these morphologies each react differently to applied confining stresses (Fig. 10). Increasing confining stress causes microfractures to progressively close, resulting in a reduction in permeability (as in Vinciguerra et al. 2005), while elliptical and circular pores show very little change with increased confining stress (Guéguen and Palciauskas 1994; Lamur et al. 2017). Microfracture closure is primarily controlled by elastic deformation, with surface roughness controlling further closure. High aspect ratio microfractures are associated with relatively high permeability at low confining pressures, but are easily closed by increased confining pressures. Conversely to what Vinciguerra et al. (2005) found, we did not find a significant difference in the sensitivity of the fracture-dominated samples to confining pressure as a function of their porosity (Fig. 11). The lowest porosity (~ 3%) tonalite and lava have similarly large reductions in permeability between confining pressures of 5 and 55 MPa as the highest porosity lava (~ 6%). Nara et al. (2011) found that samples with low aspect ratio microfractures maintained their influence on permeability even at the highest confining pressure (90 MPa).

To further investigate the effect of confining pressure on permeability, the permeability results at each pressure stage are plotted against confining pressure for all microstructure types (Fig. 10).

Pore-dominated porosity: From Fig. 10, it is apparent that increasing confining pressure has little effect on the permeability for the majority of the samples. The steepest gradient occurs between 5 and 15 MPa for all samples, with a progressive levelling off of the permeability change with confined pressure. The mixed microstructure samples show a steeper gradient than the pore-dominated samples.

Microfracture porosity: Figure 10 indicates a steeper relationship between increased confining pressure and decreased permeability when compared to the pore-dominated. For most samples, the gradient is steepest at the lower confining pressures (5–25 MPa) with a slightly shallower gradient as the confining pressure increases (25–65 MPa). The volcaniclastic sample with permeability of 9.79×10^{-21} m^2 at 55 MPa has a much steeper and more consistently steep gradient than the other samples. This suggests that microstructure continued to be closed as confining pressure increased. A possibility is that the sample is compacting non-elastically during the high confinement; however, Heap et al. (2015) report that triaxial data on ~ 30% connected porosity tuff that indicates lower porosity (< 20%) samples would likely require higher confinement than 55 MPa to initiate permanent compaction. Our preferred explanation is that these fractures in 2D were actually complex dissolution, vein precipitation, and then recrystallised pipes in 3D and thus do not simply close like microfractures. Textures in thin sections (Fig. 5c, d) support a more complex origin for these structures.

The permeability of the lowest and highest confining pressures was used to demonstrate the effect of pore structure on permeability with increased confining pressure. Figure 11 shows that while the tonalite- and microfracture-dominated volcaniclastics were formed by different processes; their porosity, permeability, and response to confining stress are similar. This suggests that similar microstructures can be found across very different lithologies. Some of the mixed microstructure volcaniclastic samples have a higher connected porosity than the pore-dominated samples and yet a lower permeability. Tuff samples in Heap et al. (2017b) showed a large range in permeability depending on grainsize and presence and absence of intergranular pore-filling cement. In general, the volcaniclastic samples form two distinct clusters, one at high connected porosity and permeability and a second at low connected porosity and permeability. However, the limited textural data provided by the small area 2D thin sections were unable to consistently distinguish distinct microstructures between these groups (e.g., NM2 1354.2 B is an outlier in Fig. 7a) and this necessitates reliance upon the sensitivity of the permeability to confining pressure to identify microstructure type.

Figure 11 indicates that for the rocks we tested, relatively low connected porosity generally correlates with samples with microfractures while relatively high connected porosity correlates with those containing pores or mixed microfractures and pores. The low connected porosity samples exhibit a relatively large decrease in permeability with increased confining pressure consistent with the closure of microfractures during elastic deformation, reducing permeability within the sample (e.g., Nara et al. 2011). The elliptical pore structure in the samples in most of the higher connected porosity samples is resistant to closure with increasing confining pressure, resulting in comparatively small decreases in permeability (e.g., Lamur et al. 2017). However, the mixed microstructure samples show a much larger decrease in permeability than the pore-dominated samples. From this observation, we can infer that elastic closure of microfractures is important even in samples with high connected porosity. Therefore, we suggest that both microfractures and pores contribute to the permeability such that the control on fluid flow is not mutually exclusive to a single microstructural type, where pores connected by fractures would also be sensitive to confinement. We also emphasize that the microfractures in these altered volcaniclastic rocks are complex with a clear history of both post-fracture dissolution and vein precipitation (Fig. 5).

Effect of increased depth on physical properties

The microstructural analysis indicates that lithology, burial depth, and hydrothermal alteration need to be considered together when interpreting microstructure. Figure 12 shows that there are large variations in permeability with depth, and when all microstructure types are plotted together, there is no systematic relationship for these rocks. The previous studies have reported systematic changes in physical properties with depth (e.g., Wyering et al. 2014) as a result of different alteration processes. At NGF, increased depth is associated with increased geothermal fluid temperatures (Boseley et al. 2012), and corresponding alteration and changes in mineralogy (e.g., Reyes 1990; Tewhey 1977; Wyering et al. 2014). The Tahorakuri samples have variations in hydrothermally deposited minerals, with the shallow samples showing relatively high zeolite content and the deep samples containing zeolite and epidote (Fig. 5). The shallow units show evidence of both dissolution, resulting in open pores, and microfracturing, dissolution, and veining (Fig. 5a, b). Samples with pore-dominated structures (Table 2) tend to have higher connected porosities (> 10%) and come from intermediate depths. This is above the silica precipitation depth (Saishu et al. 2014; Wyering et al. 2014) and porosity is the result of extensive dissolution of pumice clasts with the zeolites, epidote, and clay only partially infilling the resulting large, open pores.

Consistent with Wyering et al. (2014) and Chambefort et al. (2014) deeper samples showed that a higher proportion of the thin section contained secondary minerals and re-crystallisation by quartz and feldspar in veins, matrix, and clasts. In the deepest samples, this appeared to reduce the porosity and permeability by cementing microfractures

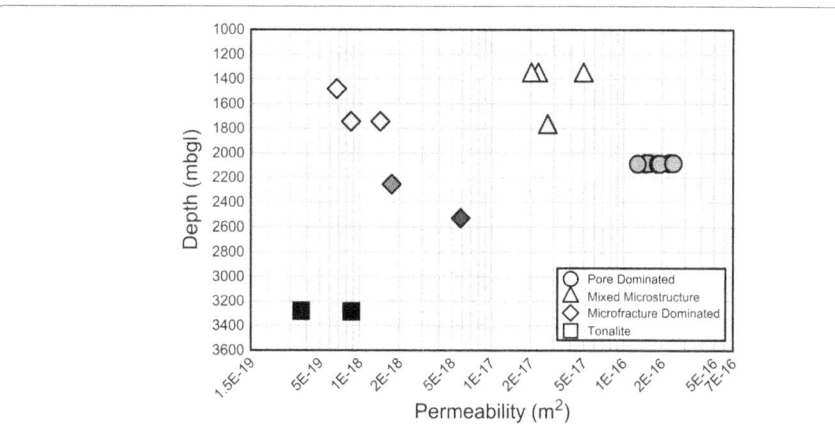

Fig. 12 Depth versus Permeability corrected for lithostatic pressure with lithologies identified. Tone darkens with depth of sample, as shown in legend in Fig. 7a

(quartz veins in Fig. 5e, f) and filling open pores. The two rock types with low connected porosities (< 8%) have similar microfracture-dominated pore structures, but come from a range of depths and their microstructure arose from two different origins: (1) the volcaniclastics may have undergone pore collapse and ductile compaction, alteration, dissolution, and re-precipitation of silica (Saishu et al. 2014) leading to a dominantly microfracture-dominated porosity (Siratovich et al. 2016; Wadsworth et al. 2016). (2) The micro-crystalline tonalite may have never formed pores and only developed fractures in response to cooling (Siratovich et al. 2015) or tectonic stresses (Davidson 2014). Figure 8 also highlights that the tonalite does not sit on the same porosity–density trendline as the volcaniclastics, suggesting that while there is evidence of alteration mineralogy in the tonalite (Fig. 5d, e), it is not as intense as for the other lithologies, contributing to its lack of pores. The identified pore structure type (i.e., pores or microfracture) plays a large role in the permeability, with microfractured samples typically having lower permeabilities that are more sensitive to confining pressure than pore-dominated samples (Fig. 11).

The textural differences in the volcaniclastic units prevent a systematic reduction in connected porosity associated with increasing burial depth and alteration. Figure 13 shows the physical properties from the Tahorakuri Formation plotted against depth (note that the values plotted are averaged from the test results). With the microstructure identified, visible groupings of physical properties within each microstructure group become apparent. For example, the microfracture-dominated samples have a relatively low permeability, fast sonic wave velocity, low connected porosity, and high dry bulk density.

Due to limited data and complex interactions between rock properties and post-deposition alteration and mineralization (e.g., Pola et al. 2014; Kanakiya et al. 2017), it is difficult to isolate the effects burial diagenesis (e.g., Wadsworth et al. 2016). In our study, these fluctuations have been attributed to the variations in primary lithological textures and microstructure as well as secondary alteration, dissolution, and re-precipitation. Other studies of burial diagenesis in geothermal systems have observed similar changes in mechanical properties with depth (Tewhey 1977; Stimac et al. 2004; Mielke

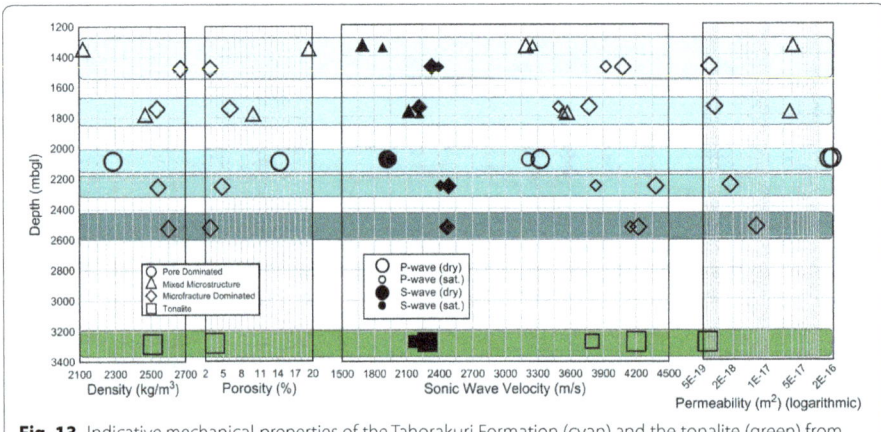

Fig. 13 Indicative mechanical properties of the Tahorakuri Formation (cyan) and the tonalite (green) from the Ngatamariki Igneous Complex arranged by depth. Colours correspond to the colours in Fig. 2

2009; Dillinger et al. 2014; Wyering et al. 2014, 2015). This shows that while the primary textures of the deposited lithologies must contribute to the effects of burial diagenesis by reducing porosity and increasing density, the hydrothermal alteration plays a much larger role in controlling the physical properties.

Conclusions

The objective of this paper was to determine the controls on the intact physical properties of a range of volcanic geothermal reservoir rocks. Because the samples were extracted from a range of depths, it was possible to perform permeability testing at confining pressure representative of the in-situ pressure conditions from which they were extracted. Microstructural analysis was performed in conjunction with the physical testing to compare the physical properties to the microstructural textures, mineralogy, and pore structure.

The main conclusions are:

- The physical properties of the tested samples reflect the microstructure types observed. Minor variations within the physical properties are attributed to variations in lithostatic stress and hydrothermal alteration processes. The volcaniclastic units show a large variation in connected porosity, dry bulk density, sonic velocity, permeability, and microstructure, attributed to the compositional range of pumice and lithic components and depositional processes resulting in vastly different primary textures.

- In general, microfracture-dominated samples have lower permeability than pore-dominated samples. A critical porosity at approximately 10–12% delineates the changeover from microfracture to pore-dominated permeability and response to confining pressure. However, few consistent correlations exist between the limited 2D thin section analysis-based quantitative microstructure shape and spatial density analysis (e.g., microfracture density) and permeability. Increased pore circularity does weakly correlate with increased permeability, likely as a result of embayments at the pore boundary that do not promote fluid flow, although we emphasize that further analysis on larger 3D volumes (as shown in Kennedy et al. 2016) is required to confirm the correlation.

- Samples with a microfracture pore structure have progressively lower permeability with increased confining pressure when compared to samples with a pore-dominated microstructure. The samples with pore-dominated pore structure show a smaller decrease in permeability with increasing confining pressure compared to the microfractured samples. All samples show the largest decrease occurring between 5 and 15 MPa.

- The non-systematic variation in the physical and mechanical properties with depth suggests that lithology, burial diagenesis, and hydrothermal alteration and dissolution all play a role in controlling the physical and mechanical properties of the reservoir rocks. In particular, the variation in the pore structure of lithologies in the Tahorakuri Formation is likely due to variations in sorting and clast density associated with depositional environments.

- The pore-dominated samples show little decrease in permeability with increased confining pressure and, depending on the effects of alteration, therefore, could retain connected porosity and permeability at great depth (> 3000 m). This then indicates that as long as pore-dominated porosity is preserved, the development of deep geothermal resources may be possible.

Our results show that matrix permeability is not simply a function of lithology, depositional environment, diagenesis, or alteration. It is the combination of all of these that leads to particular microstructure types, each of which contributes to matrix permeability differently. Certain processes tend to occur at specific pressures or temperatures, with different fluids and for different lithologies and primary textures. From a geothermal perspective, this suggests that building a permeability model requires careful geological and physical characterisation of the rock and rock mass for each unique geothermal system. This would also be the case for petroleum reservoirs, dewatering and excavations in volcanic systems or hydrothermally altered systems.

Authors' contributions
JLC performed the laboratory testing, analysis, and writing of the manuscript. PAS, JWC, MCV, and BMK conceived the project, secured the funding, selected the samples, supervised the research and analysis, and finalised the completion of the manuscript. All authors read and approved the final manuscript.

Acknowledgements
The authors wish to thank Mercury NZ Limited, Tauhara North No. 2 Trust and Te Pumautanga o Te Arawa Trust for the use of core supplied for this study. The technical staff at the University of Canterbury was invaluable for conducting the laboratory testing.

Competing interests
The authors declare that they have no competing interests.

Funding
This research was funded by Mercury NZ Limited (formerly Mighty River Power).

Appendix

See Tables 6, 7, and 8.

Table 6 Ultrasonic wave velocities and the derived dynamic elastic constants; oven dried and saturated

Sample ID	Ultrasonic wave velocity (oven dried)				Ultrasonic wave velocity (saturated)			
	P-wave velocity (m/s)	S-wave velocity (m/s)	Young modulus (GPa)	Poisson's ratio	P-wave velocity (m/s)	S-wave velocity (m/s)	Young modulus (GPa)	Poisson's ratio
NM2 1354.2 A	3132	1850	17.5	0.23	3038	1609	14.1	0.31
NM2 1354.2 B	3327	1894	19.2	0.26	2975	1769	16.3	0.23
NM2 1354.4 A	3308	1892	19.5	0.26	3568	1692	16.8	0.35
NM2 1788 A	3536	2225	28.7	0.17	3568	2122	27.3	0.23
NM2 2254.7 A	3833	2913	35.2	0.17	4381	2484	39.6	0.26
NM3 1743 A	3401	2238	28.6	0.12	3668	2192	30.0	0.22
NM3 1743 C	3580	2275	30.3	0.16	3873	2239	31.6	0.25
NM4 1477.2 A	3927	2400	37.0	0.20	4077	2330	36.5	0.26
NM8a 2525.5 B	4141	2464	39.0	0.23	4491	2543	42.8	0.26
NM8a 2525.5 C	4149	2488	38.9	0.22	3953	2432	36.5	0.20
NM8a 3280 C	3714	2146	29.0	0.25	3969	2346	34.2	0.23
NM8a 3284.7 C	3883	2180	30.2	0.27	4434	2234	33.2	0.33
NM11 2083 A	3212	1989	21.4	0.19	3287	1881	20.3	0.26
NM11 2083 B	3124	1935	20.1	0.19	3395	1872	20.3	0.28
NM11 2083 C	3270	2057	22.6	0.17	3366	1959	21.8	0.24
NM11 2083.34 A	3186	1887	19.6	0.23	3233	1855	19.3	0.25
NM11 2083.34 B	3147	1947	20.5	0.19	3288	1961	21.4	0.22
NM11 2087.4 A	3284	1967	22.2	0.22	3348	1990	22.8	0.23
NM11 2087.4 B	3175	1914	20.4	0.21	3364	1904	21.0	0.26
NM11 2087.4 C	3319	1884	20.6	0.26	3297	1902	20.8	0.25
NM11 2087.4 D	3182	1968	21.4	0.19	3340	1963	22.1	0.24

Table 7 True permeability at confining pressures from 5 to 65 MPa using Klinkenberg correction

Sample ID	Confining pressures (MPa)						
	5	15	25	35	45	55	65
NM2 1354.2 A	7.80E−17	4.94E−17	3.83E−17	3.39E−17	3.11E−17	2.78E−17	2.63E−17
NM2 1354.2 B	5.26E−17	2.29E−17	2.23E−17	1.91E−17	1.71E−17	1.62E−17	1.52E−17
NM2 1354.4 A	3.87E−17	2.04E−17	1.57E−17	1.41E−17	1.30E−17	1.24E−17	1.18E−17
NM2 1788 A	3.46E−17	2.70E−17	2.36E−17	2.24E−17	2.22E−17	2.12E−17	2.07E−17
NM2 2254.7 A	5.16E−18	2.57E−18	1.76E−18	1.35E−18	9.91E−19	6.91E−19	6.12E−19
NM3 1743 A	6.17E−19	8.62E−19	4.98E−19	3.62E−19	2.57E−19	1.70E−19	1.30E−19
NM3 1743 C	2.95E−18	1.43E−18	9.61E−19	7.17E−19	5.63E−19	4.58E−19	4.29E−16
NM4 1477.2 A	1.58E−18	6.76E−19	1.32E−19	6.16E−20	2.27E−20	9.79E−21	–
NM8a 2525.5 C	2.01E−17	8.15E−18	5.80E−18	4.70E−18	3.98E−18	3.27E−18	2.68E−18
NM8a 3280	2.41E−18	1.29E−18	6.59E−19	3.63E−19	1.40E−19	1.27E−19	–
NM8a 3284.1 A	1.86E−16	1.66E−16	1.60E−16	1.57E−16	1.52E−16	1.51E−16	1.48E−16
NM8a 3284.1 B	1.83E−16	1.64E−16	1.55E−16	1.51E−16	1.47E−16	1.45E−16	1.42E−16
NM8a 3284.1 C	1.68E−16	1.51E−16	1.49E−16	1.41E−16	1.39E−16	1.40E−16	–
NM8a 3284.7 C	5.98E−18	3.02E−18	1.33E−18	8.74E−19	5.45E−19	3.76E−19	3.15E−19
NM11 2083.34 A	2.80E−16	2.55E−16	2.43E−16	2.34E−16	2.33E−16	2.27E−16	2.23E−16
NM11 2083.34 B	2.63E−16	2.39E−16	2.26E−16	2.19E−16	2.13E−16	2.08E−16	2.05E−16
NM11 2087.4 A	1.54E−16	1.38E−16	1.29E−16	1.27E−16	1.23E−16	1.21E−16	1.20E−16
NM11 2087.4 B	2.25E−16	2.01E−16	1.92E−16	1.88E−16	1.84E−16	1.81E−16	1.78E−16
NM11 2087.4 C	2.23E−16	1.95E−16	1.85E−16	1.79E−16	1.72E−16	1.72E−16	1.70E−16
NM11 2087.4 D	1.81E−16	1.56E−16	1.48E−16	1.45E−16	1.42E−16	1.30E−16	–

−, not tested

Table 8 Bulk lithostatic and hydrostatic stress applied to each formation

Bulk stress applied by lithostatic pressure and hydrostatic pressure (MPa)

Formation	NM2		NM3		NM4		NM8a		NM11	
	Litho.	Hydro.	Litho.	Hydro.	Litho.	Hydro.	Litho.	Hydro.	Litho.	Hydro.
Oruanui	1.6	1.1	1.6	1.1	0.64	0.44	0.74	0.51	0.78	0.54
Huka falls Fm	1.6	1.5	1.6	1.5	1.5	1.5	1.7	1.7	1.8	1.7
Rhyolite	2.9	1.6	5.4	3.0	4.3	2.4	1.3	0.74	2.9	1.6
Waiora Fm	0.73	0.34			4.5	2.1	1.8	0.83	3.9	0.18
Whakamaru group ignimbrite	4.0	2.0	2.1	1.0	3.6	1.8	6.5	3.2	1.1	5.6
Tahorakuri sedimentary succession	7.0	3.6	6.2	3.2	2.3	1.2	6.9	3.5	4.8	2.5
Tahorakuri quartz-rhyolite	1.6	6.9	4.1	1.8	4.1	1.8				
Tahorakuri pyroclastic succession	25	11	17	7.3	23	9.6	38	16	28	12
Tahorakuri volcaniclastic andesite breccia	1.3	0.54					2.5	1.0		
Tahorakuri volcaniclastic andesite breccia	3.3	1.3								
Porphyritic microdiorite–diorite							3.9	1.4		
Quartz-bearing diorite							5.8	2.1		
Tonalite							11	4.5		

References

Bai H, Pecher IA, Adam L, Field B. Possible link between weak bottom simulating reflections and gas hydrate systems in fractures and macropores of fine-grained sediments: results from the Hikurangi margin, New Zealand. Mar Pet Geol. 2016;71:225–37.

Barton N. Rock quality, seismic velocity, attenuation and anisotropy. London: Taylor & Francis; 2007.

Bibby HM, Caldwell TG, Davey FJ, Webb TH. Geophysical evidence on the structure of the Taupo Volcanic Zone and its hydrothermal circulation. J Volcanol Geotherm Res. 1995;68(1–3):29–58.

Bignall G. Ngatamariki geothermal field geoscience overview. GNS Sci Consult Rep. 2009;94:41.

Boseley C, Bignall G, Rae AJ, Chambefort I, Lewis B. Stratigraphy and hydrothermal alteration encountered by monitor wells completed at Ngatamariki and Orakei Korako in 2011. In: Proceedings, New Zealand geothermal workshop. 2012. p. 8.

Brace WF, Walsh JB, Frangos WT. Permeability of granite under high pressure. J Volcanol Geotherm Res. 1968;73(6):2225–36.

Cant JL. Matrix permeability of reservoir rocks, Ngatamariki geothermal field, Taupo Volcanic Zone, New Zealand. M.Sc. thesis, University of Canterbury, Christchurch, New Zealand. 2015.

Chambefort I, Lewis B, Wilson CJN, Rae AJ, Coutts C, Bignall G, Ireland TR. Stratigraphy and structure of the Ngatamariki geothermal system from new zircon U–Pb geochronology: implications for Taupo Volcanic Zone evolution. J Volcanol Geotherm Res. 2014;274:51–70.

Colombier M, Wadsworth FB, Gurioli L, Scheu B, Kueppers U, Di Muro A, Dingwell DB. The evolution of pore connectivity in volcanic rocks. Earth Planet Sci Lett. 2017;462:99–109.

Coutts C. Revision of Ngatamariki stratigraphy (unpublished Mighty River Power report). 2013.

Davidson J. The effect of fractures on fluid flow in geothermal systems, Taupo Volcanic Zone, New Zealand. Ph.D. thesis, University of Canterbury. 2014.

Dillinger A, Huddlestone-Holmes CR, Zwingmann H, Ricard L, Esteban L. Impacts of diagenesis on reservoir quality in a sedimentary geothermal play. In: Proceedings, thirty-ninth workshop on geothermal reservoir engineering, Stanford University, Stanford, California. 2014.

Eastwood AA. The Tahorakuri Formation: investigating the early evolution of the Taupo Volcanic Zone in buried volcanic rocks at Ngatamariki and Rotokawa geothermal fields. M.Sc. thesis, University of Canterbury, Christchurch, New Zealand. 2013.

Farquharson IJ, Heap MJ, Varley N, Baud P, Reuschlé T. Permeability and porosity relationships of edifice-forming andesites: a combined field and laboratory study. J Volcanol Geotherm Res. 2015;297:52–68.

Farrell NJC, Healy D. Anisotropic pore fabrics in faulted porous sandstones. J Struct Geol. 2017. https://doi.org/10.1016/j.jsg.2017.09.010.

Géraud Y. Variations of connected porosity and inferred permeability in a thermally cracked granite. Geophys Res Lett. 1994;21(11):979–82.

Glassley WE. Subsurface fluid flow. Geotherm Energy: CRC Press; 2010. p. 51–67.

Guéguen Y, Palciauskas V. Introduciton to the physics of rocks. Princeton: Princeton University Press; 1994.

Heap MJ, Kennedy BM. Exploring the scale-dependent permeability of fractured andesite. Earth Planet Sci Lett. 2016;447:139–50.

Heap MJ, Mollo S, Vinciguerra S, Lavallée Y, Hess KU, Dingwell DB, Baud P, Iezzi G. Thermal weakening of the carbonate basement under Mt. Etna volcano (Italy): implications for volcano instability. J Volcanol Geotherm Res. 2013;250:42–60.

Heap MJ, Lavallee Y, Petrakova L, Baud P, Reuschle T, Varley NR, Dingwell DB. Microstructural controls on the physical and mechanical properties of edifice-forming andesites at Volcan de Colima, Mexico. J Geophys Res Solid Earth. 2014;119(B4):2925–63.

Heap MJ, Kennedy BM, Pernin N, Jacquemard L, Baud P, Farquharson J, Scheu B, Lavallée Y, Gilg HA, Letham-Brake M, Mayer K, Jolly AD, Reuschlé T, Dingwell DB. Mechanical behavior and failure modes in the Whakaari (White Island volcano) hydrothermal system, New Zealand. J Volcanol Geotherm Res. 2015;295:26 42.

Heap MJ, Kennedy BM, Farquharson JI, Ashworth J, Mayer K, Letham-Brake M, Reuschlé T, Gilg HA, Scheu B, Lavallée Y, Siratovich P, Cole J, Jolly AD, Baud P, Dingwell DB. A multidisciplinary approach to quantify the permeability of the Whakaari/White Island volcanic hydrothermal system (Taupo Volcanic Zone, New Zealand). J Volcanol Geotherm Res. 2017a;332:88–108.

Heap MJ, Kushnir ARL, Gilg HA, Wadsworth FB, Reuschlé T, Baud P. Microstructural and petrophysical properties of the Permo-Triassic sandstones (Buntsandstein) from the Soultz-sous-Forêts geothermal site (France). Geotherm Energy. 2017b. https://doi.org/10.1186/s40517-017-0085-9.

Heard HC, Page L. Elastic moduli, thermal expansion, and inferred permeability of two granites to 350 °C and 55 megapascals. J Geophys Res Solid Earth. 1982;87(B11):9340–8.

Hurst AW, Bibby HM, Robinson R. Earthquake focal mechanisms in the central Taupo Volcanic Zone and their relation to faulting and deformation. NZ J Geol Geophys. 2002;45(4):527–36.

Jafari A, Babadagli T. Effective fracture network permeability of geothermal reservoirs. Geothermics. 2011;40(1):25–38.

Kanakiya S, Adam L, Esteban L, Rowe M, Shane P. Dissolution and secondary mineral precipitation in basalts due to reactions with carbonic acid. J Geophys Res Solid Earth. 2017;122(6):4312–27.

Kennedy BM, Wadsworth FB, Schipper CI, Jellinek MJ, Vasseur J, von Aulock FW, Hess K-U, Russell JK, Lavallée Y, Dingwell DB. Surface tension and gas escape from magma. Earth Planet Sci Lett. 2016;433:116–24.

Khandelwal M. Correlating P-wave velocity with the physico-mechanical properties of different rocks. Pure Appl Geophys. 2013;170(4):507–14.

Klinkenberg LJ. The permeability of porous media to liquids and gases. API drilling and production practice. New York: American Petroleum Institute; 1941. p. 200–13.

Klug C, Cashman KV. Permeability development in vesiculating magmas: implications for fragmentation. Bull Volcanol. 1996;58:87–100.

Lamur A, Kendrick JE, Eggertsson GH, Wall RJ, Ashworth JD, Lavallée Y. The permeability of fractured rocks in pressurised volcanic and geothermal systems. Sci Rep. 2017;7(1):6173.

Lane SJ, Gilbert JS. Fluid motions in volcanic conduits: a source of seismic and acoustic signals. London: Geological Society Publishing House; 2008.

Lewis DW, McConchie D. Practical sedimentology. 2nd ed. New York: Chapman & Hall; 1994.

MBIE. Energy in New Zealand, 2016 calendar year edition. Ministry of business, innovation and employment. Crown Copyright 2017. Wellington, New Zealand. 2017. ISSN 2324-5319.

Mielke P. Properties of the reservoir rocks in the geothermal field of Wairakei, New Zealand. Ph.D. thesis, Technische Universität Darmstadt. 2009.

Mueller S, Scheu B, Spieler O, Dingwell DB. Permeability control on magma fragmentation. Geology. 2008;36:399–402.

Mueller S, Scheu B, Kueppers U, Spieler O, Richard D, Dingwell DB. The porosity of pyroclasts as an indicator of volcanic explosivity. J Volcanol Geotherm Res. 2011;203(3):168–74.

Murphy H, Huang C, Dash Z, Zyvoloski G, White A. Semianalytical solutions for fluid flow in rock joints with pressure-dependent openings. Water Resour Res. 2004;40(12):W12506.

Nara Y, Meredith PG, Yoneda T, Kaneko K. Influence of macro-fractures and micro-fractures on permeability and elastic wave velocities in basalt at elevated pressure. Tectonophysics. 2011;503(1–2):52–9.

Olalla C, Hernandez LE, Rodriguez-Losada JA, Perucho Á, González-Gallego J. Volcanic rock mechanics: rock mechanics and geo-engineering in volcanic environments. Boca Raton: CRC Press; 2010.

Peacock DCP, Anderson MW, Rotevatn A, Sanderson DJ, Tavarnelli E. The interdisciplinary use of "overpressure". J Volcanol Geotherm Res. 2017;341:1–5.

Pola A, Crosta GB, Fusi N, Castellanza R. General characterization of the mechanical behaviour of different volcanic rocks with respect to alteration. Eng Geol. 2014;169:1–13.

Reyes AG. Petrology of Philippine geothermal systems and the application of alteration mineralogy to their assessment. J Volcanol Geotherm Res. 1990;43(1–4):279–309.

Ruddy I, Andersen MA, Pattillo PD, Bishlawi M, Foged N. Rock compressibility, compaction, and subsidence in a high-porosity chalk reservoir: a case study of Valhall field. J Pet Technol. 1989;41(07):741–6.

Rust AC, Cashman KV. Permeability of vesicular silicic magma: inertial and hysteresis effects. Earth Planet Sci Lett. 2004;228(1–2):93–107.

Saar MO, Manga M. Permeability-porosity relationship in vesicular basalts. Geophys Res Lett. 1999;26:111–4.

Saishu H, Okamoto A, Tsuchiya N. The significance of silica precipitation on the formation of the permeable–imperme-able boundary within Earth's crust. Terra Nova. 2014;26:253–9.

Shea T, Houghton BF, Gurioli L, Cashman KV, Hammer JE, Hobden BJ. Textural studies of vesicles in volcanic rocks: an integrated methodology. J Volcanol Geotherm Res. 2010;190(3):271–89.

Siratovich PA. Thermal stimulation of the Rotokawa andesite: a laboratory approach. Ph.D. thesis, University of Canterbury, Christchurch, New Zealand. 2014.

Siratovich PA, Heap MJ, Villeneuve MC, Cole JW, Reuschlé T. Physical property relationships of the Rotokawa Andesite, a significant geothermal reservoir rock in the Taupo Volcanic Zone, New Zealand. Geotherm Energy. 2014;2:10.

Siratovich PA, Villeneuve MC, Cole JW, Kennedy BM, Bégué F. Saturated heating and quenching of three crustal rocks and implications for thermal stimulation of permeability in geothermal reservoirs. Intern J Rock Mech Min Sci 2015;80:265–80.

Siratovich P, Heap MJ, Villeneuve MC, Cole JW, Kennedy BM, Davidson J, Reuschle T. Mechanical behaviour of the Rotokawa Andesites (New Zealand): insight into permeability evolution and stress-induced behaviour in an actively utilised geothermal reservoir. Geothermics. 2016;64:163–79.

Stimac JA, Powell TS, Golla GU. Porosity and permeability of the Tiwi geothermal field, Philippines, based on continuous and spot core measurements. Geothermics. 2004;33(1–2):87–107.

Tewhey JD. Geologic characteristics of a portion of the Salton Sea geothermal field. Davis: Geothermal Resource Council; 1977.

Ulusay R, Hudson JA. The complete ISRM suggested methods for rock characterization, testing and monitoring: 1974–2006. In: Ulusay R, Hudson J, editors. Commission on testing methods, International Society of Rock Mechanics. Ankara: IRSM Turkish National Group; 2007.

Underwood EE. Stereology, or the quantitative evaluation of microstructures. J Microsc. 1969;89(2):161–80.

Vasconcelos G, Lourenço PB, Alves CAS, Pamplona J. Ultrasonic evaluation of the physical and mechanical properties of granites. Ultrasonics. 2008;48(5):453–66.

Vinciguerra S, Trovato C, Meredith PG, Benson PM. Relating seismic velocities, thermal cracking and permeability in Mt. Etna and Iceland basalts. Int J Rock Mech Min Sci. 2005;42(7):900–10.

Wadsworth FB, Vasseur J, Scheu B, Kendrick JE, Lavallée Y, Dingwell DB. Universal scaling of fluid permeability during volcanic welding and sediment diagenesis. Geology. 2016;44:219–22.

Watson A. Geothermal engineering: fundamentals and applications. 1st ed. Dordrecht: Springer; 2013.

Wilson CJN, Rowland JV. The volcanic, magmatic and tectonic setting of the Taupo Volcanic Zone, New Zealand, reviewed from a geothermal perspective. Geothermics. 2016;59:168–87.

Wilson CJN, Houghton BF, McWilliams MO, Lanphere MA, Weaver SD, Briggs RM. Volcanic and structural evolution of Taupo Volcanic Zone, New Zealand: a review. J Volcanol Geotherm Res. 1995;68(1–3):1–28.

Wright HMN, Cashman KV, Gottesfeld EH, Roberts JJ. Pore structure of volcanic clasts: measurements of permeability and electrical conductivity. Earth Planet Sci Lett. 2009;280:93–104.

Wu XY, Baud P, T-f Wong. Micromechanics of compressive failure and spatial evolution of anisotropic damage in Darley Dale sandstone. Int J Rock Mech Min Sci. 2000;37(1):143–60.

Wyering LD. The influence of hydrothermal alteration and lithology on rock properties from different geothermal fields with relation to drilling. Ph.D. thesis, University of Canterbury, Christchurch, New Zealand. 2014.

Wyering LD, Villeneuve MC, Wallis IC, Siratovich PA, Kennedy BM, Gravley DM, Cant JL. Mechanical and physical properties of hydrothermally altered rocks, Taupo Volcanic Zone, New Zealand. J Volcanol Geotherm Res. 2014;288:76–93.

10

Criteria and geological setting for the generic geothermal underground research laboratory, GEOLAB

Eva Schill[1*], Jörg Meixner[2], Carola Meller[2], Manuel Grimm[2], Jens C. Grimmer[2], Ingrid Stober[2] and Thomas Kohl[2]

*Correspondence:
eva.schill@kit.edu
[1] Institute for Nuclear
Waste Disposal, Karlsruhe
Institute of Technology,
Hermann-von-Helmholtz-Platz
1, 76128 Karlsruhe, Germany
Full list of author information
is available at the end of the
article

Abstract

High flow rate injection and related hydromechanical interaction are the most important factors in reservoir development of Enhanced Geothermal Systems (EGS). GeoLaB, a new generic geothermal underground research laboratory (URL), is proposed for controlled high flow rate experiments (CHFE) to address limited comprehension of coupled processes connected to EGS reservoir flow conditions. As analogue for typical EGS development, CHFE require specific hydromechanical conditions including a connected fracture network in crystalline basement rock, sufficient hydraulic fracture transmissivities, a strike-slip to normal faulting tectonic regime, controllable hydraulic boundary conditions, and hydrothermal alteration fracture fillings that improve conditions for hydromechanical interaction. With the aim to identify most appropriate areas for future site selection, four criteria have been established based on the EGS reference site of Soultz. Two URLs in crystalline basement worldwide approximate the requirements of a new generic GeoLaB and may be used for accompanying experimentation. Besides favourable geological, hydraulic, and stress conditions, the vicinity to long-term EGS production favours the southern Black Forest as potential region for GeoLaB. Therefore, an exemplary site assessment has been carried out at "Wilhelminenstollen" in the southern Black Forest (Germany). New remote sensing, hydrochemical, and geophysical analyses as well as reactivation potential, and stress modelling were added to complement existing geological and hydrogeological information. At this site, reactivation potential analysis reveals two local maxima prone for shear reactivation as strike-slip faults. The highest lineament density is observed for the N110°E strike direction that is associated with both slip and dilation tendency maxima. Clay minerals occur in fractures and the matrix. Local, partly water-bearing fractures, when partly filled with ore minerals, were connected to veins in the tunnel using shallow geophysical methods. Hydrochemical data reveal infiltration of the tunnel water from at least 500 m above the tunnel. The results suggest a crystalline basement with a fracture network that is regionally connected and water-conducting. Hydraulic conductivity in the southern Black Forest granite is estimated to amount to about $4.5 \cdot 10^{-8}$ m s^{-1} at 500 m depth. The hydraulic boundary conditions exclude unknown drainage. Analyses of the influence of topography on orientation and magnitude of the maximum stress indicate a minimum overburden of about 500 m for regional reactivation to be valid. In conclusion, the southern Black Forest and in particular "Wilhelminenstollen" offers favourable condition for CHFE. Final decision on the GeoLaB site is to be drawn from forthcoming exploration wells.

Keywords: Geothermal underground research laboratory, Fractured crystalline basement, Black forest, GeoLaB

Background

Most geothermal resources worldwide are found in fractured crystalline basement and may be approached using enhanced geothermal systems (EGS) technology (e.g. MIT 2006). This technology dates back to the 1970s, when first hot dry rock (HDR) projects led to the design of artificial sub-surface heat exchangers in crystalline basement rock. The outcome of the Fenton Hill HDR experiment (1970–1995, Los Alamos National Laboratory, USA) was the identification of the hydraulic challenges related to reservoir creation (Brown 2009). Reservoir testing and development revealed that the characteristics of the crystalline basement appear to be highly variable. Some of the related questions have been addressed along the EGS learning curve, particularly at the Soultz-sous-Forêts EGS site and during its follow-up projects. With regard to productivity enhancement of fractured granitic basement, different injection schemes and operation modes have been tested and validated with respect to their efficiency (Nami et al. 2008; Schill et al. 2015). Cyclic injection along with production from a second well has been found to be most effective for reservoir development (Schindler et al. 2010; Schill et al. 2015). Due to perceptible induced seismicity, however, the deep reservoir at Soultz has mainly been treated chemically in the late stage of the development phase. During operation of the Soultz power plant, occurrence of induced seismicity was reduced in frequency and magnitude by distributing the injected volume over different injection points and limiting the overall flow rate (Cuenot et al. 2011). Still, high flowrate injection is a crucial issue for large-scale industrial development and economic viability of EGS.

As such, current EGS development for controlled enhancement of the reservoir is hindered by fundamental understanding of processes occurring at high volumes and rates of flow in fractured reservoir rocks. Additional complexity results among others from high differential stresses, the related mechanical behaviour, and the chemical response of the fluid. Although experimental progress has been made, fundamental constitutive laws of flow and thermo-hydraulic-mechanical-chemical (THMC) coupling presently are not being considered adequately in reservoir engineering and plant operation in fractured environments. It is, therefore, most important to carry out large-scale in situ tests under hydrogeological conditions and comprehensively study these aspects (Freeze and Javandel 2008).

Necessity for real-scale experiments

Brady and Brown (2006) detail five inherent complexities in rock mechanical investigation: (1) fracture mechanics, (2) scale effects, (3) tensile strength, (4) fluid interaction, and (5) alteration. Although being established already 20 years ago these are still most important mechanics topics and remain key aspects in characterising the mechanical influence in a reservoir. They involve especially the bias in evaluation of laboratory and field data in terms of the natural complexity of heterogeneity of the host rock but also of the complexity flow pathways. Out of a large assessment, two aspects may be picked out to describe the need for research in a reservoir simulator.

Scale effects are a well-known limitation in describing mechanical processes in subsurface rock since they may exhibit tremendous changes in strength when varying the sample sizes under laboratory conditions. Peak strength of rock, a key parameter separating the linear from the failure regime (i.e. involving possible seismicity), is highly dependent of rock sample size. Barton et al. (1985) have identified variabilities by more than 50 % by direct shear test with smaller sample sizes involving considerably higher peak strength than larger samples.

The origin of wing-crack mechanisms and their evolution within a shear zone during a stimulation as proposed by Jung (2013) addresses another challenge in reservoir simulation rarely performed in rock mechanics due to not existent opportunities. It involves the creation of new void spaces by the injection of large fluid volumes. Identical mechanisms to wing-cracks in structural geology are proposed when injecting large amounts of fluid in host rock. Under a given stress regime, wing-cracks within a shear zone could connect consecutively in a shear zone until forming tensile fractures.

Besides mechanical interaction, hydraulic effects from high reservoir flow rates need to be studies more in detail to investigate the range of validity for the typically assumed Darcy flow and fluid dynamics need to be carefully evaluated. Zimmerman and Yeo (2000) have shown that the Stokes equations for laminar flow leave their range of validity from $Re > 10$ and new Navier–Stokes solvers need to be applied. In a heterogeneous network with tube-like structures at the interface of fractures (Wennberg et al. 2016) these conditions are easily met and apparent transmissivity decreases with higher flowrate. Until now, there are ample theoretical and numerical considerations of the flow behaviour in fractured rock; however, little was done to quantify these effects in laboratories or under real field conditions (Kohl et al. 1997).

The implications on hydraulic field and rupture mechanisms are obvious and can be easily extended to thermal and chemical impacts. Therewith, injection in fractured reservoirs can include a high variety of interactions induced from hydraulic fields (Gaucher et al. 2015), known as THMC for thermal, hydraulic, mechanical, and chemical interaction (Kolditz et al. 2012). It is most important to address these questions during large-scale observation to calibrate coupled mechanical models also from real experimental testing under reservoir condition. Providing the specific experimental background going beyond the expertise from nuclear research and applying new probabilistic based approaches are important perspectives for reservoir engineering. Careful analyses of probabilistic response spectra together with geological structural analysis resulting in new hydraulic calibration parameters of fracture network models under variable flow rate can only be performed in a 3D environment where processes are monitored in space and time. In this context, underground research laboratories (URLs) are best suited to monitor and quantify the interrelated processes during large volume injection in fractured rock by large-scale in situ observational methods to complement present-day laboratory-scale experiments. Such experimental facilities for application and monitoring of experiments should address key aspects of a safe and economically efficient use of geothermal energy.

Experiments in a geothermal URL

In-line with related geo-disciplines, such as nuclear waste disposal research, we propose a specific URL for geothermal purposes, which is to be located in a typical EGS environment to conduct fundamental experiments in space and time (i.e. 4D). As pointed out by NEA (2013) for nuclear waste disposal research, URLs offer an excellent opportunity to integrate multiple disciplines (e.g. geology, hydrology, and engineering), build technical teams, and gain practical experience that will be invaluable for the next generation of researchers. Decision-making requires practical demonstrations of key technical elements to prove the robustness of the proposed concept as well as to establish confidence in the technology. URLs play an important and multi-faceted role in these scientific assessments and demonstrations, since they provide a realistic environment for characterising and testing the selected technical approaches and materials. In areas of operational safety, acquiring geological information on the repository scale, and constructional and operational feasibility, only URLs can provide reliable in situ data. URLs can also provide tangible benefits in enhancing participation by the general scientific community and confidence of both technical and non-technical stakeholders. The many successful URLs operated for other research purposes reflect their high value.

The 4D controlled high flow rate experiments (CHFE) in a geothermal URL may address the validity of fundamental flow and hydro-mechanical laws in fractured rock, i.e. on surfaces of variable roughness and tortuosity, and shall provide new insights into the relevant petrophysical, hydrogeological, and mechanical processes in space and time. A typical experiment may include fluid injection into an appropriate fracture zone underneath the URL (Fig. 1). Measurement devices in observation wells and on the surface enable full 3D process monitoring and can be adapted to different experiments and scientific purposes. Furthermore, the URL may provide access to a modified surface for further investigation. A schematic setting of the planned geothermal URL, GeoLaB, is shown in Fig. 1. Our concept of such an infrastructure facility includes a 1–2 km long tunnel with 3–4 side caverns, from which the CHFEs and other key experiments can be performed (Fig. 1b).

In order to prepare the site selection for GeoLaB, in this study, we establish criteria for a generic geothermal URL for CHFE. Applying these criteria to existing URLs, we demonstrate the necessity for a new generic URL. Based on these results, the Southern Black Forest appears to reveal appropriate conditions and is selected for further investigation. Furthermore, the study includes a review of suitable existing mines in the Black Forest. Since it is crucial to install the GeoLaB in an environment close to hydraulic reservoir conditions, the detailed planning concept needs to follow the tectonic setting in the adjacent rock. Especially, the structure of the target fault zone imposes boundary

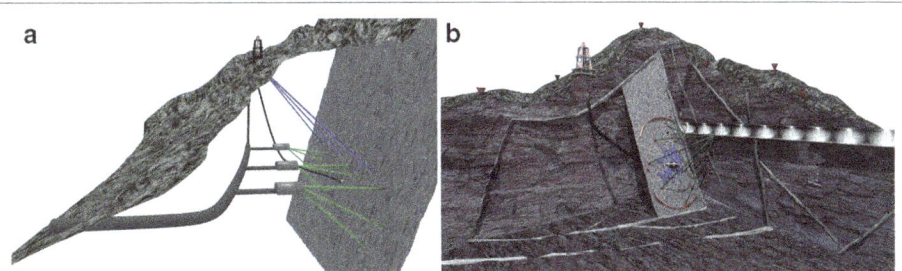

Fig. 1 a Possible layout of the planned geothermal URL (not to scale), GeoLaB, with an access tunnel to a fracture zone, three experimental caverns, and exploration wells (*blue*) and observation wells (*green*) from the tunnel into the fracture zone. In addition to the monitoring stations (*red cones*), the main injection well (*black*) for controlled high-flow experiments (CHFE) is reached from surface boreholes targeting the fracture zones next to experimental caverns. **b** Possible CHFE with high flow rate injection from the surface into a fracture zone (*blue volume*) and related mechanical deformation indicated by *red ring* segments

conditions on URL construction and the installation of tools to achieve the scientific targets. In this study, we describe an exploration procedure aiming at the characterisation of a fault zone in a target rock that is predestined for geothermal production in central Europe. In this respect, a case exploration study is carried out at "Wilhelminenstollen" in the southern Black Forest, assuming an extension of existing mining structures for construction of a geothermal URL. Based on the results the necessary exploration steps towards a test site in the Variscan crystalline rock are detailed.

Criteria for a generic geothermal URL

In analogy to the procedure chosen in nuclear waste disposal research, it is distinguished between "generic" scientific investigations in analogue URLs and "site-specific" geothermal URLs linked to a real operation (NEA 2013). In geothermal research, site-specific investigations are to be conducted under real conditions in the deep subsurface using tomographic approaches between geothermal wells (e.g. DOE FORGE project, http://www.energy.gov/eere/forge). Complementary, GeoLaB is intended to be developed as generic URL to gain general experience with respect to geological, physical, and chemical processes under high flow rate conditions, including model testing and verification of investigation and measurement techniques and to identify process interactions by means of experiments. Generic URLs are to be installed in the host rock type, but not in the particular geological formation. Although closely mimicking hydraulic or geological reservoir conditions, generic geothermal URLs are not able to reproduce parameters, such as temperature, stress magnitude, and geochemistry in a specifically selected host rock formation.

A geothermal URL designed for CHFE is associated with specific requirements on hydraulic and geological properties of typical EGS reservoirs and, hence, on the GeoLaB site. EGS technology is best studied in the Upper Rhine Graben (URG). This accounts for the Soultz project, the unique proto-type of EGS, in particular, since nowhere else worldwide so many hydromechanical stimulation experiments of different type have been accomplished. Lithologic, hydraulic, and stress conditions in a geothermal URL should thus be comparable to conditions encountered there, but also be transferrable to the crystalline basement of Central Europe and worldwide. Note that URG's host rocks are exposed in the adjacent low mountain ranges. Against this background, we aim at installing GeoLaB at a site approximating Soultz condition to reduce the number of degrees of freedom when comparing results of CHFE to reservoir experiments. Furthermore, since only few developed EGS sites exist worldwide, the definition of a typical EGS environment is based on the theoretical considerations of Garnish (2002) and the Soultz-sous-Forêts EGS project.

EGS reference Soultz-sous-Forêts

The Soultz site is located in the southern part of the URG's thermal anomalies (Baillieux et al. 2013). The reservoir in Soultz comprises two different granite varieties:

- a porphyritic K-feldspar monzogranite (from 1420 to 4700 m) with a highly altered and fractured intermediate section (between about 2700 and 3900 m) having a fracture densities of up to 2.86 m^{-1} for fractures with an acoustic aperture of >1–2 mm is observed in the upper part (Genter and Castaing 1997; Dezayes et al. 2005). A correlation between alteration zones and permeable fractures as well as an increased tendency to shear during stimulation was found for this section (e.g. Evans et al. 2005).

- In the lower part (from 4700 to 5000 m) a two-mica granite with fracture densities of up to 1.97 m^{-1} is observed (Dezayes et al. 2005).

In the granitic basement at the Soultz site, three reservoir levels have been developed at about 2000, 3500, and 5000 m depth. The different reservoirs fulfil most of the EGS criteria (Table 1) as defined by Garnish (2002). The upper reservoir extending from the top of the basement at 1420 m depth down to about 2000 m includes an approximately 100 m thick alteration zone at its top. This zone is mainly characterised by precipitated hematite, low magnetic susceptibility (Rummel and König 1991), high electric conductivity (Geiermann and Schill 2010), and high values of heat production of up to 7 μW m^{-3} (Pribnow 2000). As regards reservoir engineering, at the Soultz site, about 35 major injection and production experiments have been carried out, during which different types of hydraulic and chemical stimulations were tested and evaluated for their effectiveness (Schill et al. 2015). Thus, Soultz can be considered a reference EGS site.

The production temperatures are higher than 140 °C, the stimulated rock volume is larger than 2·10^8 m^3, hydraulic impedance is lower than 0.1 MPa kg^{-1} s^{-1} (in the wells GPK1 and GPK2), and the produced fluid can be re-injected to the complete extent. The effective heat exchange area has not been estimated. The crucial flow rate of 50–100 L s^{-1} was not reached in Soultz. On the contrary, it had to be reduced to about 15 L s^{-1} in 2013 (Schill et al. 2015). Overall hydraulic conductivity in the non-stimulated reservoirs ranges from 1.2·10^{-9} m s^{-1} in the upper to 2·10^{-8} m s^{-1} in the intermediate to <2·10^{-9} m s^{-1} in the deep reservoir. Separating the most permeable fracture at 3490 m depth from the test section reduces the hydraulic conductivity by one order of magnitude to about 1·10^{-9} m s^{-1} in the intermediate reservoir (Jung et al. 1995). Hydraulic conductivities in the fractured zone between 2850 and 3100 m were determined to be

Table 1 Requirements to be met by EGS according to Garnish (2002) and reservoir parameters of the Soultz-sous-Forêts EGS

	EGS requirement	Soultz 3.5 km reservoir	Soultz 5 km reservoir
Flow rate (L s^{-1})	50–100	24	23
Mean wellhead fluid temperature (°C)	150–200	140	157.5
Effective heat exchange area	>2·10^6 m^2	N/A	N/A
Rock volume (m^3)	>2·10^8	About 7·10^8	About 2.7·10^9
Hydraulic impedance	<0.1 MPa L^{-1} s^{-1}	0.06 MPa L^{-1} s^{-1} (GPK1, injection)	0.1 MPa L^{-1} s^{-1} (GPK1, injection)
		0.06 MPa L^{-1} s^{-1} (GPK2, production)	0.05 MPa L^{-1} s^{-1} (GPK2, production)
			0.25 MPa L^{-1} s^{-1} (GPK3, injection)
			0.25 MPa L^{-1} s^{-1} (GPK4, production)
Water loss at the surface	<10 %	0 % (Total reinjection)	0 % (Total reinjection)

Data from the 3500 and 5000 m reservoirs were obtained in 1997 and 2011, respectively (modified after Schill et al. 2015)

up to $6 \cdot 10^{-8}$ m s^{-1} (Sausse et al. 2006). The lower limit of hydraulic conductivity of the matrix is $>10^{-11}$ m s^{-1} (Rummel and König 1991).

During reservoir development, the factors determining the effectiveness of hydraulic stimulation are the injection volume, downhole pressure, flow rate, injection scheme, and the ambient stress field. The stress field at the Soultz site was determined using drilling-induced tensile fractures and borehole breakouts (Valley and Evans 2007) as well as additional seismic information (Cornet et al. 2007). Mean orientations of the maximum principal stress component (S_{Hmax}) of $169 \pm 14°$ and 175 ± 10 obtained by means of the respective methods. The stress field at Soultz is characterised by a transition from normal faulting (NF) to the strike slip regime (SS) at a depth of about 3200 m. In the NF regime of the upper and intermediate reservoirs, maximum slip tendency coincides with the orientation of S_{Hmax}, while maxima are forecast at conjugated angles of $30°$ to S_{Hmax} in the SS regime of the intermediate and deep reservoirs. With respect to the reactivation potential, these structural trends are most favourably oriented in the ambient state of stress (Fig. 2). Hence, they are supposed to control subsurface fluid flow and associated reservoir processes in Soultz.

The significance of such regional forecast for reservoir processes was verified by observations made during different hydraulic stimulation experiments. For example, the 00JUN30 stimulation of GPK2 well below the NF/SS transition zone (open hole section: 4402–5026 m) in June/July 2000 caused an overall seismic cloud comprising about 7200 single events. Maximum orientation of this cloud is oriented parallel to S_{Hmax} (Hettkamp et al. 2004). Although stress re-orientation close to fault zones is not taken into account, the distribution of the failure plane orientations (Fig. 3) follows the distribution expected on the basis of slip and dilation tendency analyses.

The geothermal reservoir at the Soultz site has experienced significant fracturing and complex alteration since Variscan times. The resulting distinct fracture networks provided pathways for hydrothermal fluids, thus enabling fluid-rock interactions between geothermal brines and reservoir rocks. Pervasive and localised vein alterations can be distinguished in the reservoir rocks. An early propylitic, weak iso-chemical alteration of the matrix without a change of the granite texture incorporates partial transformation of biotite and hornblende into chlorite, the replacement of plagioclase by illite, and the formation of hydrogarnet and epidote within the granite (Genter et al. 2000). A second alteration includes strong chemical and textural changes mainly within and in the immediate vicinity of fractures (Fig. 4). Alteration ranges from moderate to extreme grades (e.g. Meller et al. 2014). Typically, a brecciated and cataclased zone is observed between the protolith and a central, often-sealed quartz vein. Here, primary silicates, such as feldspar and biotite, were transformed into clay minerals, with the original texture of the granite being destroyed by shearing. The precipitation in veins contains secondary quartz, barite, illite, carbonates, iron oxides, and, locally, mixed-layers of illite–smectite and chlorite-smectite (Genter et al. 2000; Schleicher et al. 2006). In some altered zones, quartz, biotite, plagioclase, and hornblende are dissolved completely (Ledésert et al. 2010).

The average porosity of the fresh granite is <1 % (Genter et al. 2000). For hydrothermally altered samples, they determined porosities of 1.7–25 % (often higher than those of breccia and microbreccia) using mercury porosity measurements. The porosity of

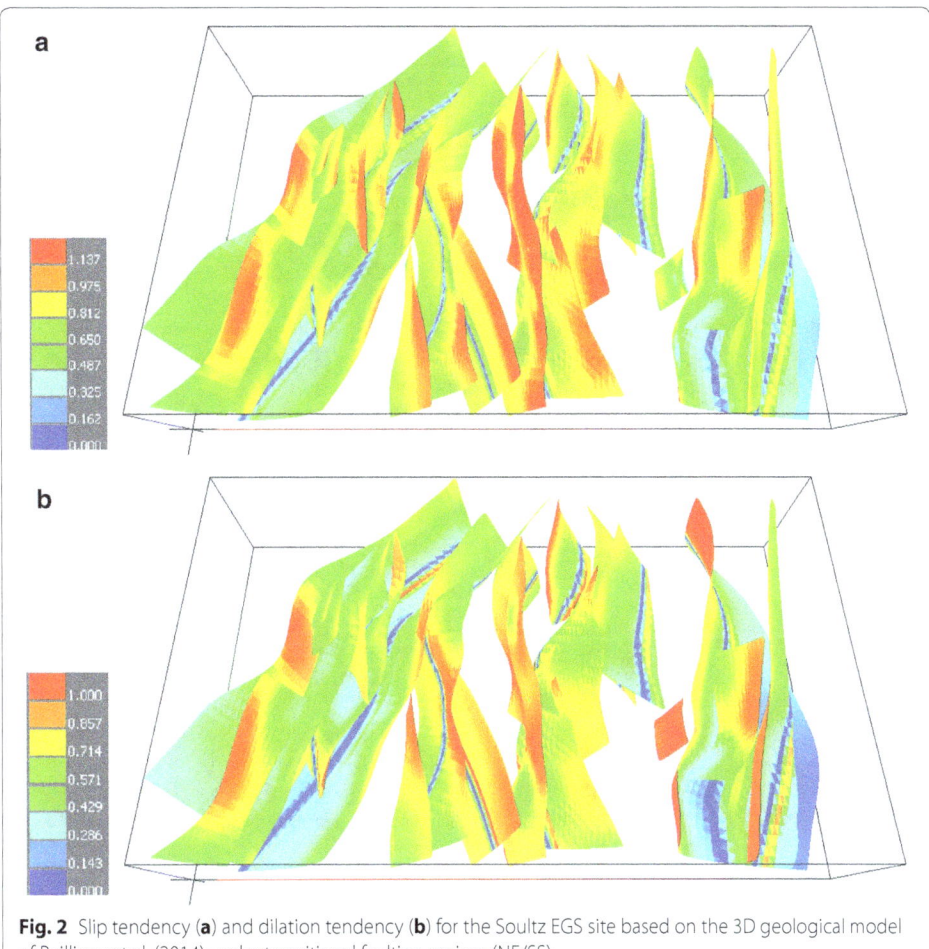

Fig. 2 Slip tendency (**a**) and dilation tendency (**b**) for the Soultz EGS site based on the 3D geological model of Baillieux et al. (2014) under transitional faulting regime (NF/SS)

altered plagioclase, which makes up to 40 % of the rock volume (Ledésert et al. 1999), is between 20 and 27 %.

Fracture apertures were determined using different techniques. Electric apertures obtained from azimuthal resistivity images range from about 1 μm to 1 mm (Sausse and Genter 2005). Apertures for the upper reservoir were determined from three different core samples (Sausse 2002). Strongly altered granite with chlorite sealing (from 1557.8 m depth) revealed a mean aperture of 0.448 ± 0.101 mm. Weakly altered granite with hematite sealing (from 1789.6 m depth) was found to have a mean aperture of 0.518 ± 0.257 mm and unaltered granite with drilling-induced fractures (from 2075.46 m depth) had a mean aperture of 0.748 ± 0.101 mm. An upper limit of apertures is given by the width of the fault-related damage and alteration zones (Massart et al. 2010). For the largest group of fractures with a mean lateral extension of about 2700 m, a mean width of damage/alteration zone of about 6 m of (with a maximum width of 13 m) was determined by analysing UBI logs. With decreasing fault size, the relation between extension and width follows a power-law distribution.

Mechanical weakening of rock in hydrothermal alteration zones and the influence of hydrothermally altered fluid pathways are significant for hydro-mechanical processes (e.g. Meller and Kohl 2014; Sausse 2002; Zoback et al. 2012). E-moduli of hydrothermally

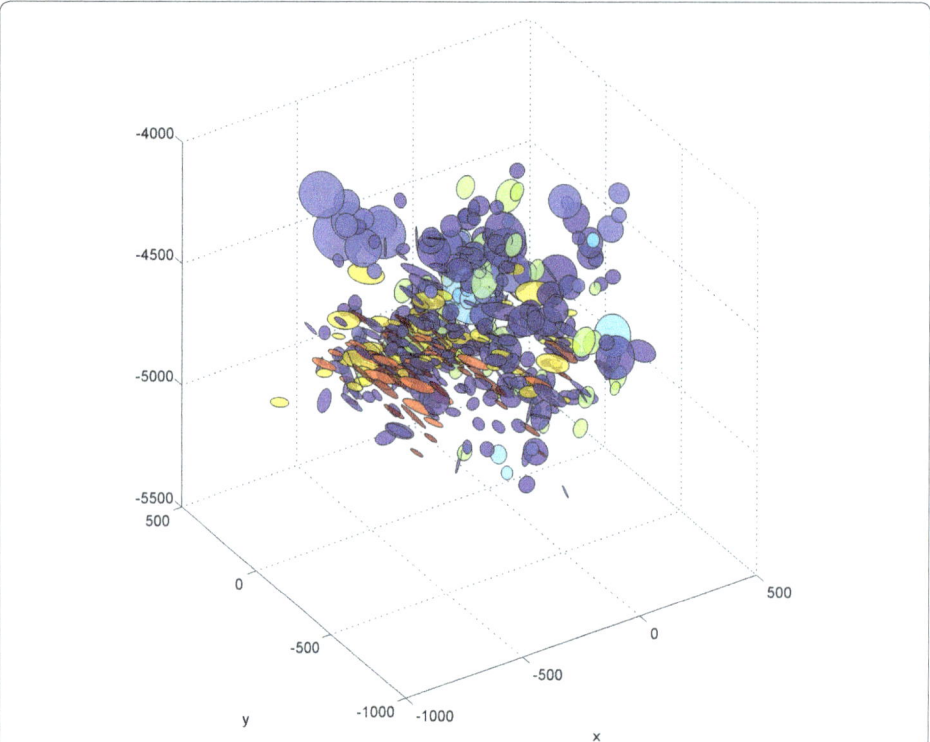

Fig. 3 Orientation and size of the failure planes during the stimulation 00JUN30 of the GPK2 open hole section from 4431 to 5084 m including 7200 registered and 715 located events, with focal plane solution, modified after Schoenball et al. (2012). *Colour* indicates different fracture families

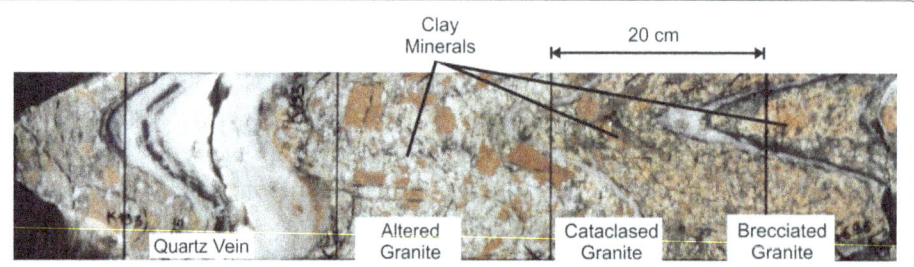

Fig. 4 EPS1 core K195 at 2156 m depth (*top*) showing the zonation of an altered fault. Two consecutive horizontal lines represent 20 cm. A quartz vein is often sealing the centre of a fracture and is surrounded by an alteration halo, characterised by an assemblage of secondary minerals formed during hydrothermal alteration. The adjacent brecciated and cataclased zones are highly porous and permeable zones (Meller et al. 2014)

altered granite samples, for example, are reduced from about 55 to 40 GPa compared to fresh rock samples (Valley and Evans 2007). Furthermore, a correlation between the occurrence of aseismic creep on fractures and the clay content (alteration grade) of fractures and between the cumulative occurrence of borehole breakouts and a high density of clay-filled fractures was observed. Moreover, the maximum magnitude of seismic events was observed to decrease with increasing clay content inside fractures. Evans et al. (2005) demonstrated that hydrothermally altered fractures can be enhanced best during stimulation. Thus, target candidates for geothermal exploration are fractures and faults, which are (1) optimally oriented in the ambient stress state, (2) critically stressed for shear reactivation, and (3) show increased fluid conductivities due to alteration processes.

Selection criteria for a geothermal URL

Considering the relevant observations at Soultz and the controllable and specific boundary conditions required for the planned CHFEs, a geothermal URL focusing on EGS should fulfil the following four selection criteria:

1. Complex geological boundary conditions are to be avoided. A rather homogenous crystalline matrix with a high density of connected and highly permeable natural fractures, i.e. a fracture network, provides optimal conditions for CHFE. In the engineered volume of Soultz with a vertical reservoir extension of about 3500 m, maximum mean fracture length and damage zone size are of about 2700 and 60 m, respectively (Massart et al. 2010). A realistic rock volume developed by an URL is $\ll 2{\cdot}10^8$ m^3 (Table 1), which is much smaller than the defined EGS reservoir size suggested by Garnish (2002). Hence, we may consider fracture densities on the order of 2–3 m^{-1} with major fault and fracture zone lengths on the order of <400–500 m (Sausse et al. 2010). The width of these zones (including damage and alteration zones) is estimated to be on the order of up to <1–2 m (Massart et al. 2010). The lower limit of the overall hydraulic conductivity is set by the initial hydraulic conductivities of the Soultz reservoirs, which range from about 10^{-9}–10^{-8} m s^{-1} and the lower bound of matrix conductivity of about >10^{-11} m s^{-1}. Considering enhancement by stimulation (factor 10), we set this criterion to a hydraulic conductivity of >10^{-7}–10^{-6} m s^{-1}. This is in agreement with the permeability estimated by Kohl et al. (2000) for the fracture zones at Soultz.

2. Controllable hydraulic boundary conditions cause CHFEs to act on natural fractures or matrix, only. This is a pre-requisite for the validity of the CHFEs and the follow-up numerical models. Extensive drainage to surrounding known or unknown adits that may occur in areas of historic mining activities must be excluded.

3. Hydrothermal alteration products in the matrix and along the faults are to cover the illite to smectite range. Clay minerals are known to reduce the mechanical/frictional strength of fractures, i.e. the pressure required to cause shear during hydraulic stimulation. The contribution of mechanically induced rock alteration should be small, because the profound disintegration of large volumes of rock (e.g. grus) results in hydraulic conditions, which are no longer controllable (e.g. in terms of discrete hydraulic pathways, discrimination between matrix and fault, often no symmetry of the fault structure).

4. In order to carry out CHFEs, fractures that are favourably oriented for reactivation in the ambient stress field are an intrinsic pre-requisite of a geothermal URL. Since the stress field orientation across the European crystalline basement is highly variable, no specific stress field can be chosen, but a favourable reactivation potential has to be considered instead. Since the geothermal resources occur in normal or strike-slip regimes rather than thrust fault regimes, we may consider the latter as less favourable. Non-representative, atypical stress perturbation resulting from, e.g. the

glacial rebound should be avoided. In mountain regions, significant perturbations of the regional stress field pattern by local topography need to be taken into account. To avoid such additional variation in the stress field that biases CHFEs with respect to reservoir condition, for the URL design, it is strived for a maximum variation in magnitude of the principal stress of <10 %. Depending on the topography of a potential URL site, a minimum depth below terrain level has to be considered.

URLs existing in the crystalline basement

A number of URLs worldwide have been installed in crystalline basement rock (Fig. 5; Table 2). Except for the Josef URL, which comprises a number of different lithologies also of sedimentary origin, the host rocks exhibit only few lithological changes across the individual URLs. Host rocks range from gneissic (e.g. in the Mine Reiche Zeche) to granitic basement. The fractures at the Lindau Test Site and the Mine Reiche Zeche are commonly filled with ore minerals (Himmelsbach et al. 1998; Bayer 1998).

In the following section, suitability of the existing URLs for CHFEs will be discussed using the above four selection criteria. A compilation of the criteria for the URLs is given in Table 3. It should be noted that (1) the URL Lac du Bonnet is dismantled, (2) Onkalo is a site-specific URL for nuclear waste disposal and (3) the two URLs Sudbury and Sanford are not suitable for CHFE due to their sensitive instrumentation. All four URLs are, therefore, excluded from further consideration.

Criterion 1 is the crystalline lithology, including a well-connected fracture network and a high fracture transmissivity. Based on the minimum hydraulic conductivity range and the expected fracture apertures given above, for CHFEs, URLs with fracture transmissivities in the range of $>10^{-4}\,\mathrm{m}^2\,\mathrm{s}^{-1}$ may be take into account. In Europe, such condition are found in the Lindau Test Site and the Äspö Hard Rock Laboratory (HRL) and partly in the Mine Reiche Zeche (Table 2). Hydraulic testing revealed well-connected fracture networks at Äspö HRL and the Black Forest, in which the Lindau Test site is situated (Stanfors et al. 1999; Stober and Bucher 2014). At Grimsel, for instance, alpine tension fractures may have larger apertures and, hence, transmissivity, but are typically

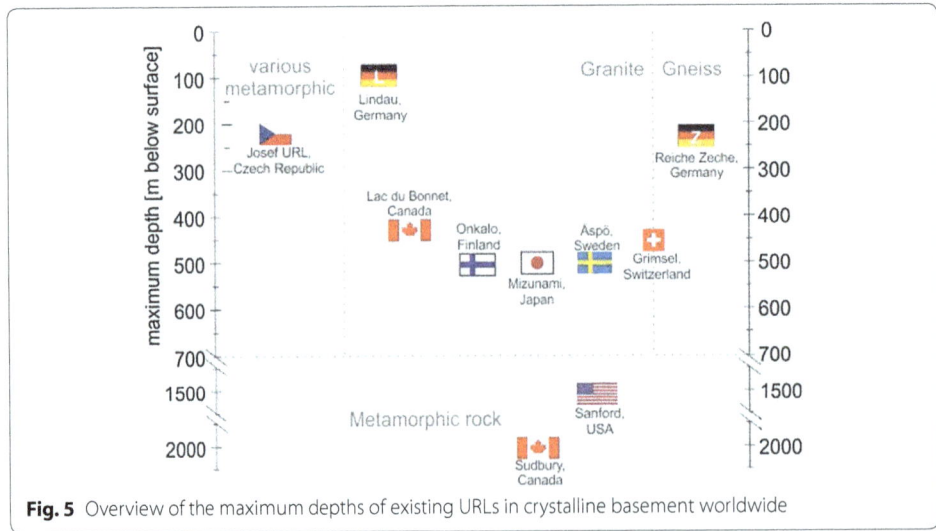

Fig. 5 Overview of the maximum depths of existing URLs in crystalline basement worldwide

Table 2 Overview of general hydraulic conditions and secondary fracture fillings in selected URLs in the crystalline basement worldwide (e.g. Himmelsbach et al. 2003; Stanfors et al. 1999; Himmelsbach et al. 1998; Komulainen et al. 2014; Davison 1984; Kumazaki et al. 2003)

Name	Hydraulic properties	Secondary minerals in fractures
Grimsel Test Site, Switzerland	MHC: 10^{-12}–10^{-11} m s^{-1}	Albite, epidote, muscovite, mica, chlorite, calcite
	FT: up to 10^{-10}–10^{-5} m^2 s^{-1}	CM: <10 % mainly illite
Äspö Hard Rock Laboratory, Sweden	MHC: >10^{-12} m s^{-1}	Chlorite, epidote, zoisite, albite, calcite, fluorite, sericite, and zeolites, hematite
	FT: up to >10^{-4} m^2 s^{-1}	CM: illite, mixed-layer clay
	Overall pumping rate: about 20 L s^{-1}	
Onkalo, Finland	FT (depth <85 m): up to >10^{-6} m^2 s^{-1}	Fe-sulphides, pyrrhotite, pyrite, and calcite
	FT (depth >85 m): up to >10^{-7} m^2 s^{-1}	CM: illite, smectite-group, kaolinite
AECL URL, Lac du Bonnet, Canada	MHC: <10^{-11} m s^{-1}	Chlorite, iron oxides, carbonates, epidote
	FT: up to >10^{-3} m^2 s^{-1}	CM: present (type N/A)
Mizunami URL, Japan	HC (fresh Toki granite): >10^{-7} m s^{-1}	Quartz, plagioclase, K-feldspar and biotite, iron oxides, chlorite, calcite
	HC (altered Toki granite): >10^{-8} m s^{-1}	CM: kaolinite, montmorillomite
Josef URL, Czech Republic	HC (overall): 2–5·10^{-9} m s^{-1}	Ore (gold,...), quartz, calcite, barite
		CM: n/a
Mine Reiche Zeche, Freiberg, Germany	MHC: <10^{-12} m s^{-1} FT: up to 10^{-8}–10^{-4} m^2 s^{-1}	Ore: sulphides, quartz, fluorite, sulphates, barite, feldspars, pyrite CM: n/a
Lindau test site, Black Forest, Germany	HC (fractured ore dike): 5·10^{-6}–10^{-4} m s^{-1}	Quartz, fluorite, barite, ore
	HC (fractured granite): 5·10^{-10}–10^{-8} m s^{-1}	CM: kaolinite
	FT: up to >10^{-6}–5·10^{-4} m^2 s^{-1}	

HC hydraulic conductivity, *MHC* matrix HC, *FT* fracture transmissivity, *CM* clay minerals

Table 3 Criteria match of existing URLs in the crystalline basement worldwide

Name	Criterion 1:		Criterion 2:	Criterion 3:	Criterion 4:	
	Fractured and rather homogenous crystalline matrix	Fracture transmissivity > 10^{-4} m^2 s^{-1}	Controllable hydraulic boundary condition	Comparable alteration (clay minerals)	Suitable stress condition	Sufficient overburden
Grimsel Test Site	Yes	No	Partly	No	No	Yes
Äspö Hard Rock Laboratory	Yes	Yes	Yes	Yes	Limit	Yes
Mizunami URL	Yes	Limit	Yes	No	No	Yes
Josef URL	No	No	Yes	No	Yes	Limit
Mine Reiche Zeche	Yes	Limit	No	Yes	Yes	Limit
Lindau Test Site	Yes	Yes	Yes	N/A (No)	Yes	No

Limit at the limit of the criteria. *Partly* the hydraulic boundary conditions are generally controllable, but due to storage lakes above the URL, special equipment may be needed

short or poorly connected (Himmelsbach et al. 2003). Geothermal energy development activities at Josef URL and Grimsel Test Site focus on HDR or multi-frac technology aiming at engineering naturally impermeable crystalline rock. These approaches represent the opposite end-member of EGS technology are complementary to CHFE.

Criterion 2, controllable hydraulic boundary conditions, further reduces the number of URLs suitable for CHFEs. At the Mine Reiche Zeche, 800 years of continuous mining activities resulted in a dense network of adits causing complex and non-controllable boundary conditions for CHFEs. It should be mentioned that topography-driven fluid circulation at the Lindau Test Site and influence from the Baltic Sea at the Äspö HRL lead to a certain complexity of natural boundary conditions. In the latter case, this mainly concerns the chemical composition of the fluid.

Criterion 3, hydrothermal alteration in the illite–smectite range, is fulfilled at the Äspö HRL. At the Lindau Test Site, only kaolinite is found as clay mineral. Interestingly, clay minerals occur preferably in NW–SE striking fractures (Himmelsbach, pers. comm.). However, hydrothermal alteration has not been studied in detail at the Lindau Test Site. In the Black Forest, alteration often resembles clay mineral assemblage at the Soultz site. For example, 60–90 wt% of illite, smectite, and chlorite are observed in altered granitic rocks of the southern Black Forest batholith (Brockamp et al. 2015).

Criterion 4 refers to suitable regional and homogenous stress conditions across the URL. Near the Äspö HRL, various stress measurements were made in the tunnel and nearby wells using different measurement techniques. The results are summarised, e.g. by Hakami (2003). Large scattering of S_{Hmax} orientation and principal stress magnitudes is observed. The scattering occurs across the entire depth range down to 800 m and appears to be independent of the measurement technique. Thus, regional and in situ stress states are not fully constrained. However, a mean S_{Hmax} orientation of N119°E–N138°E is observed in the SE Sweden area (Stephansson et al. 1991) and linear stress–depth relations predict a transitional stress state between thrust and strike-slip faulting with $S_v \approx S_{hmin} \ll S_{Hmax}$ (Hakami and Min 2009; Klee and Rummel 2002). Hence, fault and fracture reactivation may occur on three differently oriented failure planes. If S_{hmin} is the smallest principal stress magnitude, a strike-slip faulting regime is supposed to reactivate vertical structures at two conjugated 30° angles around S_{Hmax}. If S_v is smallest, a thrust faulting regime is expected to reactivate sub-horizontal structures with a dip of about 30°. All three fracture orientations are predominantly observed at Äspö (Rehn et al. 1997). Favourably oriented and critically stressed structures additionally show increased fracture conductivities. In the borehole KAS06, for example, water-conducting fractures strike around 132° and are aligned almost parallel to S_{Hmax} (Sehlstedt and Strahle 1991).

At the Lindau Test Site, application of a slip- and dilation tendency analysis on the resulting fault pattern reveals the high reactivation potentials for the N–S striking ore dike as strike-slip fault and for NW–SE striking structures as normal faults within the ambient stress field (see below). However, strong perturbation of the stress field inside the existing is expected due to a steep slope topography in combination with a low overburden of only 90 m. This is considered insufficient for providing rather topography-independent stress conditions in the tunnel.

It can be concluded that the Äspö HRL and Lindau Test Site approximate best criteria for CHFEs. While at the Lindau Test Site insufficient overburden represents a criteria of exclusion for the existing design of the tunnel, at Äspö HRL the general stress situation is unsatisfactorily described and furthermore includes most likely thrust faulting. Both URLs may be used in a qualified sense for complementary experiments. In agreement with these findings and the criteria for GeoLaB, we have carried out further studies in the southern Black Forest for the following reasons:

1) The basement outcropping in the Black Forest represents to a large extent the lithology of the geothermal host rock in the URG.
2) Appropriate hydrothermal alteration is widely observed in the Black Forest (Brockamp et al. 2015).
3) Extended fracture networks are indicated (Stober and Bucher 2014).
4) Apart from overburden the Lindau Test Site indicates the general suitability of the southern Black Forest. Hydraulic condition fixed in criterion 1 are achieved (Himmelsbach et al. 1998) and the regional stress field represents or at least approximates conditions of major EGS projects in the URG.

Geology and hydrogeology of the black forest

Geology of the Black Forest

From N to S, the Black Forest is subdivided into four zones: the Baden-Baden Zone (BBZ), Central Black Forest Gneiss (CSGC), and Southern Black Forest Granite Complexes (SSGC) that are separated by the Badenweiler-Lenzkirch Zone (BLZ) (Fig. 6). The low-grade metamorphic rocks of BBZ comprise schists, meta-greywackes, and marbles and are separated from the medium to high-grade metamorphic rocks of the CSGC by a major NE–SW striking dextral-transpressive ductile shear zone. Shearing occurred between 335 Ma and the intrusion of an about 325 Ma old biotite-muscovite granite that was cataclastically deformed during sinistral brittle reactivation (Eisbacher et al. 1989; Hess et al. 2000; Kober et al. 2004; Wickert et al. 1990). The CSGC essentially consists of migmatitic and, locally, mylonitic biotite-plagioclase-gneisses hosting irregularly shaped and sized bodies of amphibolites and eclogites, serpentinised peridotites, and granulites (Kalt and Altherr 1996; Marschall et al. 2003). It was intruded at about 330–325 Ma by batholiths of different size and numerous granitic to porphyritic dykes (Hess et al. 2000). The BLZ is an about 5–10 km wide metasedimentary-metavolanic succession that includes low-grade to non-metamorphic Ordovician to late Carboniferous sedimentary and volcanic rocks (e.g. Krecher and Behrmann 2007). It was overthrust by the CSGC along the dextral transpressive N- to NW-dipping, mylonitic-cataclastic Todtnau thrust (Krohe and Eisbacher 1988; Eisbacher et al. 1989). The southern boundary of BLZ is a WNW-trending shear zone (e.g. Krecher and Behrmann 2007). The SSGC was intruded by the large South Black Forest Batholith between 334 and 328 Ma (Schaltegger 2000; Todt 1976).

Major brittle, commonly mineralised fault zones are observed in abandoned mines (e.g. Werner and Dennert 2004) and quarries and—on a regional scale—at the basement-cover interfaces. During late Variscan times, i.e. from about 310 to 330 Ma, deep and high-level crystalline basement rocks cooled to below 300 °C in the Moldanubian

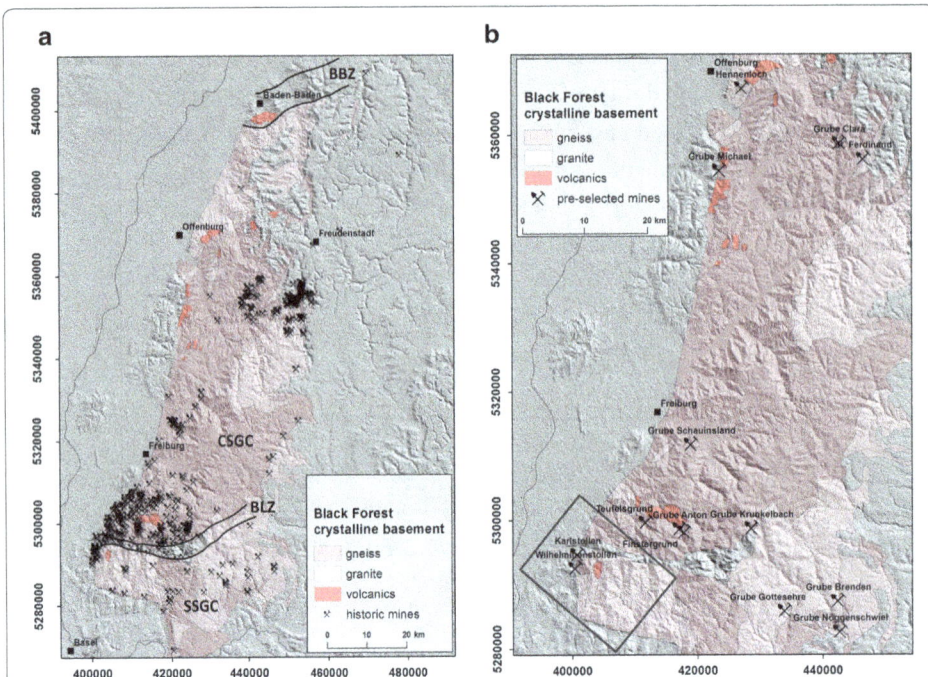

Fig. 6 a Lithological and geological units of the Black Forest, including historic major mining activities. **b** Pre-selected historic mines in the southern and central Black Forest. *Rectangle* indicates area of investigation (Fig. 9). *BBZ* Baden-Baden Zone, *CSGC* Central Black Forest Gneiss Complex, *BLZ* Badenweiler-Lenzkirch Zone, *SSGC* Southern Black Forest Granite Complex

zone of the Black Forest and, hence, reached upper crustal brittle levels, with retrograde mylonitic shear zones being transformed into cataclastic fault zones (e.g. Eisbacher et al. 1989; Grimmer et al. 2016).

The crystalline basement rocks of the Black Forest display important late- to post-Variscan alteration and mineralisation features (e.g. Simon 1990; Zuther and Brockamp 1988). The Southern Black Forest batholith has significantly reduced ∂^{18}O-values compared to its magmatic equilibrium values and to the 325 Ma old granites of the Northern Black Forest Batholith (Hess et al. 2000; Schaltegger 2000; Simon 1990; Hoefs and Emmermann 1983). On the regional scale, the directions, timing, and relative proportions of descending low ∂^{18}O meteoric paleofluids versus ascending or recycled and internally buffered high ∂^{18}O paleofluids are largely unknown. On the local scale, detailed geochemical studies of post-Variscan mineralised veins indicate a mixture of ascending and descending fluids during mineral formation in fault zones (e.g. Bons et al. 2014). Radiometric dating of minerals originating from hydrothermal alteration of crystalline basement rocks and Permo-Triassic cover rock successions as well as of minerals originating from fault-related, mineralised veins reveal multiple and complex post-Variscan phases of pervasive and localised fluid-rock interactions (Brockamp et al. 2015; Zuther and Brockamp 1988; Glodny and Grauert 2009).

Hydrogeology of the Black Forest

Features of brittle deformation, such as faults and joints, are the principal fluid-conducting structures in crystalline basement rock. In the Black Forest and URG area, hydraulic tests in boreholes down to 5 km depth revealed hydraulic conductivity values that range over nine orders of magnitude from 10^{-13} to 10^{-4} m s^{-1} with a log mean of $7 \cdot 10^{-6}$ m s^{-1} over 175 tests (Stober and Bucher 2007). The large variance observed at shallow depth decreases rapidly to a range between 10^{-8} and 10^{-6} m s^{-1} at about 1 km depth. A characteristic value of 10^{-8} m s^{-1} is observed at 4 km depth. This decrease appears to be pronounced in gneissic basement (Stober 1995). Thus, mean hydraulic conductivities of $1 \cdot 10^{-6}$ m s^{-1} in fractured granite are generally higher than in fractured gneiss ($5 \cdot 10^{-8}$ m s^{-1}; Fig. 7).

Several hydraulic test data indicated the influence of faults or fault zones acting as distant hydraulic boundaries. Hydraulic conductivities were lower than the values of the undisturbed crystalline basement, retrieved from the radial flow period (Stober and Bucher 2014). This permeability contrast was attributed to low permeabilities within the core of the fault. Investigation of water table fluctuations due to earth tides, of drawdown data from long-term pumping tests, and of hydrochemical and thermal data from long-term observations of deep circulating systems (natural thermal springs) shows that the open water-conducting fractures and pore spaces form an interconnected network in the crystalline basement with characteristic hydraulic properties (e.g. Stober and Bucher 2014). Thus, the crystalline basement can be regarded as an aquifer in hydraulic terms.

Bucher and Stober (2000) and Pearson et al. (1991) summarised the hydrochemical characterisation of crystalline fluids in the Black Forest and adjacent areas. Three major types of groundwater can be distinguished (Fig. 8). Besides near-surface fresh type 1 water (Fig. 8a), CO_2-rich mineral type 2 water (Fig. 8b) is predominantly occurring in the central Black Forest (e.g. Bad Peterstal, Bad Griesbach, Bad Teinach), whereas saline thermal type 3 water (Fig. 8c) is found mostly in the northern and southern Black Forest (e.g. Wildbad, Bad Liebenzell, Baden-Baden, Bad Säckingen, Zurzach). The latter results from the mixing of surface freshwater, saltwater, and a water–rock reaction component from an up to about 3–5 km deep reservoir. CO_2-rich mineral water in the central Black

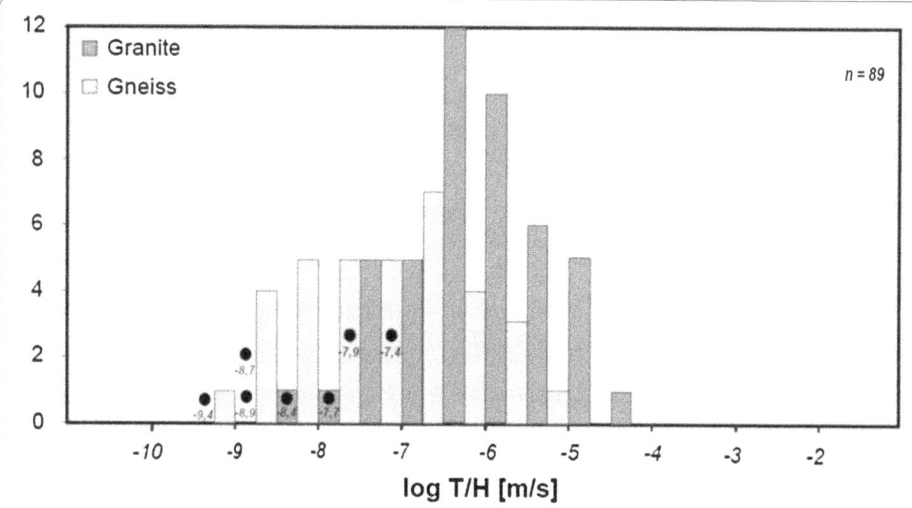

Fig. 7 Hydraulic conductivities from 89 hydraulic tests in the granitic and gneissic basement of the Black Forest, modified after Stober (1995). Wells with mixed lithology, granite, and gneiss, are not considered. *Black dots* additional hydraulic conductivities of deep wells in the URG for comparison

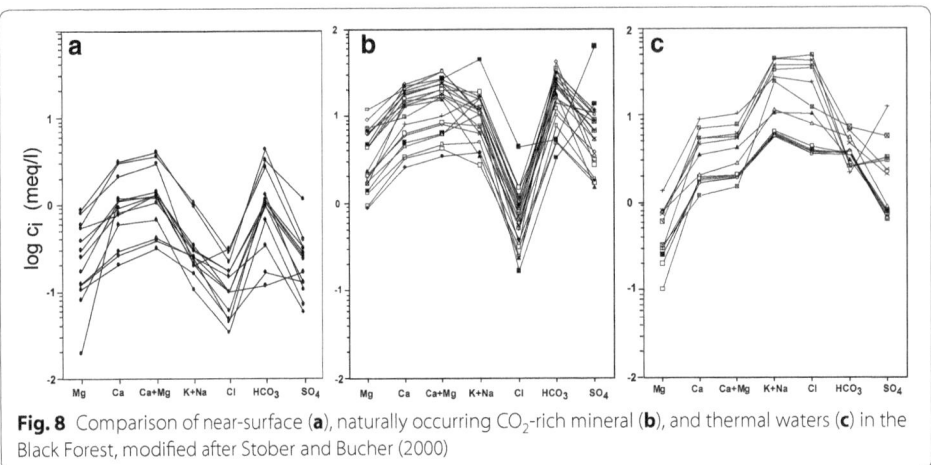

Fig. 8 Comparison of near-surface (**a**), naturally occurring CO_2-rich mineral (**b**), and thermal waters (**c**) in the Black Forest, modified after Stober and Bucher (2000)

Forest is of low salinity and its chemical composition results from the reaction of CO_2-rich water with the crystalline rock matrix at relatively shallow depth (<500 m) exclusively. The main water components are Ca, Na, and HCO_3 (Fig. 8b). In types 1 and 2, the total of dissolved solids (TDS) is enhanced. The type 1 groundwater resembles main water components of type 2, but has a significantly lower TDS and is poor in CO_2. It develops due to interaction of rainwater with crystalline basement rocks. Thus, a distinct stratification in hydrochemistry with increasing depth results: TDS increases and the water type changes from Ca-(Na)-HCO_3 to Na–Cl rich waters.

Topography-driven deep circulation systems in the Black Forest occur in granites. Highly permeable fracture and fault zones in granites are used as ascent channels and flow paths by deep hot saline crystalline basement waters. Although spatially closely associated, the saline deep waters and the CO_2-rich mineral waters are hydraulically and chemically unconnected (e.g. Bucher and Stober 2000).

Preliminary criteria validation in the southern Black Forest

An exploration concept has been established to pre-characterise a suitable site for Geo-LaB in the Black Forest from the surface. It aims at indicating the subsurface condition that are relevant to CHFE in the run-up of large exploration including exploration wells. The concept is based on the recommendations for characterising, modelling, and monitoring fractured rock sites of the National Academies of Sciences (2015). It includes geometric characterisation using lineament mapping techniques, geomagnetic, geoelectric, georadar measurements, geological mapping. In the absence of boreholes, no geophysical logging or hydraulic testing is envisaged at this stage. In addition, a hydrochemical exploration is included. The concept was applied and tested along the Wilhelminenstollen, an exploration tunnel selected from about 700 abandoned mines in the Black Forest according to the selection criteria above. The study is to be updated upon the continuation of the project.

Structural setting of the southern Black Forest

In addition to the geological and structural mappings summarised in the Geological Map of Baden-Württemberg (1:50,000, LGRB, 2015), lineaments of the southern Black forest were mapped using high-resolution digital elevation models with up to 5×5 m resolution. Compared to earlier studies (Franzke et al. 2003), a higher lineament density in the southern Black Forest and predominant (S)SW, (W)NW, and N trending strike directions are observed (Fig. 9). These directions are also observed in the geothermal reservoir at Riehen and Basel in the URG (Meixner et al. 2016). From stress measurements in the entire Black Forest a strike-slip regime with a S_{Hmax} orientation of N140°E $\pm$ 10° was obtained.

The resulting fault and stress field models were used for slip- and dilation tendency analyses (Fig. 9). The reactivation potential reveals two local maxima. Faults that strike about N110°E and N170°E are prone for shear reactivation as dextral and sinistral strike-slip faults, respectively. Dilation is most likely to occur in directions parallel to S_{Hmax}. Highest lineament density is observed for the N110°E strike direction that exhibits both slip and dilation tendency maxima.

Possible sites for GeoLaB in the southern Black Forest

Generally, there are two main approaches to constructing a geothermal URL in the southern Black Forest: using and connecting to an existing tunnel or mine or constructing a new tunnel. Existing tunnels or mines may provide the necessary infrastructure, e.g. road access to the location. Moreover, rock properties, the fracture network, and hydraulic conditions can be investigated in situ before major investments are committed.

Of the more than 700 historic mines in the Black Forest (Steen 2004; Werner and Dennert 2004), 15 sites were pre-selected according to the size and accessibility of the tunnel (Fig. 6). To evaluate their suitability for CHFE, the lithology and overburden were analysed from literature. Fracture transmissivity can only be inferred occasionally from the water budget in the mines. Commonly, information on hydrothermal alteration is lacking. Regional stress conditions in the Black Forest are generally suitable for CHFEs. Controllable hydraulic boundary conditions is one of the most important criteria for CHFEs in pre-existing tunnel systems.

To avoid complex and unknown tunnel systems, we selected a single exploration tunnel, the Wilhelminenstollen near Badenweiler, as potential GeoLaB site. Wilheminenstollen is documented well and led to no further mining activity. Furthermore, it fulfils a number of other criteria such as lithology, open fracture zones, and clay minerals comparable to the Soultz site. For these reasons, it was chosen as a case study site to test an assessment procedure for a geothermal URL in crystalline basement rock. Geological, hydrogeological, and geophysical methods have been applied to indicate the suitability of the Wilhelminenstollen for an URL.

Geological setting at the Wilhelminenstollen

The 521 m long, E–W trending Wilhelminenstollen is located SE of Badenweiler in the SSGC (Fig. 10). Together with the earlier constructed Sehringer Stollen, which is located approximately 50 m higher, it was installed around 1920 at Sehringen to explore the mineralised veins in the Blauen granite and was abandoned a few years later. The tunnel entrance is located immediately east of the Black Forest escarpment fault. Today, about 37,500 m^3 of drinking water per year are collected from open fractures in the tunnel.

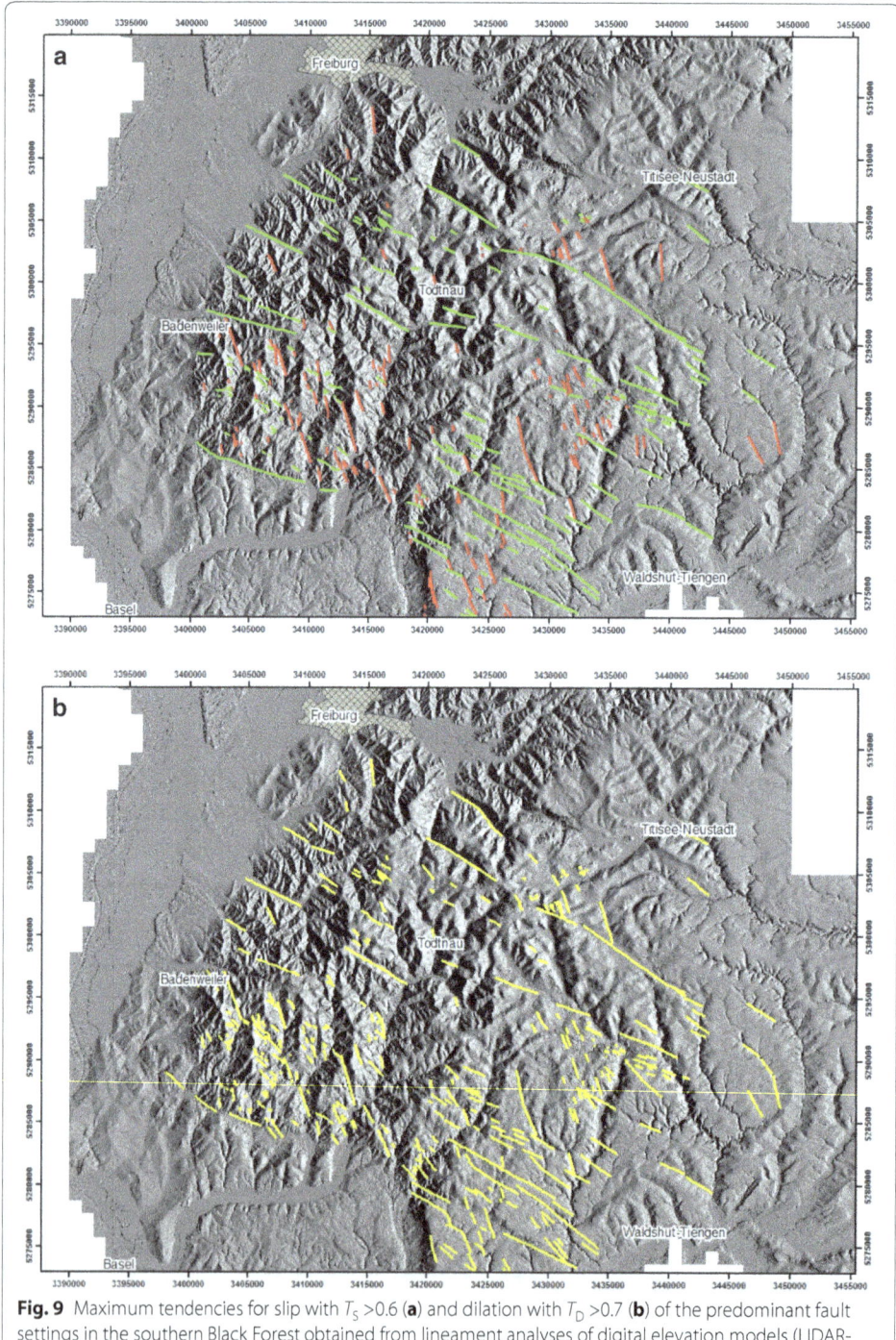

Fig. 9 Maximum tendencies for slip with T_S >0.6 (**a**) and dilation with T_D >0.7 (**b**) of the predominant fault settings in the southern Black Forest obtained from lineament analyses of digital elevation models (LIDAR-DEM5). *Red/green* sinistral/dextral reactivation

Structurally, three major features laterally frame the study area. To the north, segmented E–W striking normal faults separate the SSGC from the Badenweiler-Lenzkirch-Zone. East of the Mt. Blauen, the N to NNW striking Schweighof fault zone and the N–S trending graben structure of Marzell juxtapose the SSGC granites against Permian volcanics. To the west, the NNE striking Black Forest escarpment fault marks the structural

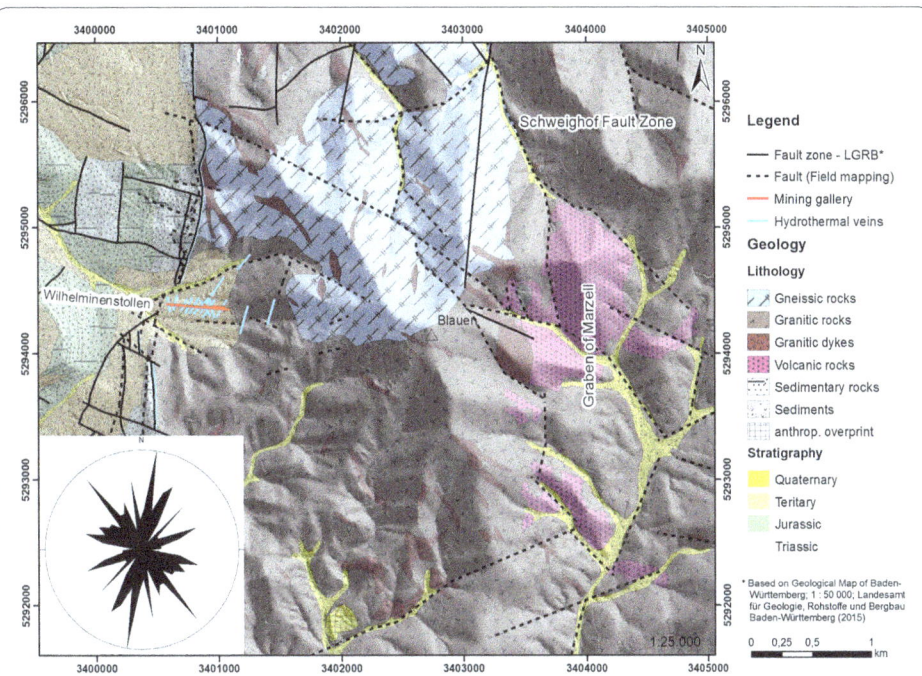

Fig. 10 Geological setting in the surroundings of the Wilhelminenstollen, including fault zones, as obtained from the geological mapping and remote sensing. Measurements of fracture geometries indicate several predominant strike directions with two peaks in NNE-SSE and NW–SE direction (see inlet). *Inlet* symmetric distribution statistics of orientation of joint strike directions ($n = 127$) in the Blauen granite around the Wilhelminenstollen display predominant NNE-and NW-striking joints and subordinate WSW- and WNW-striking joints

border between the exposed crystalline basement rocks and the Mesozoic cover. It is segmented by NE and SE striking transfer faults which accommodate an E–W extension in the form of oblique normal faults with a sinistral and dextral a strike-slip component. NW to NNW and NE to SW trending faults resulted from post-orogenic extensional tectonics during Variscan orogeny, but show polyphase reactivations in the Mesozoic and Cenozoic (Huber and Huber-Aleffi 1984; Schumacher 2002).

Two different petrologic units, the Blauen granite and the Wiese-Wehra formation, are exposed in the study area. The medium- to fine-grained biotite-type Blauen granite was linked to the about 328–333 Ma old Malsburg granite (Sawatzki et al. 2003). The second petrologic unit mainly consists of gneissic rocks commonly designated as diatexites of the Wiese-Wehra formation.

NNE striking faults and fracture zones are often filled by cataclastic material or by hydrothermally and mechanically altered secondary minerals, such as hematite, Fe-hydroxide, clay minerals, barite, or quartz. Especially the NNE striking escarpment fault zone is accompanied by several hydrothermal veins. These deposits are fault- and fracture-hosted veins and seem to be directly linked to brittle deformation processes. Formation ages indicate hydrothermal activity in the Black Forest area since Variscan times, with a peak in the Jurassic (Bons et al. 2014). But NNE alignment of the veins close to the escarpment fault may also indicate correlation of Cenozoic rifting and origin of hydrothermal circulation along this fault. Breccia textures of the veins also reflect repeated cycles of fault reactivation and mineral precipitation. The veins almost constantly strike NNE-SSW with varying dip angles between about 45–80°. Interestingly, most of the veins dip to the E, opposite to the escarpment fault, which was active during URG formation. This may indicate that hydrothermal circulation and mineral precipitation took place at both conjugated shear angles during active normal faulting. The epigenetic ore deposits in the study area mainly consist of lead-bearing quartz-barite veins and were formed by mixture of fluids of different origin (Bons et al. 2014; Kneer 2006). The distribution of the inflow of water into the tunnel and the high Ba concentration of 442.3 µg L^{-1} indicate a clear link between the barite-filled fractures and fluid pathways.

X-ray diffraction (XRD) analyses of powders of fracture fillings from the Black Forest escarpment fault and a fracture in the front part of the Wilhelminenstollen revealed an asymmetric peak at 2θ (CuKalpha) = 10 Å characterised by tailing towards lower 2θ values. This indicates the presence of mica/illite together with an illite–smectite interstratification in the alteration products. Ethylene glycol treatment confirmed the presence of the swellable interstratification. No indication of free smectite was found.

Fault characterisation

Due to the lack of geological outcrops in the southern Black Forest, the identification of faults from the surface requires geophysical measurements. In a first reconnaissance study, the faults and veins near the Wilhelminenstollen were identified successfully using electric resistivity measurements. Geo-electric measurements were conducted using a 4-point Light High Power with 60 electrodes (LMG) and Geo-Test (Geophysik Dr. Rauen). Dipole-dipole and pole-dipole geometries with an electrode offset of 0.5–45 m were used. Geographic coordinates were acquired using a Trimble R4 with Sapos-correction. In addition, geomagnetic and georadar data have been acquired (not shown). Compared to geoelectric measurements, they are less conclusive.

First, a calibration profile was measured across the Black Forest escarpment fault exposed along the road to Sehringen (CP, Fig. 11). The fault zone dips towards WNW with about 60°. As indicated by the geological map, the exact location and the number of branches of the targeted fault zone are a matter of discussion. Both, electric resistivity and phase shift indicate an anomaly between profile meters 5 and 10 (Fig. 12). The anomalies are observed to an off-set of >10 m corresponding to a depth of approximately 5 m. There is some minor indication of a smaller anomaly around profile meters 20 in the apparent resistivity.

Hydrothermal veins partly filled with quartz and Mg-, Pb-ore minerals and barite within the Wilhelminenstollen mostly dip to the ESE or ENE with about 45–60 °C. Here, they are mapped from the surface in variable quality using electric resistivity (Fig. 13). Correlation of the near-surface anomalies and the hydrothermal veins in the tunnel is possible when considering an E–W offset due to the dip of the structures. Two low resistivity anomalies in the western part of P4 coincide well with the shifted position of veins in the tunnel. These two veins are Pb-bearing structures. The vein at profile meter 102 is not detected by resistivity measurements. This may be attributed to the fact that this vein is filled by quartz and Mg only. The easternmost veins at the edge of the profile (profile meters 149.5 and 151.5) are not covered by the profile.

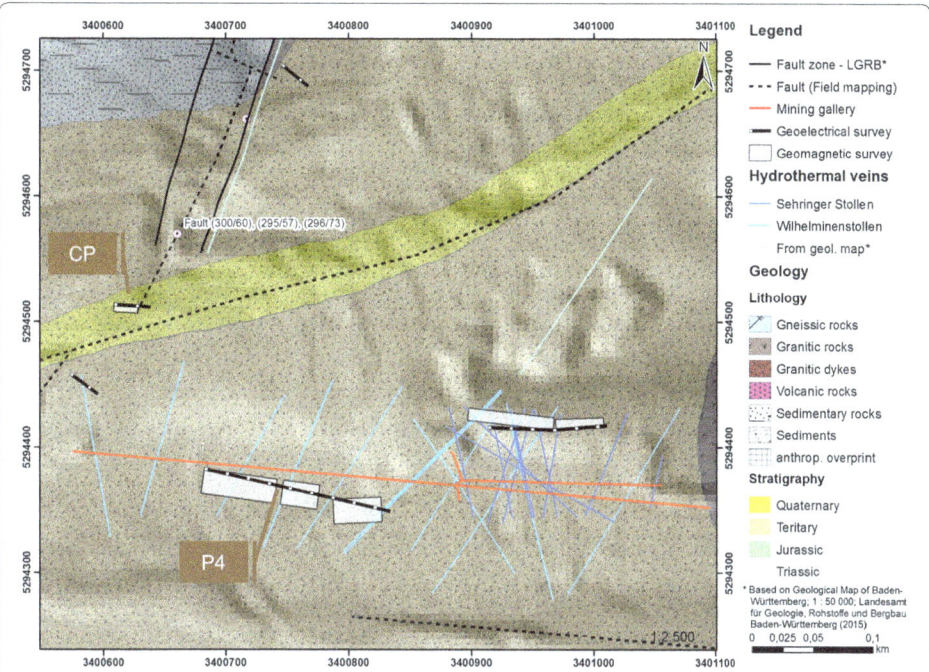

Fig. 11 Distribution of faults near the Wilhelminenstollen and veins (*blue lines*) observed in the Wilhelminen- and Sehringer Stollen (*red lines*) and extrapolated to the geophysical measurement areas (geo-magnetics) and profiles (geo-electrics). *CP* calibration profile (Fig. 13)

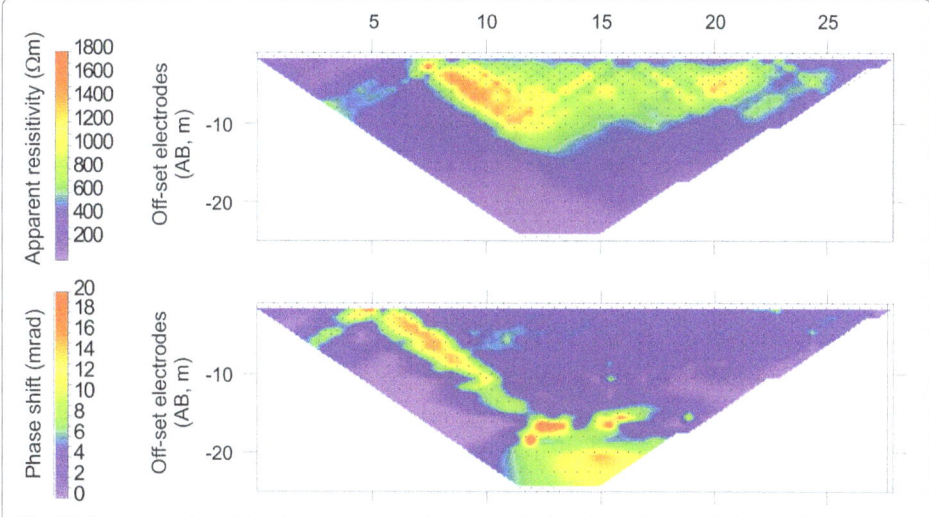

Fig. 12 Pseudo-section of the electric resistivity distribution (*top*) and the phase shift (*bottom*) from dipole–dipole measurements along the NW–SE calibration profile (CP) across the Black Forest escarpment fault at Sehringen (Fig. 11)

Hydrogeology

Hydraulic tests performed in the granitic basement of the southern Black Forest suggest hydraulic conductivities of the order of $4.5 \cdot 10^{-8}$ m s^{-1} at 500 m depth. Fluids from the near-surface granitic basement exhibit similar Ca- or Na-HCO$_3$-dominated chemical characteristics (Fig. 14). Near-surface waters generally are of low mineralisation. TDS increases with increasing depth, as indicated by the three analyses from Bad Säckingen.

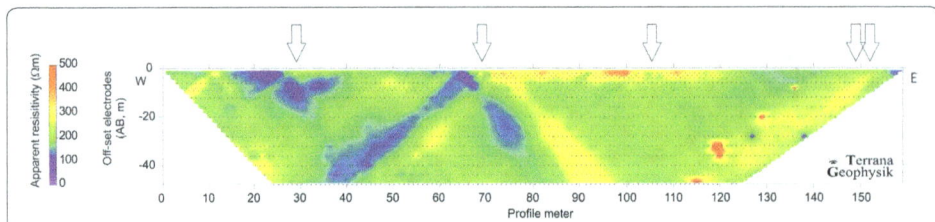

Fig. 13 Electric resistivity distribution along profile P4 (Fig. 11). *Arrows* indicate the extrapolation of the hydrothermal veins mapped in the Wilhelminenstollen to the surface assuming a constant dip between 45° and 60°

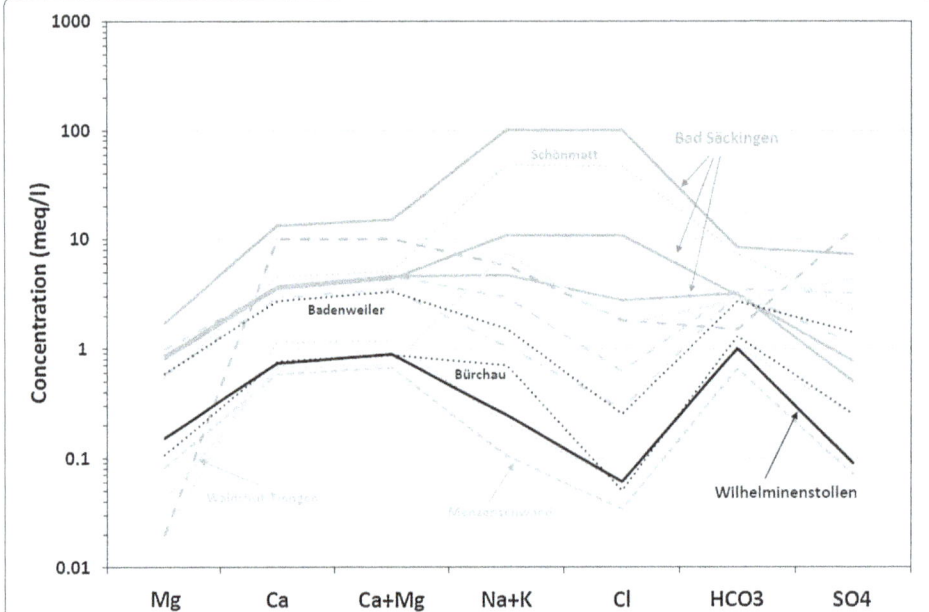

Fig. 14 Chemical composition of selected water samples from the southern Black Forest reveals increasing ion-concentration and water type changes from near-surface Ca-HCO$_3$-type (e.g. Wilhelminenstollen, Menzenschwand, Bürchau) (Rolker et al. 2015 and this study) to an Na–Cl water-type (different depth sections in Bad Säckingen and Schönmatt, data from Stober (1995), Stober and Bucher (2014), and Rolker et al. (2015)

The deepest well (600 m) shows the highest TDS and is dominated by Na–Cl, whereas both TDS and NaCl decrease at shallower depth. Near-surface waters (<100 m), in general, possess low TDS values and water chemistry is dominated by Ca and HCO$_3$ (e.g. Menzenschwand, Wilhelminenstollen). However, also thermal spring waters often have a near-surface water component (Fig. 14, e.g. Badenweiler, Bürchau). In contrast to Bürchau, the water from the Wilhelminenstollen has significantly lower Na and SO$_4$ concentrations. In Badenweiler and Bürchau, low mineralisation and maximum sulphate equilibrium temperatures of about 60–80 °C indicate a small potential for water–rock interaction only (Rolker et al. 2015).

The relationship between δ^{18}O and δ^{2}H of thermal springs from different sites in the southern Black Forest is shown in Fig. 15. The isotope ratio follows the global meteoric water line. The values from Bürchau and Badenweiler show differences between sampling campaigns 1 and 2. The values from Wilhelminenstollen (sampling campaign 3, this study) lie between the above values. A shift towards higher δ^{18}O values in campaign

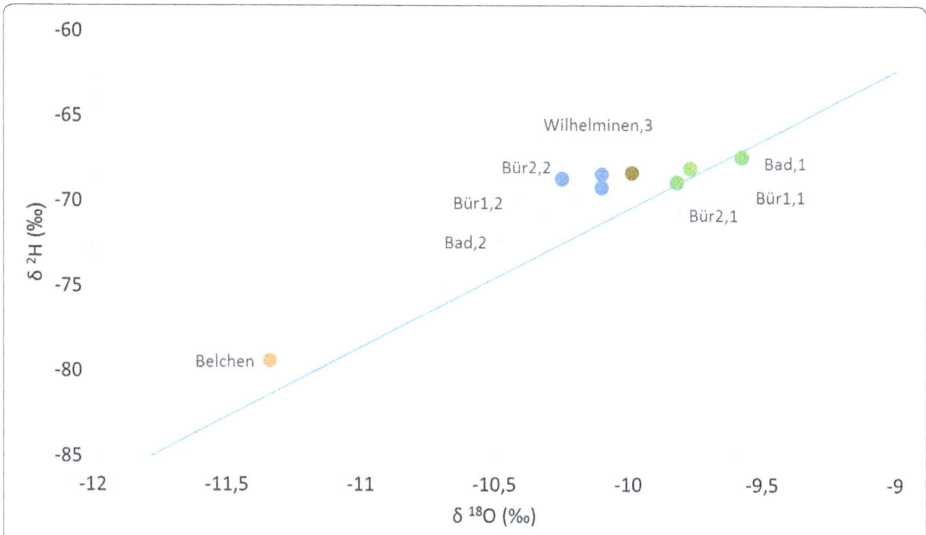

Fig. 15 H/O-isotope distribution of the Wilhelminenstollen water (this study) in comparison with data from adjacent waters of the southern Black Forest from Badenweiler and Bürchau (Rolker et al. 2015)

2 was attributed to a higher contribution of rainwater. The difference in $\delta^{18}O$ and δ^2H values between the Mt. Belchen and the sites of Bürchau, Badenweiler, and Wilhelminenstollen along the meteoric water line can be attributed to altitude effects, since there is a difference in altitude of several hundred metres between the sites and Mt. Belchen. Following earlier studies of Pearson et al. (1991), a local relationship is identified, including the southern Black Forest in the area of Bürchau. For Mt. Belchen, the best fit between this relationship and true infiltration elevation was obtained from $\delta^{18}O$ values. This gives a minimum infiltration altitude for the Wilhelminenstollen of about 730 masl. The hydrochemical data on Wilhelminenstollen, its minimum infiltration altitude, and the topographic situation result in a locally controlled hydrogeological framework with a maximum catchment area of 0.5 km².

Stress model for the topographic condition

Local stress in mountain areas is affected by the topography, the rheological properties of the rock, and the ambient tectonic state of stress. Geometry and mass of the topography may cause local reorientation and lateral variation of the magnitude of the principal stress directions (Miller and Dunne 1996; Pan et al. 1995; Savage and Morin 2002). Since GeoLaB aims at approximating reservoir condition best, the URL design strives for rather topography-independent conditions of the stress field.

Rock type and ambient stress state are rather homogenous across the planned GeoLaB tunnel and hence to solely assess the influence of topography on the stress field distribution near the Wilhelminenstollen, a finite element was generated for an area of 19 × 20 km around the Wilhelminenstollen (Fig. 6b) with a vertical extension from the topography down to 5 km b.s.l. (Fig. 16). First-order geomechanical modelling using ABAQUS© covers gravitational loading of the 3D model only. The resulting orientation and magnitude of the principal stress component are shown in Fig. 16.

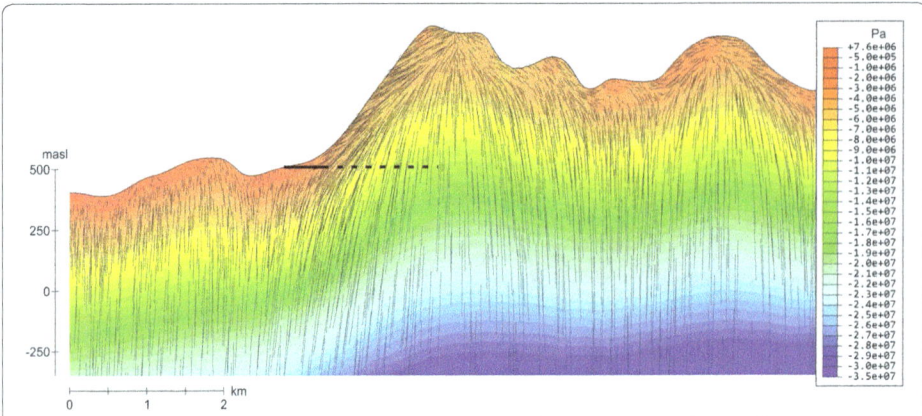

Fig. 16 Stress field perturbation by the topography near the Wilhelminenstollen obtained from geo-mechanical modelling using ABAQUS© on a 19 × 20 × 5 km finite element model. Orientation (*black lines*) and magnitude (*colour signature*) of the maximum principal stress component are shown (positive/negative magnitudes: tensile/compressive stress). *Black line* extension of the Wilhelminenstollen, *dashed line* possible extension to 2 km length

Close to the surface, both, orientation and magnitude of the maximum principal stress component are strongly influenced by the topography. At steep slopes and shallow depth (first 100–200 m), the maximum principal stress component significantly deviates from vertical orientation due to topographic stresses. Here, the vertical component of the stress tensor is no longer a principal stress. As a result, failure conditions and fracture patterns in these areas deviate significantly from regional trends. Consequently, regional stress field models are no longer applicable to predict of the reactivation potential of faults and fractures. Subvertical orientation of the maximum principal stress component is observed below ridges and plateaus and at considerable depth beneath strong topographic gradients only. Note that the modelled local stress field changes significantly, when superposing on the gravitational stress field the ambient compressive far-field tectonic stress. At this early planning stage of GeoLaB, this more comprehensive model hampers evaluation of the stress perturbations that are purely driven by topography, but it will be considered for the design of the tunnel.

The onset of the Wilhelminenstollen is located at the foot of Mt. Blauen. With its current length of about 500 m, it is located entirely in the critical near-surface zone of highly perturbed stress states. For the specific setting of the Wilhelminenstollen, a tunnel length of >1.5 km is suggested when considering the criteria of stress perturbation of <1 MPa and approaching vertical orientation of the maximum principal stress direction for gravity. This results in a minimum overburden of about 500 m.

Conclusion

Particular aspects of geothermal characterisation and operation can only be obtained through access to the underground environment. Engineering of geothermal production from fractured crystalline rocks using flow rates of more than 30 L s^{-1} requires knowledge of geological, physical, and chemical processes that can be gained partly on the downhole field scale and partly by large-scale research laboratory facilities.

As was pointed out by the NEA (2013), URLs provide important technical knowledge and increase confidence in the process of facility siting and design, engineering support, and evaluation of safety. Clearly, such a generic URL will not have all real reservoir conditions (such as temperature, magnitude of stress components, and chemical composition of water), but confidence in the suitability of a potential geological environment and engineering feasibility can be gained by a verification of individual processes and operation concepts. URLs offer an unparalleled opportunity to demonstrate the engineering concept of EGS and instil confidence in the wide range of stakeholders. A transparent underground experimentation and operation programme of an URL platform will yield specific information that also is of direct relevance to authorities assessing the risk of underground utilisation. In this context, a geothermal URL represents a worldwide unique research installation with specific environmental settings being required for EGS development.

Four criteria have been established based on EGS experience and applied to existing generic URLs. These are a relatively homogenous crystalline basement matrix with a well-connected fracture network with fracture transmissivities in the range of $>10^{-4}$ m^2 s^{-1}. Fractures should be characterised by hydrothermal alteration including illite and smectite among others. The local stress field is characterised by strike-slip (to normal faulting) reactivation and fractures are favourably oriented. Furthermore, the stress field is rather homogeneous with a variation in magnitude <1 MPa across the envisaged experimental volume. Obviously, a geothermal URL will not match reservoir conditions in terms of temperature, rock-water interaction, and the stress field. CHFEs that address these parameters will rather study gradients of these fields.

Application of the four criteria to existing generic URLs in the crystalline basement worldwide reveals that apart from the inferred stress field, the Äspö HRL, and apart from overburden, the Lindau Test Site, provide favourable conditions for CHFEs. These URLs may be used to carry out specific accompanying or complementary experiments.

Typical deep geothermal reservoir rocks of the URG are exposed in the Black Forest. In general, tectonic, geological, and hydrogeological conditions in the southern Black Forest match the established criteria for CHFEs in a geothermal URL. In the Wilhelminenstollen, controllable hydraulic boundary conditions are valid. An exploration concept to pre-characterise suitable GeoLaB sites in the Black Forest from the surface has been tested successfully. It involves exploration from the surface and aims at a first assessment of the established criteria in the run-up of large exploration including exploration wells. In this respect, the Wilhelminenstollen area is characterised as follows:

- A dense fracture network inferred from lineament analyses is completed by observations in the existing tunnel and resistivity measurements.
- Alteration zones with clay minerals or ore mineralisation in fractures are characterised well by electric resistivity.
- Hydraulically active fractures and fault zones in a regional network are identified at several sites in the Black Forest. In the absence of wells, local transmissivities cannot be acquired, but a typical natural in-flow into the tunnel of several L s^{-1} can be estimated. Increasing TDS and higher NaCl concentration are expected.

- Low influence of topography-induced stress variations requires a minimum overburden of about 500 m.
- Regional reactivation potential is highest for large differential stresses. At Wilhelminenstollen, they are maximum for fractures and faults oriented N110°E and N170°E.

The analysis performed has provided key steps for developing a generic URL in the southern Black Forest. In first order, most of the criteria are fulfilled in the study area. For a final decision on the technical feasibility of GeoLaB at Wilhelminenstollen as well as the conceptualisation and optimisation of the design, a large geophysical exploration campaign as well as exploration wells including hydraulic testing is necessary.

Authors' contributions
ES reviewed the condition of the Soultz EGS site and the exisitng URLs and carried out the fluid chemical analyses and the geophysical surveys, JM carried out the remote sensing and slip- and dilation tendency analyses, CM contributed to the review of the Soultz EGS site and the existing URLs, MG reviewed the exiting mines in Black Forest and performed the numerical simulation on the influence of topography, JG reviewed the geological condition of the Black Forest, IS reviewed the hydrogeological condition and TK summarized the scientific concept of GeoLaB. All authors established the criteria. All authors read and approved the final manuscript.

Author details
[1] Institute for Nuclear Waste Disposal, Karlsruhe Institute of Technology, Hermann-von-Helmholtz-Platz 1, 76128 Karlsruhe, Germany. [2] Geothermal Research Group, Institute of Applied Geosciences, Karlsruhe Institute of Technology, Adenauerring 20b, 76131 Karlsruhe, Germany.

Acknowledgements
This study was carried out in the framework of the HGF portfolio project "Geoenergy" under the Helmholtz topic "Geothermal Energy Systems". We would like to thank I. Blechschmidt (Nagra, Switzerland), R. Christiansson (SKB, Sweden) and J. Pacovsky (Centre of Experimental Geotechnics, Czech Republic) for detailed information on their URLs. F. Nitschke and E. Eiche (KIT, Germany) kindly measured and evaluated the hydrochemical data of Wilhelminenstollen. G. Markl (University of Tübingen, Germany) kindly commented on the condition of the pre-selected mines in the southern Black Forest. M. Waldhör and A. Patzelt (Terrana Geophysik, Germany) are acknowledged for the geophysical field measurements. A. Steudel (KIT, Germany) kindly provided analyses on clay minerals.

Competing interests
The authors declare that they have no competing interests.

References
Baillieux P, Schill E, Abdelfettah Y, Dezayes C. Possible natural fluid pathways from gravity pseudo-tomography in the geothermal fields of Northern Alsace (Upper Rhine Graben). Geotherm Energy J. 2014;2(16):1–14.

Baillieux P, Schill E, Edel JB, Mauris G. Localization of temperature anomalies in the Upper Rhine Graben: insights from geophysics and neotectonic activity. Int Geol Rev. 2013;55(14):1744–62.

Barton N, Bandis S, Bakhtar K. Strength, deformation and conductivity coupling of rock joints. Int J Rock Mech Min Sci Geomech. 1985;22(3):121–40. doi:10.1016/0148-9062(85)93227-9.

Bayer M. Die Himmelfahrt Fundgrube. Ein Führer durch das Lehr—und Besucherbergwerk der TU Bergakademie Freiberg. Freiberg: TU Bergakademie Freiberg; 1998.

Bons PD, Fusswinkel T, Gomez-Rivas E, Markl G, Wagner T, Walter B. Fluid mixing from below in unconformity-related hydrothermal ore deposits. Geology. 2014;42(12):1035–8. doi:10.1130/g35708.1.

Brady BHG, Brown ET. Rock mechanics: for underground mining. Springer Science & Business Media 2013.

Brockamp O, Schlegel A, Wemmer K. Complex hydrothermal alteration and illite K-Ar ages in Upper Visean molasse sediments and magmatic rocks of the Variscan Badenweiler-Lenzkirch suture zone, Black Forest, Germany. Int J Earth Sci. 2015;104(3):683–702.

Brown DW. Hot dry rock geothermal energy: important lessons from Fenton Hill. In: Proceedings of the thirty-fourth workshop on geothermal reservoir engineering. Stanford: Stanford University; 2009.

Bucher K, Stober I. The composition of groundwater in the continental crystalline crust. In: Stober I, Bucher K, editors. Hydrogeology in crystalline rocks. Dordrecht: KLUWER academic Publishers; 2000. p. 141–76.

Cornet F, Bérard T, Bourouis S. How close to failure is a granite rock mass at a 5 km depth? Int J Rock Mech Min Sci. 2007;44(1):47–66.

Cuenot N, Frogneux M, Dorbath C, Calo M. Induced microseismic activity during recent circulation tests at the EGS site of Soultz-sous-Forêts (France). In: Proceedings of the 36th workshop on geothermal reservoir engineering. 2011.

Davison CC. Monitoring hydrogeological conditions in fractured rock at the site of Canada's Underground Research Laboratory. Ground Water Monit Rem. 1984;4(4):95–102. doi:10.1111/j.1745-6592.1984.tb00899.x.

Dezayes C, Genter A, Hooijkaas GR. Deep-seated geology and fracture system of the EGS Soultz reservoir (France) based on recent 5 km depth boreholes. Proceedings, paper 1612 pdf. 2005.

Eisbacher GH, Lüschen E, Wickert F. Crustal-scale thrusting and extension in the Hercynian Schwarzwald and Vosges, central Europe. Tectonics. 1989;8(1):1–21. doi:10.1029/TC008i001p00001.

Evans KF, Genter A, Sausse J. Permeability creation and damage due to massive fluid injections into granite at 3.5 km at Soultz: 2. Critical stress and fracture strength. J Geophys Res. 2005;110:B04204. doi:10.1029/2004JB003169.

Franzke HJ, Werner W, Wetzel H-U. Die Anwendung von Satellitenbilddaten zur tektonischen Analyse des Schwarzwalds und des angrenzenden Oberrheingrabens. Jh Landesamt für Geologie Rohstoffe und Bergbau Baden-Württemberg LGRB. 2003;39:25–54.

Freeze RA, Javandel I. An interview with Paul Witherspoon, distinguished hydrogeologist from the USA. Hydrogeol J. 2008;16:811–5.

Garnish J. European activities in Hot Dry Rock research. Paper presented at the open meeting on enhanced geothermal systems. Reno; 2002.

Gaucher E, Schoenball M, Heidbach O, Zang A, Fokker P, van Wees J-D, Kohl T. Induced seismicity in geothermal reservoirs: a review of forecasting approaches. Renew Sustain Energy Rev. 2015;52:1473–90. doi:10.1016/j.rser.2015.08.026.

Geiermann J, Schill E. 2-D Magnetotellurics at the geothermal site at Soultz-sous-Forêts: resistivity distribution to about 3000 m depth. Comptes Rendus Geoscience. 2010;342:587–99.

Genter A, Castaing C. Scale effects in the fracturing of granite. Comptes Rendus De L Academie Des Sciences Serie Ii Fascicule a-Sciences De La Terre Et Des Planetes. 1997;325(6):439–45.

Genter A, Traineau H, Ledésert B, Bourgine B, Gentier S. Over 10 years of geological investigations withing the HDR Soultz Project, France. In: World Geothermal Congress. 2000. p. 3706–12.

Glodny J, Grauert B. Evolution of a hydrothermal fluid-rock interaction system as recorded by Sr isotopes: a case study from the Schwarzwald, SW Germany. Mineral Petrol. 2009;95(3–4):163–78.

Grimmer JC, Ritter JRR, Eisbacher GH, Fielitz W. The late Variscan control on the location and asymmetry of the Upper Rhine Graben. Int J Earth Sci. 2016. doi:10.1007/s00531-016-1336-x.

Hakami E, Min K-B. Modelling of the state of stress, preliminary site description Laxemar subarea—version 1.2. Report vol R-06-17. Stockholm: Svensk Kärnbränslehantering; 2009.

Hakami H. Äspö Hard Rock Laboratory, update of the rock mechanical model 2002. International progress report vol IPR-03-37. Stockholm: Svensk Kärnbränslehantering; 2003.

Hess JC, Hanel M, Arnold M, Gaiser A, Prowatke S, Stadler S, Kober B. Variscan magmatism at the northern margin of the Moldanubian Vosges and Schwarzwald I. Ages of intrusion and cooling history. Eur J Mineral. 2000;12:79.

Hettkamp T, Baumgärtner J, Baria R, Gerard A, Gandy T, Michelet S, Teza D. Electricity production from hot rocks. In: Proceedings, 29th workshop on geothermal reservoir engineering, California: Stanford University; 2004. p. 26–8.

Himmelsbach T, Hötzl H, Maloszewski P. Solute transport processes in a highly permeable fault zone of Lindau Fractured Rock Test Site (Germany). Ground Water. 1998;36(5):792–800. doi:10.1111/j.1745-6584.1998.tb02197.x.

Himmelsbach T, Shao H, Wieczorek W, Flach D, Schuster K, Alheid H-J, Liou T-S, Bartlakowski J, Krekeler T. Effective field parameter EFP (Grimsel). Wettingen: Nagra; 2003.

Hoefs J, Emmermann R. The oxygen isotope composition of Hercynian granites and pre-Hercynian gneisses from the Schwarzwald, SW Germany. Contrib Miner Petrol. 1983;83:320–9.

Huber M, Huber-Aleffi A. Das Kristallin des Südschwarzwaldes: NTB 84-30. NTB. 1984.

Jung R. EGS—goodbye or back to the future 95. Effective and sustainable hydraulic fracturing. Rijeka: InTech; 2013.

Jung R, Willis-Richard J, Nicholls J, Bertozzi A, Heinemann B. Evaluation of hydraulic tests at Soultz-sous-Forêts, European HDR Site. In: Proceedings of the world geothermal congress. 1995. p. 2671–76.

Kalt A, Altherr R. Metamorphic evolution of garnet-spinel peridotites from the Variscan Schwarzwald (Germany). Geol Rundsch. 1996;85(2):211–24.

Klee G, Rummel F. Rock stress measurements in the Äspö HRL, hydraulic fracturing in boreholes KA2599G01 and KF0093A01, vol. IPR-02-02. Stockholm: Svensk Kärnbränslehantering AB; 2002.

Kneer S. Geochemische Untersuchungen der Gangartmineralisationen im Bergbaurevier Badenweiler-Sehringen. Diploma Thesis, Eberhard Karls Universität. 2006.

Kober B, Kalt A, Hanel M, Pidgeon RT. SHRIMP dating of zircons from high-grade metasediments of the Schwarzwald/SW-Germany and implications for the evolution of the Moldanubian basement. Contrib Miner Petrol. 2004;147(3):330–45.

Kohl T, Bächler D, Rybach L. Steps towards a comprehensive thermo-hydraulic analysis of the HDR test site Soultz-sous-Forêts. In: Proceedings World Geothermal Congress. 2000. p. 2671–76.

Kohl T, Evans K, Hopkirk R, Jung R, Rybach L. Observation and simulation of non-Darcian flow transients in fractured rock. Water Resour Res. 1997;33(3):407–18.

Kolditz O, Görke U-J, Shao H, Wang W. Thermo-hydro-mechanical-chemical processes in porous media. Berlin: Springer; 2012.

Komulainen J, Hurmerinta E, Pöllänen J. Monitoring measurements by the difference flow methods during the year 2011, drillholes OL-KR41, -KR42, KR45, -KR46 and ONK-KR13. Eurajoki: Posiva; 2014.

Krecher M, Behrmann JH. Tectonics of the Vosges (NE France) and the Schwarzwald (SW Germany): evidence from Devonian-Carboniferous active margin basins and their deformation. Geotecton Res. 2007;95(1):61–86.

Krohe A, Eisbacher GH. Oblique crustal detachment in the Variscan Schwarzwald, southwestern Germany. Geol Rundsch. 1988;77(1):25–43. doi:10.1007/bf01848674.

Kumazaki N, Ikeda K, Goto J, Mukai K, Iwatsuki T, Furue R. Synthesis of the shallow borehole investigations at the MIU construction site. Tech. Rep. TN7400. Ibaraki: Japan Nuclear Cycle Development Institute; 2003.

Ledésert B, Berger G, Meunier A, Genter A, Bouchet A. Diagenetic-type reactions related to hydrothermal alteration in the Soultz-sous-Forets granite, France. Eur J Mineral. 1999;11(4):731–41. doi:10.1127/ejm/11/4/0731.

Ledésert B, Hebert R, Genter A, Bartier D, Clauer N, Grall C. Fractures, hydrothermal alterations and permeability in the Soultz enhanced geothermal system. C R Geosci. 2010;342(7):607–15.

Marschall HR, Kalt A, Hanel M. P-T evolution of a Variscan lower-crustal segment: a study of granulites from the Schwarzwald, Germany. J Petrol. 2003;44(2):227–53.

Massart B, Paillet M, Henrion V, Sausse J, Dezayes C, Genter A, Bisset A. Fracture characterization and stochastic modeling of the granitic basement in the HDR Soultz Project (France). In: World Geothermal Congress. 2010.

Meixner J, Schill E, Grimmer JC, Gaucher E, Kohl T, Klingler P. Structural control of geothermal reservoirs in extensional tectonic settings: an example from the Upper Rhine Graben. J Struct Geol. 2016;82:1–15.

Meller C, Kohl T. The significance of hydrothermal alteration zones for the mechanical behavior of a geothermal reservoir. Geothermal Energy. 2014;2(12):21. doi:10.1186/s40517-014-0012-2.

Meller C, Kontny A, Kohl T. Identification and characterization of hydrothermally altered zones in granite by combining synthetic clay content logs with magnetic mineralogical investigations of drilled rock cuttings. Geophys J Int. 2014;199(1):465–79. doi:10.1093/gji/ggu278.

Miller DJ, Dunne T. Topographic perturbations of regional stresses and consequent bedrock fracturing. J Geophys Res B Solide Earth. 1996;101(B11):25523–36.

MIT. The future of geothermal energy, impact of enhanced geothermal systems (EGS) on the United States in the 21st century. Cambridge: Massachusetts Institute of Technology; 2006.

Nami P, Schellschmidt R, Schindler M, Tischner T. Chemical stimulation operations for reservoir development of the deep crystalline HDR/EGS system at Soultz-sous-Forêts (France). In: Proceedings, 32nd workshop on geothermal reservoir engineering. California: Stanford University; 2008. p. 28–30.

National Academies of Sciences E and Medicine. Characterization, modeling, monitoring, and remediation of fractured rock. Washington: The National Academies Press; 2015.

Nea NEA. Underground research laboratories (URL). Radioactive waste management. OECD/NEA: Paris; 2013.

Pan E, Amadei B, Savage WZ. Gravitational and tectonic stresses in anisotropic rock with irregular topography. Int J Rock Mech Min Sci Geomech. 1995;32(3):201–14. doi:10.1016/0148-9062(94)00046-6.

Pearson F, Balderer W, Loosli H, Lehmann B, Matter A, Peters T, Schmassmann H, Gautschi RU. Applied isotope hydrogeology: a case study in northern Switzerland. Amsterdam: Elsevier; 1991.

Pribnow DFC. The deep thermal regime in Soultz and implications for fluid flow. GGA Report. Hannover: GGA Institut; 2000.

Rehn I, Gustafson G, Stanfors R, Wikberg P. ÄSPÖ HRL—Geoscientific evaluation 1997/5. Models based on site characterization. Technical report, vol TR-97-06. Stockholm: Svensk Kärnbränslehantering AB; 1997.

Rolker J, Schill E, Stober I, Schneider J, Neumann T, Kohl T. Hydrochemical characterisation of a major central European heat flux anomaly: the Bürchau geothermal spring system, Southern Black Forest, Germany. Geotherm Energy. 2015;3(1):1–18.

Rummel F, König E. Density, ultrasonic velocities and magnetic susceptibility measurements on the core material from borehole EPS1 at Soultz-sous-Forêts. Internal report, vol 8. Bochum: Ruhr-Universität; 1991.

Sausse J. Hydromechanical properties and alteration of natural fracture surfaces in the Soultz granite (Bas-Rhin, France). Tectonophysics. 2002;348(1–3):169–85. doi:10.1016/s0040-1951(01)00255-4.

Sausse J, Dezayes C, Dorbath L, Genter A, Place J. 3D model of fracture zones at Soultz-sous-Forêts based on geological data, image logs, induced microseismicity and vertical seismic profiles. C R Geosci. 2010;342(7):531–45.

Sausse J, Fourar M, Genter A. Permeability and alteration within the Soultz granite inferred from geophysical and flow log analysis. Geothermics. 2006;35(5–6):544–60. doi:10.1016/j.geothermics.2006.07.003.

Sausse J, Genter A. Types of permeable fractures in granite. Geol Soc London Spec Publ. 2005;240(1):1–14. doi:10.1144/gsl.sp.2005.240.01.01.

Savage WZ, Morin RH. Topographic stress perturbations in southern Davis Mountains, west Texas 1. Polarity reversal of principal stresses. J Geophys Res B Solid Earth. 2002;107(12):5–15.

Sawatzki G, Hann HP, Groschopf R, Villinger E. Badenweiler-Lenzkirch-Zone (Südschwarzwald): Erläuterungen mit Hinweisen für Exkursionen. 1. Ausg. zur 2., überarb. Ausg. d. Geolog. Karte edn. Landesamt für Geologie. Freiburg: Rohstoffe und Bergbau Baden-Württemberg; 2003.

Schaltegger U. U-Pb geochronology of the Southern Black Forest Batholith (Central Variscan Belt): timing of exhumation and granite emplacement. Int J Earth Sci. 2000;88(4):814–28. doi:10.1007/s005310050308.

Schill E, Genter A, Kohl T, Cuenot N. Enhancement of productivity in the Soultz EGS site by 20 hydraulic and chemical stimulation experiments and long-term circulation. In: Proceedings World Geothermal Congress 2015, 19–25 April 2015. Melbourne; 2015.

Schindler M, Baumgärtner J, Gandy T, Hauffe P, Hettkamp T, Menzel H, Penzkofer P, Teza D, Tischner T, Wahl G. Successful hydraulic stimulation techniques for electric power production in the Upper Rhine Graben, Central Europe. In: World Geothermal Congress, 25–29 April 2010, Bali; 2010.

Schleicher A, Warr L, Kober B, Laverret E, Clauer N. Episodic mineralization of hydrothermal illite in the Soultz-sous-Forêts granite (Upper Rhine Graben, France). Contrib Miner Petrol. 2006;152(3):349–64. doi:10.1007/s00410-006-0110-7.

Schoenball M, Baujard C, Kohl T, Dorbath L. The role of triggering by static stress transfer during geothermal reservoir stimulation. J Geophys Res Solid Earth. 2012;117(B9):B09307. doi:10.1029/2012jb009304.

Schumacher ME. Upper Rhine Graben: role of preexisting structures during rift evolution. Tectonics. 2002;21(1):1006. doi: 10.1029/2001tc900022.

Sehlstedt S, Strahle A. Identification of water conductive oriented fractures in the boreholes KAS02 and KAS06, vol. PR 25-91-11. Stockholm: Svensk Kärnbränslehantering AB; 1991.

Simon K. Hydrothermal alteration of Variscan granites, southern Schwarzwald, Federal Republic of Germany. Contrib Miner Petrol. 1990;105:177–96.

Stanfors R, Rhén I, Tullborg E-L, Wikberg P. Overview of geological and hydrogeological conditions of the Äspö hard rock laboratory site. Appl Geochem. 1999;14(7):819–34. doi:10.1016/s0883-2927(99)00022-0.

Steen H. Geschichte des modernen Bergbaus im Schwarzwald: Eine detaillierte Zusammenstellung der Bergbauaktivitäten von 1890 bis zum Jahr 2000. 1st ed. Norderstedt: Books on Demand GmbH; 2004.

Stephansson O, Ljunggren C, Jing L. Stress measurements and tectonic implications for Fennoscandia. Tectonophysics. 1991;189(1):317–22.

Stober I. Die Wasserführung des kristallinen Grundgebirges. Ferdinand Enke Verlag:191. 1995.

Stober I, Bucher K. Herkunft der Salinität in Tiefenwässern des Grundgebirges–unter besonderer Berücksichtigung der Kristallinwässer des Schwarzwaldes. Grundwasser. 2000;5(3):125–40.

Stober I, Bucher K. Hydraulic properties of the crystalline basement. Hydrogeol J. 2007;15(2):213–24. doi:10.1007/s10040-006-0094-4.

Stober I, Bucher K. Hydraulic conductivity of fractured upper crust: Insights from hydraulic tests in boreholes and fluid-rock interaction in crystalline basement rocks. 2014. Geofluids:n/a-n/a. doi:10.1111/gfl.12104.

Todt W. Zirkon U/Pb-Alter des Malsburg-Granits vom Südschwarzwald. J Mineral Geochem. 1976;12:532–44.

Valley B, Evans KF. Stress state at Soultz-sous-Forêts to 5 km depth from wellbore failure and hydraulic observations. In: 32nd workshop on geothermal reservoir engineering. 2007.

Wennberg OP, Casini G, Jonoud S, Peacock DC. The characteristics of open fractures in carbonate reservoirs and their impact on fluid flow: a discussion. Pet Geosci. 2016. doi:10.1144/petgeo2015-003.

Werner W, Dennert V. Lagerstätten und Bergbau im Schwarzwald. Freiburg: LGRB; 2004.

Wickert F, Altherr R, Deutsch M. Polyphase Variscan tectonics and metamorphism along a segment of the Saxo-thuringian-Moldanubian boundary: the Baden-Baden Zone, northern Schwarzwald (F.R.G.). Geol Rundsch. 1990;79(3):627–47.

Zimmerman RW, Yeo I-W. Fluid flow in rock fractures: From the Navier-Stokes equations to the cubic law. In: Faybishenko B, Witherspoon PA, Benson S, editors. Dynamics of fluids in fractured rock, vol. 122. Washington: American Geophysical Union; 2000. p. 213–24.

Zoback MD, Kohli A, Das I, McClure MW. The importance of slow slip on faults during hydraulic fracturing stimulation of shale gas reservoirs. Society of Petroleum Engineers, SPE-155476-MS, SPE. 2012.

Zuther M, Brockamp O. The fossil geothermal system of the Baden-Baden trough (northern Black Forest, F.R. Germany). Chem Geol. 1988;71:337–53.

Permissions

List of Contributors

Tatyana Plaksina
Department of Petroleum Engineering, Texas A&M University, 907 Richardson Hall, College Station, TX 77843, USA

Christopher White
Department of Earth and Environmental Sciences, University of Tulane, Room 200 Blessey Hall, New Orleans, LA 70118, USA

Felina Schütz and Ernst Huenges
Section 6.2 Geothermal Energy Systems, Helmholtz Centre Potsdam-GFZ German Research Centre for Geoscience, Telegrafenberg, 14473 Potsdam, Germany

Gerd Winterleitner
Section 6.2 Geothermal Energy Systems, Helmholtz Centre Potsdam-GFZ German Research Centre for Geoscience, Telegrafenberg, 14473 Potsdam, Germany
Department for Earth and Environmental Sciences, University of Potsdam, Karl-Liebknecht Straße 24-25, Golm, Germany

Kamal Kumar Agrawal and Ghanshyam Das Agrawal
Mechanical Engineering Department, Malaviya National Institute of Technology, Jaipur 302017, India

Mayank Bhardwaj
Department of Renewable Energy, Rajasthan Technical University, Kota 304010, India

Vikas Bansal and Rohit Misra
Mechanical Engineering Department, Government Engineering College, Ajmer, Ajmer 305001, India

M. Peter-Borie and A. Loschetter
BRGM, 3 av. C. Guillemin, BP36009, 45060 Orléans Cedex 2, France

I. A. Merciu and G. Kampfer
Equinor ASA, Research and Technology, Rotvoll, Norway

O. Sigurdsson
HsOrka, Svartsengi, 240 Grindavík, Iceland

Bernard Giroux and Lyal B. Harris
Centre – Eau Terre Environnement, Institut national de la recherche scientifique, 490 rue de la Couronne, Québec, QC G1K 9A9, Canada

Hejuan Liu
Centre – Eau Terre Environnement, Institut national de la recherche scientifique, 490 rue de la Couronne, Québec, QC G1K 9A9, Canada
State Key Laboratory of Geomechanics and Geotechnical Engineering, Institute of Rock and Soil Mechanics, Chinese Academy of Sciences, Wuhan 430071, China

Steve M. Quenette
Monash eResearch Centre, Monash University, Clayton, VIC 3800, Australia
School of Mathematical Sciences, Monash University, Clayton, VIC 3800, Australia

John Mansour
School of Mathematical Sciences, Monash University, Clayton, VIC 3800, Australia

Marco Viccaro
Dipartimento di Scienze Biologiche Geologiche e Ambientali – Sezione di Scienze della Terra, Università degli Studi di Catania, Corso Italia 57, 95129 Catania, Italy
Istituto Nazionale di Geofisica e Vulcanologia – Sezione di Catania, Osservatorio Etneo, Piazza Roma 2, 95125 Catania, Italy

Esmail Ansari and Richard Hughes
Louisiana State University, 70803 Baton Rouge, LA, USA

Dietmar Kuhn
Institute of Nuclear and Energy Technology, Karlsruhe Institute of Technology, Herrmann-von Helmholtz-Platz 1, 76344 Eggenstein-Leopoldshafen, Germany

Pia Orywall
Institute of Nuclear and Energy Technology, Karlsruhe Institute of Technology, Herrmann-von Helmholtz-Platz 1, 76344 Eggenstein-Leopoldshafen, Germany
RBS wave GmbH, Postfach 311508, 70475 Stuttgart, Germany

Kirsten Drüppel
Institute of Applied Geoscience-Division of Mineralogy and Petrology, Karlsruhe Institute of Technology, Kaiserstrasse 12, 76131 Karlsruhe, Germany

Michael Zimmermann
Institute of Catalysis Research and Technology, Karlsruhe Institute of Technology, Herrmann-von Helmholtz-Platz 1, 76344 Eggenstein-Leopoldshafen, Germany

Elisabeth Eiche
Institute of Applied Geoscience-Aquatic Geochemistry Division, Karlsruhe Institute of Technology, Kaiserstrasse 12, 76131 Karlsruhe, Germany

J. L. Cant, P. A. Siratovich, J. W. Cole, M. C. Villeneuve and B. M. Kennedy
Department of Geological Sciences, University of Canterbury, Christchurch 8140, New Zealand

Eva Schill
Institute for Nuclear Waste Disposal, Karlsruhe Institute of Technology, Hermann-von-Helmholtz-Platz 1, 76128 Karlsruhe, Germany

Jörg Meixner, Carola Meller, Manuel Grimm, Jens C. Grimmer, Ingrid Stober and Thomas Kohl
Geothermal Research Group, Institute of Applied Geosciences, Karlsruhe Institute of Technology, Adenauerring 20b, 76131 Karlsruhe, Germany

Jean Schmittbuhl, Olivier Lengliné and François Cornet
EOST, Université de Strasbourg/CNRS, 5 rue René Descartes, 67000 Strasbourg, France

Nicolas Cuenot
GEIE EMC, Route de Soultz-sous-Forêts, 67250 Kutzenhausen, France

Albert Genter
ES-Géothermie, 3, Chemin du Gaz, 67500 Haguenau, France

Thomas Kohl
Institute of Applied Geoscience-Division Geothermics, Karlsruhe Institute of Technology, Kaiserstrasse 12, 76131 Karlsruhe, Germany

Index

www.ingramcontent.com/pod-product-compliance
Lightning Source LLC
Chambersburg PA
CBHW080407190526
45161CB00003B/159